The Near-surface Layer of the Ocean

The Near-surface Layer of the Ocean

K. N. FEDOROV and A. I. GINSBURG

Translated by M. Rosenberg

///VSP///
Utrecht, The Netherlands

VSP BV
P.O. Box 346
3700 AH Zeist
The Netherlands

First published in Russian 1988

ISBN 90-6764-136-7

CIP-DATA KONINKLIJKE BIBLIOTHEEK, DEN HAAG

The near-surface layer of the ocean / K. N. Fedorov,
A. I. Ginsburg.—Utrecht: VSP.—Ill.
Vert. van: Pripoverchnostnyj sloj okeana.—Moskva:
Gidrometeoizdat, 1988.—with index, lit. opg.
Subject heading: oceanography

Phototypeset at Thomson Press (India) Ltd., New Delhi
Printed in The Netherlands by Koninklijke Wöhrmann B.V., Zutphen.

Contents

v

Foreword

The structure and variability of the main physical fields in the thin near-surface water layer, responding rapidly to all short-term and local effects of the atmosphere and the radiant energy flux from the sun, are discussed. The greatest attention is given to the anomalous phenomena on the ocean surface, the thermal boundary layer and its association with convection, the freshening role of precipitation, the diurnal thermocline and the cyclic nature of related processes, the specific features of the near-surface currents, and certain problems of the effectiveness of remote sensing of the ocean by the aerospace technique. The results of specialized measurements in the ocean are used in the analysis, including data obtained by the authors, as well as carefully selected satellite information.

This book is intended for oceanographers–physicists with long-term experience of participating in sea expeditions, as well as for marine meteorologists, specialists in ocean–atmosphere interactions, and for those who use satellite information on the ocean for scientific and practical purposes.

K. N. FEDOROV
A. I. GINSBURG

Preface

The development of satellite methods in recent years, besides its direct contribution to our knowledge of the ocean and to the modern arsenal of research means in oceanography, has also had a somewhat unexpected indirect effect. It has attracted the attention of scientists to the surface of the ocean. Many phenomena, known to seamen from time immemorial (e.g. current rips), were practically discovered anew and were subjected to detailed investigation and description from a new point of view. Everything relating to the state of the ocean surface and to the anomalies of this state, i.e. everything that could exert an effect in one way or another on the formation of surface radiation that should be recorded from space, became the object of special intensive studies. It can be stated without any exaggeration that the ocean surface would certainly not have attracted such close attention if it were not for such a strong stimulus as the possibility of viewing the ocean from space, and, what is still more important, the possibility of placing measuring instruments in orbit.

However, everything occurring on the ocean surface is associated in one way or another with the underlying water layers, even when the surface phenomena are induced by the direct effect of the atmosphere. This is true, first of all, in regard to the anomalous states of the surface. Hence, it is impossible to confine oneself only to the ocean surface when studying these phenomena. It is no coincidence that interest in the ocean surface gradually turned into a detailed study of the ocean near-surface water layer. In this case, the book deals not with the upper quasi-homogeneous layer (UQL) or the upper active layer (UAL) of the ocean, but with the water layer which, in the majority of cases, is thinner and responds faster to all short-term and local effects of the atmosphere and direct solar irradiance. Its lower boundary is determined by quite different considerations to those in the cases of the UQL and UAL. It is connected in the majority of cases with the position of the diurnal (daily) thermocline that develops as a result of diurnal solar heating, and is a most important element of the near-surface layer.

We have made an attempt in this monograph to state in one book everything new on the near-surface layer, which has been learned in 10 years of intensive studies. The lack of previous knowledge in this field was due to the fact that a standard series of bathometers, the first of which was rarely placed on a cable less than 1 m from the surface, was the basic means of measurement in physical oceanography and then in oceanology. The measured horizon 0 m was then ascribed to this bathometer. As a result, the most interesting, by its characteristics, 1–2 m thick ocean layer, adjoining from below the sea surface, was completely lost from the sphere of studies. Other methods and means of measurements were necessary to fill this gap. Laboratory experiments were also necessary alongside the new field measurements.

Simply the wish to illustrate in the greatest detail and more logically the earlier unknown aspects of the variability of the ocean near-surface layer hydrophysical fields predetermined the contents of this book and the distribution of material in the chapters. One of the difficulties that the authors encountered arose when distributing the explanatory material between the general introduction to the book and the introductory sections in the individual chapters. The novelty of the subject required a considerable volume of explanation. It was absolutely necessary to consider in this case the degree of knowledge of the potential reader. It was decided to orientate the book towards the oceanographer–physicist, having practical experience in sea expeditions in various latitudes and under various weather conditions, and knowing well the problems of sea meteorology and the ocean–atmosphere interaction. Accordingly, the most general data on the near-surface layer and its links with atmospheric and other forces are presented in the Introduction, while the more concrete explanatory information is given at the beginning of the respective chapters. It was thus possible to avoid repetition. We also hope that specialists–theoreticians will find in this book material and incentive to formulate new problems and to develop new concepts and numerical models. The choice of concrete problems, encompassing the contents of the individual chapters and sections, serves this objective. We wanted, first of all, to present in detail the problem of the thermal boundary layer in water, which is often called the 'skin-layer' or 'cold surface film', and its close association with convection. Special attention is given to the problem of salinization of the boundary layer during evaporation. The diurnal thermocline, having a cyclic nature and associated with diurnal solar heating, was another important objective of description. The effects of cyclicity are transmitted due to the diurnal thermocline to practically all the processes occurring in the near-surface layer of the ocean, including convection. The latter does not occur beyond the limits of a thin near-surface layer of the ocean in temperate and low latitudes during daytime under intensive solar heating and in the absence of a strong wind, and it can encompass the whole UQL only during the night. Questions on the cyclicity of the convection process, the hierarchy of its scales, and its space structure under various conditions are discussed by us in the light of a series of new facts obtained from observations in the ocean. It is important to point out that convection of the near-surface layer of the ocean in winter at high latitudes penetrates to great depths and represents an effective mechanism for the formation of deep waters. The characteristics of the near-surface layer forming during the summer and autumn become determinative in regard to the processes of formation of deep waters and their properties, which are then preserved for a long time. It is possible to speak in this context of the climate-forming role of the near-surface layer of the ocean.

The range of problems chosen by us for discussion was also motivated by the wish to show the physical association between the phenomena on the surface and the processes occurring in the deep layers of the ocean, which is important for the development of remote methods of studying the ocean. In this respect, the near-surface layer plays an essential role of the basic transient link. The useful signal, which can be then 'read' by means of remote receiving equipment, recording one or other radiation from the ocean surface, is formed and converted just there. Though many publications have been dedicated to remote methods of studying the ocean by means of satellites, the role of the near-surface layer has not been discussed in this context. Little is known of concrete scientific results already obtained on the basis of satellite data. At the same time, observations and measurements from satellites have already provided much new information on the character of motion and variability of waters nearby the ocean surface. Special sections in Chapter 5 of this book are dedicated to recently discovered non-stationary ordered forms of motion in the near-surface layer generated by localized impulse forces. Certain methodological problems dealing with the influence of the near-surface layer on the information content of remote methods are discussed in the last chapter.

On the whole, the sum of new knowledge obtained in recent years on the physics of the ocean near-surface layer, its variability, and the dynamics of its waters seems to us to be substantial and important. Nevertheless, the book is by no means an exhaustive handbook on problems dealing with the near-surface layer and the surface of the ocean. To note, there are well-known publications [40, 58, 72, 73, 240, etc.] on certain similar questions and especially closely-related problems so that it would simply be inexpedient to duplicate them here. On the other hand, we found it necessary to show what new problems have arisen due to the necessity of considering the specific features of the near-surface layer in spheres where wide-known traditional approaches were formed a long time ago. One of these spheres is the description of UQL evolution in one-dimensional models; another is the spectral approach to analysing anomalous states of the sea surface.

Because of the extent of our own interests and owing to the limited volume of the book and inexpediency of duplicating recently published monographs it was decided to leave aside practically fully interesting hydro-optical aspects of the problem, not to discuss in detail the problem on the physical nature of Langmuir circulations considered in refs. [73, 89], and to discuss only very briefly in Section 3.3.7 problems dealing with surface-active and other substances on the ocean surface. When considering the large number of various flows (heat, moisture, salts, gases, kinetic energy, etc.) through the free surface of the ocean, we included in the analysis only those (see Chapters 3 and 4) that form the thermal balance of the near-surface layer. The others are discussed only as much as required for examining the physical processes occurring in this layer. We have not discussed here at all the problems of global distribution of various flows on the surface of the World Ocean because the given problem has been critically analysed in detail recently by Woods [240]. Neither have we included in the description the states of the near-surface layer in the case of strong winds and storms. The point is that under intensive wind–wave mixing the thickness of the near-surface layer of the ocean coincides with the UQL thickness, while the thermal boundary layer ceases to exist as an integral whole. In this case, free convection no longer plays an essential role. In a certain sense, the near-surface layer is manifested most vividly in calm weather and under light and moderate winds (force 1–4). It is under these conditions that the space–time variability of the hydrophysical fields in this layer is so strong and characteristic that it gives grounds for isolating it as an individual object of investigation with a variety of physical regimes inherent only in it. Therefore, it is quite natural that the series of problems discussed in this monograph has been predetermined greatly by the processes occurring in the presence of a relatively calm ocean surface. It is no coincidence that the most interesting anomalies in the sea surface state are observed under these conditions, e.g. thermal inhomogeneities of calm weather, current rips, choppy water, traces of internal waves, frontal convergences, Langmuir cells, etc., and the developing diurnal thermocline mentioned above.

The book is intended, first of all, for specialists in the physics of the sea and hydrometeorologists who are familiar with the results of field hydrophysical measurements in the ocean and with the problems of the ocean–atmosphere interaction in the local synoptic and global scales. It is also expected that the material contained in this book will assist in formulating new theoretical concepts, in improving numerical models of the ocean and its interactions of various scales with the atmosphere, and in developing methods of remote sensing of the ocean from space. Correspondingly, the book will be of interest to specialists in hydromechanics and numerical simulation, as well as to representatives of various natural scientific, marine, and engineering professions who work in the sphere of developing satellite methods. The book may be useful for students and postgraduates of physical oceanography.

The idea of writing the book arose while discussing at a meeting of the Scientific Council of the P. P. Shirshov Institute of Oceanology of the USSR Academy of Sciences the basic results of field and laboratory measurements made by us. The problems of the variety of regimes of the near-surface layer hydrophysical fields were also discussed at the 'Atmosphere–ocean–space' seminar of Academician G. I. Marchuk in 1981.

A substantial part of the studies, whose results form the basis of this monograph, was carried out by us at the P. P. Shirshov Institute of Oceanology of the USSR Academy of Sciences and in some cases with the participation of workers of the Department of Experimental and Space Oceanology. Most valuable was the assistance of our colleagues, namely, V. L. Vlasov, P. A. Derevshchikov, S. N. Dikarev, M. V. Yemelyanov, A. G. Zatsepin, A. S. Kazmin, A. G. Kostyanoi, N. P. Kuzmina, A. G. Ostrovsky, A. M. Pavlov, L. I. Piterbarg, V. B. Rodionov, and V. E. Sklyarov. A great deal of valuable material was obtained as a result of the tireless inventiveness and assistance of V. T. Paka, Director of the Atlantic Department of the Institute, and his colleagues. V. B. Radionov very kindly calculated the coefficients of correlation between the ocean surface temperature field and the temperature fields at different deep horizons for a series of large-scale surveys (see Section 6.1). Section 6.2 was written together with L. I. Piterbarg. Extremely useful was the discussion of Section 6.3 with I. A. Leikin. We express our sincere gratitude to all of them, as well as to everyone else who participated directly or indirectly in discussing the material included in this book.

Chapter 1

Object and methods of the investigation

1.1. Object of the investigation and its basic characteristics

We consider the near-surface layer of the ocean as a thin layer with a thickness ranging from several metres to one or several tens of metres adjoining the free surface of the ocean directly from below and including this surface, the thermal boundary layer,* the diurnal (or daily) thermocline, and all the stratifications that arise because of higher frequency processes and non-periodic atmospheric effects. The near-surface layer absorbs solar radiation directly and reacts to heating, cooling, evaporation, and freshening due to direct local effects of the atmosphere as well as to near-surface water exchange between neighbouring areas. Conditions are created in this layer for the development and suppression of convection and turbulence. All this causes increased space and time variability of the main hydrophysical fields, whose statistical characteristics often surpass the variability indices of the same fields in the underlying layers of the ocean, with the exception of variability characteristics induced by internal waves on fixed horizons of the seasonal and main thermocline. The near-surface layer of the ocean with its increased concentration and specific composition of suspended matter forms the colour of the ocean, i.e. the spectral composition of radiation from the ocean. The lower boundary of the near-surface layer of the ocean is considered to be the lower boundary of the diurnal (daily) thermocline developing during daylight under relatively calm conditions of the sea. This boundary is not fixed by depth. By the time it has reached maximum development [about 1600–1800 local solar time (LST)], the near-surface layer usually contains $\simeq 90$–95% of the daily solar heat absorbed by the ocean. The dynamics of the layer, as we will see below (see Chapter 5), is characterized mainly by non-stationary movements of various nature. The thickness of the near-surface layer coincides with that of the upper quasi-homogeneous layer (UQL) under intensive wind–wave and convective mixing.

   The novelty of the oceanic near-surface layer concept is associated with the gradual introduction of new measuring methods into oceanological practice. These methods include continuous vertical sounding with high resolution, recording when towing in near-surface horizons, use of strings of sensors, and special sounding and measuring devices that do not disturb the structure of

---

* We consider the thermal boundary layer hereafter (see Chapter 2) as the layer associated with molecular exchange of heat through the water surface.

the thin water layer near the free surface, and use of remote sensing methods for measuring the ocean surface temperature.

Formerly, when carrying out standard hydrological measurements in the sea, the upper bathometer was always placed at the horizon of 1 m. In principle, nobody ever measured accurately the distance from the sea surface to the first bathometer, and, what is more, it could not be measured accurately because the length of the bathometer proper was comparable to this distance. The bathometer was usually sunk into the water 'with a reserve' so that it would not be exposed when held under swell and waves, and it was placed as close as possible to the surface when the sea was calm. In any case, this horizon was always designated as 0 m in tables of standard oceanographic data. It was then assumed that the vertical temperature and salinity distribution in a layer several metres thick near the ocean surface was homogeneous or changed very little and so the difference of the actual horizon of measurements from 0 m could be neglected. This erroneous assumption had formed greatly under the influence of judging variability on the basis of averaged climatic information.

Oceanologists have known for a long time that the ocean UQL is neither everywhere nor always homogeneous, or even quasi-homogeneous (by the vertical and the horizontal) [74, 110, 223]. Partly with the above circumstance the difference is associated between the notion 'sea surface temperature' (SST, or OST for the ocean surface temperature), relating to the 'thin surface layer of seawater from several micrometres to 1–2 cm', and the notion 'sea surface layer temperature' (SSLT), which means 'averaged by 1–2 min temperature of the seawater upper layer with a maximum thickness 1 m at the site of measurement', introduced in 1977 in the 'Instructions on hydrological work in oceans and seas' [88]. However, this difference did not reflect fully the specific features of the thermal structure and the variability of what we call here the near-surface layer of the ocean. It is associated mainly with the obvious physical difference between the radiation temperature of the surface in the infra-red (IR) range of the spectrum and the thermodynamic temperature of the interior, as well as with the presence of a thin thermal boundary layer near the ocean surface (see Chapter 3). The greatest variability of the thermal structure, which is not considered by the difference between OST (SST) and SSLT, is observed in the near-surface layer during intensive solar heating in calm or light-wind weather [19, 109], and is associated mainly with volume absorption of solar energy, evaporative heat losses, convection, modulation of the near-surface layer by the internal waves, and the peculiarities of near-surface salinity stratification. The high small-scale variability of salinity in the near-surface layer is determined mainly by the long lasting consequences of its freshening by heavy rains. In these cases, horizontal differences in the salinity of the order of 1‰ per kilometre occur in the absence of intensive wind mixing in the upper 1 m layer of the ocean [24]. Considerable seasonal and inter-year anomalies in the salinity of the essentially greater space scale may occur owing to anomalous discharge of rivers or the melting of polar ice [176].

The origin of hydrostatically stable thermal or salinity stratification close to the surface generates the so-called 'blocking effect', which prevents the propagation of wind–wave and convective turbulent energy from the surface downward into the thickness of the UQL. Stable stratification suppresses the earlier developed turbulence and can completely prevent the development of shear instability or convection within the UQL. Since the variability of the thermal structure in the near-surface layer of the ocean is characterized most often by a clearly expressed diurnal cycle, the intensity of turbulence in the thickness of the UQL should also experience the diurnal cycle, which is confirmed by observations [81]. However, salinity stratification is not only less subjected to diurnal variations, but it can also disturb greatly the diurnal cycle of the thermal structure variations, producing in the low latitudes long-term overheating of the near-surface layer which is preserved in the course of natural synoptic periods. The inter-year negative anomalies in the salinity of the near-surface layer may induce complete termination of winter convection in the high latitudes (e.g. in the Labrador Sea [176]) which may disturb the natural process of deep-water formation for a long time.

The high degree of variability of the near-surface characteristics is neither an anomalous nor an exotic phenomenon. It is typical during the whole year for the entire inter-trade wind zone of the World Ocean; for the vast subtropical areas, where the trade wind velocity drops to 5 m/s; and in the temperate latitudes in summer. In total, the regions where such variability can be observed make up more than 40% of the World Ocean area. Regions such as the Sargasso Sea in the Atlantic, the western part of the Pacific and the northern part of the Indian Ocean are most typical from the point of view of increased variability of the thermohaline structure of the oceanic near-surface layer.

Space–time scales of the variability of the near-surface layer characteristics raise a series of methodological problems for scientists associated, for example, with the comparability of results of hydrophysical measurements made near the ocean surface by different methods [22] (see also Chapter 3).

Let us now consider the most important factual data on the temperature and salinity variability in the near-surface layer of the ocean in order to understand better all the aspects of the problem discussed above.

1.2. Basic types of vertical thermal structure of the near-surface layer of the ocean and several characteristics of the temperature variability in it

Three basic types of vertical thermal structure of the near-surface layer of the ocean are, in principle, possible, according to current knowledge [25, 102, 239]. They reflect the fundamentally different types of strictly local (one-dimensional) energy exchange with the atmosphere. All the other types are, in essence, greater or smaller deviations from the basic types, which are due to the specific features of the development prehistory and the local interaction with the atmosphere, and the influence of the three-dimensional character of the processes forming the structure of the near-surface layer (circulation cells, advection, convection, etc.). The three basic types stated are shown in Figs 1.1a–1.1c, where the main

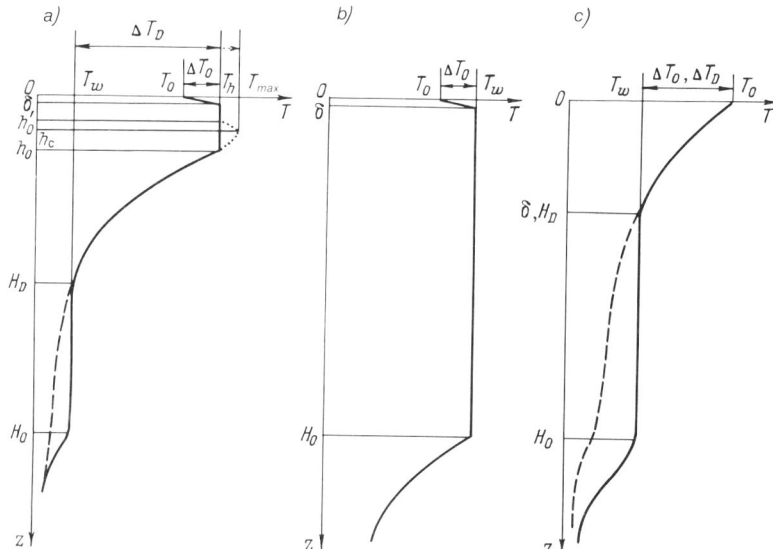

Figure 1.1. The three basic types of vertical thermal structure of the near-surface layer of the ocean. (a) Type I: fully developed diurnal vertical temperature profile at heat transfer from the surface and volume absorption of solar radiation near by the surface in almost calm, sunny weather. (b) Type II: well-expressed upper quasi-homogeneous layer at nocturnal convection or intensive wind–wave mixing. (c) Type III: near-surface thermocline which is a consequence of a heat flux from the atmosphere into the ocean through a free surface. All the other symbols are given in the text.

letter symbols are given for a number of parameters and will be used in the text below.

Type I (Fig. 1.1a): Fully developed diurnal vertical temperature profile at negative sum of q_0 of latent and sensible heat fluxes between the water and air, and effective long-wave radiation on the free water surface ($z = 0$) in clear, sunny light-wind weather (force 1–3); the diurnal mixed layer is well expressed and below it is the strongly pronounced diurnal thermocline which formed against the background of the seasonal and/or nocturnal UQL as a result of wind mixing or convection.

Type II (Fig. 1.1b): Well-expressed UQL at negative thermal balance of the free surface, intensive convection (e.g. during the night) or at a wind of force 4 or more.

Type III (Fig. 1.1c): Near-surface thermocline which is the consequence of positive q_0 on the surface (e.g. it is a rare case when the air is considerably warmer than the water at 100% humidity). The near-surface thermocline may be of diurnal or episodic character.

If the UQL is not found in the prehistory of structures of types I and III, then a gradual decrease of temperature, i.e. transition to the seasonal or main thermocline, continues below the diurnal or near-surface thermocline (dotted line in Figs 1.1a and 1.1c). Thus, in the most general case the vertical thermal structure of the ocean contains three thermoclines in the diurnal time (under solar heating), i.e. main, seasonal, and diurnal.

The symbols used in Fig. 1.1 are as follows:

H_0 depth of the UQL lower boundary location

H_D depth of the diurnal thermocline lower boundary location, and conditional boundary of the near-surface layer of the ocean

δ thickness of the thermal boundary layer

h_0 diurnal mixed-layer lower boundary

h_0' lower boundary of convection penetration (may reach H_0 during the night)

h_c thermal compensation level

T_0 thermodynamic temperature of the free surface, designated SST (sea surface temperature) in the case of its correct restoration on the basis of remote (e.g. satellite) measurements, or its assumed value

T_w UQL temperature; if $H_0 = H_D$ (type II), it is also designated as NSLT, the near-surface layer temperature

ΔT_0 temperature difference across the thermal boundary layer

ΔT_D temperature diurnal difference across the thermocline

T_h diurnal mixed-layer temperature

T_{max} temperature of the subsurface temperature maximum observed sometimes at the level of thermal compensation

Let us consider some of the characteristics of the temperature variability in the near-surface layer of the ocean. We will show that the procedures of space–time averaging, which are inevitable when calculating generalized climatic characteristics, mask the true temperature variability in the near-surface layer of the ocean. Let us, for example, analyse the variability of the characteristic which is always adopted as T_0 in a region in the Tropical Atlantic (16°35′ N, 33°25′ W) where the hydrophysical experiment 'Polygon-70' was carried out for 7 months in 1970. The climatic background is characterized in this region by the following values* [4, 91]:

Normal annual value of $\bar{T}_0$	25.0°C
Annual cycle amplitude of T_0	3.0°C
RMS temperature deviation σ_D during the annual cycle of T_0	0.8°C
Daily amplitude of T_0	0.2–0.3°C
RMS temperature deviation σ_D in the diurnal cycle of T_0	0.1°C

The above values give the impression of a very insignificant variability of T_0 in the annual and diurnal cycles. They only confirm information well learned from text-books about the low seasonal and diurnal variability of the ocean surface thermal regime in low latitudes. It is interesting to note that the mean

---

*In this chapter, T_0 will imply not the true values of the thermodynamic temperature of the ocean surface, but the temperature of the near-by horizons taken for it. The corresponding corrections are discussed in Chapter 3.

Table 1.1
Factual mean monthly values of $\bar{T}_0$ at the Tropical Atlantic polygon area in 1970 [91]

Month:	February	March	April	May	June	July	August	September	$\bar{T}_0$ (II – IX)	$\pm \sigma$
$\bar{T}_0(°C)$:	24.4	24.6	24.3	24.5	24.5	24.6	25.3	26.2	24.8	0.64

Table 1.2
Characteristics of daily and inter-daily variations of T_0 at the Tropical Atlantic polygon area in 1970 after [1]

Observation period	No. of measurements, n	$\bar{T}_0(°C)$	$T_{0,max}$ (°C)	$T_{0,min}$ (°C)	δT_0 (max – min)	$\sigma_D(°C)$	$\sigma_M(°C)$	σ_M^2/σ_D^2
14 June–16 July	746	24.6	24.9	24.2	0.7	0.30	0.48	2.6
27 July–16 August	504	25.2	26.3	24.5	1.8	0.46	1.27	7.7
27 August–13 September	432	26.1	27.6	25.1	2.5	0.83	2.15	6.8

monthly values of $\bar{T}_0$, obtained by averaging hourly measurements made by a thermometer in a Schpindler frame from February to September 1970 (Table 1.1), are well within the generalized climatic limits.

At the same time, the factual inter-daily and daily variations greatly exceeded the climatic level by their amplitudes and statistical characteristics, especially in July–September when the trade winds started to abate and intensive heating of the near-surface layer commenced. This is demonstrated in Table 1.2, which lists the other results of processing the same hourly measurements.

It is clear from the data in Table 1.2 that the amplitude δT_0 of the inter-daily variations of T_0 exceeds the mean climatic daily amplitude from three to ten times, and the relation of the dispersion of the inter-daily oscillations σ_M^2 to the dispersion of the factual daily oscillations σ_D^2 also fluctuates from two to eight. It is interesting that the root mean square daily variability σ_D is always greater than the climatic level and increases noticeably with the heating of the near-surface layer in the course of the annual cycle.

It is necessary to note that this is the situation in the region where the trade winds rarely drop to below force 3–4. The inter-daily temperature variability is more profound in the near-surface layer in regions where calm or light-wind conditions are observed more often, as we have shown by the example of the Sargasso Sea (see, for example, ref. [102]).

In our opinion, the following are the reasons for such systematic differences between the actual variability of T_0 and its climatic indices:

(1) increased space–time variability of the basic thermodynamic characteristics of the near-surface layer of the ocean during natural synoptic periods (2–6 days) which is practically not sensed when averaged for 1 month; and

(2) Utilization of standard hydrological series data (in particular, horizon '0 m') to obtain averaged climatic characteristics.

We will now show that the differences in the measuring methods of T_0 and in the distances from the surface where its value is measured actually result in great differences in the characteristics of the variability obtained (see Table 1.3).

A comparison of the first two lines of Table 1.3 gives a view of the differences obtained when using a thermometer in a Schpindler frame and the zero horizon of the hydrological series to obtain the values $\bar{T}_0$, T_0, and σ, and the extremum values in the same region in the same period of time. In both cases, the root mean square deviations of σ with the same values of $\bar{T}_0$ differ 1.5-fold and the limits of δT_0 variations differ 2-fold. The wider range of variability is produced naturally by a thermometer in a Schpindler frame. A towed sensor (lines 3 and 4) gives a view of the space variability of T_0 in light-wind days when temperature patchiness is encountered in the near-surface layer (so-called 'thermal inhomogeneities of calm weather' [109, 114]. The dispersions (σ^2) of the T_0 fluctuations, recorded on the horizon 0.15 m during a 1–3 h period on tacks from 40 to 90 km long, are of the same order as those of the inter-daily fluctuations of T_0 by the data of horizon 0 m of the hydrological series (line 2). The emergent profiler (lines 5–7) and the 'Boomerang' probe (lines 8–11) demonstrate the vertical temperature variability in the upper 2 m, as well as the space variability in scales less than 0.5 km during a period of about 10–20 min. While solar heating can create in the afternoon (1300–1600 h) temperature differences (ΔT_D) of 0.8–2.7°C and even more in the upper 2 m layer (see Section 4.3), the fluctuations can reach 0.5–1.3°C between the maximum and minimum values in short periods of time (several minutes) at levels close to the surface. These fluctuations are most likely associated with the drift of the vessel through a temperature field which is strongly modulated in the near-surface layer by the internal waves effect [103, 109]. The full difference between the temperature extremum values in the 0–2 m layer in only 5 weeks from 18 August to 25 September 1978, according to Table 1.3, is 4.7°C, which is quite commensurable with the mean annual amplitude T_0 for many years at latitude 29° in the Sargasso Sea, and is 50% of the full range of oscillations between the summer and winter extremum values of the temperature in this region [4]. A similar high space–time variability of the temperature in the near-surface layer of the Sargasso Sea has been recorded independently and described in ref. [128], where variations of 1–1.5°C were observed in 10–15 s of the sensor drift on a horizon only 5–8 cm below the surface, as well as vertical differences of up to 3°C in the upper 2 m layer in calm, sunny weather in May 1973.

It follows from the above results that the thermal structure and space–time variability of the temperature field in the near-surface layer of the ocean are very different under various conditions of interaction and energy exchange between the ocean and the atmosphere. At least five sharply differing regimes of the near-surface layer of the ocean can be distinguished on this basis:

Table 1.3
Indices of T_0 variability in the near-surface layer obtained by various methods at different distances from the surface (POLYMODE experiment area in the Sargasso Sea, 1978; 29°N, 70°W; light-wind weather, wind force 1–3)

No.	Date	Measuring method	Horizon (m)	n	$\bar{T}_0, T_0, T_w$(°C)	±σ(°C)	$T_{0,max}$(°C)	$T_{0,min}$(°C)	δT_0(°C)	Remarks
1	18 August–3 September	Thermometer in Schpindler frame	0.10–0.15	420	28.4	0.53	30.5	26.8	3.7	Within polygon area
2	24 August	Hydrological series (0 m)	1–2	31	28.37	0.37	29.68	27.78	1.90	Same
3		Sensor towed near the surface[a]	0.15	143	28.2	0.20	28.7	27.7	1.0	2225–2354 h; l = 40 km[b]
4	20 September	Emergent profiler[a]	0.15	320	28.6	0.29	29.6	27.8	1.8	1621–1941 h; l = 90 km
5	25 September		0	4	27.3	0.33	27.65	26.92	0.73	1311–1322 h; l = 0.5 km
6			1	4	27.1	0.13	27.22	26.90	0.32	Same
7			2	4	26.9	0.04	26.95	26.84	0.11	Same
8	25 August	'Boomerang' probe[a] (vertical profile, see Fig. 1.2)	0	1	31.5	—	—	—	—	1600 h
9			1	1	29.8	—	31.1	29.8	1.3	Limits measured from float
10			2	1	28.8	—	29.2	28.7	0.5	Same
11			8–20	1	27.8	—	—	—	—	UQL temperature

[a]Details of these methods are given in Section 1.4.
[b]l = Tack length; n = number of measurements.

(1)   a regime of intensive wind–wave mixing (wind velocity 10 m above water surface $u_{10} > 8$–10 m/s);

(2)   a regime of intensive convection (nocturnal, autumn–winter, or associated with frontal downdrafts of cold air over the ocean);

(3)   a regime of Langmuir circulations (u_{10} from 3 to 10 m/s);

(4)   a regime of intensive solar heating in calm and light-wind weather (u_{10} from 0 to 3–5 m/s) with modulations of heated layer by internal waves and in the absence of the latter; and

(5)   a regime of near-surface freshening by precipitation.

The vertical temperature profiles of type I and their various modifications correspond to regimes 3–5, while profile $T(z)$ of type II always corresponds to regimes 1 and 2 (see Fig. 1.1). Profiles of type III are significantly rarer owing to the specific features of heat exchange between the ocean and the atmosphere.

The diurnal thermocline under various regimes can be at different horizons by the end of daylight, i.e. from 2–5 to 30–40 m. The least deep diurnal thermocline (from 2 to 5 m) with sharp vertical temperature gradients is usually bound with intensive freshening of the near-surface layer by precipitation. Langmuir cells and intensive wind–wave mixing can propagate the daily heating to 30–40 m, where the diurnal thermocline becomes very weak and practically merges with the seasonal one. There may be practically no diurnal thermocline at very intensive mixing and in the absence of direct solar heating. The

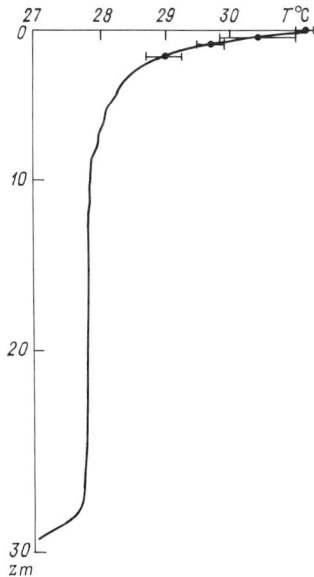

Figure 1.2. Typical vertical temperature profile under conditions of intensive solar heating in the upper 10 m layer of the Sargasso Sea according to measurements made by V. T. Paka and us (27th cruise of the research vessel *Akademik Kurchatov*, 25 August 1978, about 1600 LST). The horizontal lines near the surface show the range of temperature variability by the data of measurements from a float.

distribution of daily solar heat income, which is of the order of $21-25\,\text{MJ/m}^2$ $(500-600\,\text{cal/cm}^2)$* in temperate and low latitues [142], on layers of greatly differing thickness produces a $3°C$ range of local variability of T_0 from one regime to another. The convergent and divergent currents induced by Langmuir circulations and internal waves in the near-surface layer horizontally redistribute the heat absorbed near the surface in the light time of the day. We have observed situations when, for example, by the end of the day and owing to the effect of the internal waves the difference between the enthalpy of the ocean upper layer at two neighbouring points and at a distance of only 1 km reached the sum of solar radiation absorbed daily. In this case, as we have noted above, the amplitude of the space variability of T_0 can easily reach $2°C$.

Discrimination of the stated regimes should contribute to the development of realistic numerical models without which it is practically impossible to consider the effects of the near-surface layer of the ocean.

1.3. Variability of salinity in the near-surface layer of the ocean

The salinity variability in the near-surface layer of the ocean, unlike its temperature variability, has been hardly studied. Until recently, there has been practically no measuring equipment for obtaining information on the horizontal and vertical distributions of the salinity with a resolution of hundreds of metres and tens of centimetres, respectively. The traditional method of determining the salinity close to the surface has been, and continues to be, the taking of samples by means of a bucket from a vessel underway, or, at best, by using the water intake on the vessel, making, as a result, measurements at individual sites of the ocean. Devices and methods have been developed only recently to record salinity continuously at a depth of several metres from a vessel underway, and to obtain vertical profiles of the salinity in the near-surface layer (see Section 1.4). Unfortunately, aerospace methods of determining the salinity lack the required sensitivity (0.1 parts per thousand) which would make it possible to obtain operative information on the salinity across the entire surface area of the World Ocean, to estimate characteristic space scales of the salinity variability, to determine the average distance between the salinity fronts, and so on. The existing remote SHF methods make it possible to determine water salinity with a maximum accuracy of 1–2‰, which is comparable to the range of salinity variability in the open part of the World Ocean. However, these methods can now be used for comparatively rough measurements and estimates in coastal regions.

Despite the inadequacy of the greater part of available methods, the total amount of information accumulated today [32, 103] allows us to name the basic mechanisms determining the mesoscale variability of the salinity field in

---

*Hereafter, equivalents of heat values are also stated in CGS units to assist the reader in verifying and estimating respective temperature variations.

the ocean and the approximate ranges of this variability under the influence of each individual mechanism.

1.3.1. *Freshening of coastal ocean regions by the discharge from large rivers*

Freshening can reach very large values (up to 10–15‰), the main contrast in salinity (density) being found within a zone with a width of 100–150 m. The forming discharge fronts are usually characterized by very sharp gradients of salinity. The following examples may be cited from those given in refs [32, 103]: 0.2‰ per 1 m in the Connecticut river; 0.2–0.4‰ per 100 m in the Elbe and Weser rivers in Helgoland Bay in the North Sea; 0.8–1.0‰ per 100 m in the Hudson River in New York Bay; more than 1‰ per 1 km in the Orinoco river. Still more profound contrasts and sharper gradients are recorded at estuarine fronts. For example, variations of salinity near by the surface in Delaware Bay reached 4‰ per 1 m.

Although the greatest salinity difference in the discharge and estuarine fronts is concentrated in a comparatively narrow zone not far from the coast, the freshening influence of such large rivers as the Orinoco, Amazon, Parana, Ganges, Brahmaputra, and Irrawaddy on the structure of the upper layer is observed hundreds and even thousands of kilometres away from the river mouth areas of the coast. Traces of freshening of the ocean by the discharge of the Amazon, the largest river in the world, are found near the surface at a distance of 1000–1200 km from the delta. The waters freshened by the discharge of the Amazon frequently move away from the coast in the form of large isolated lenses 'floating' on the surface. According to available data, the salinity in these lenses can drop to 31–33‰, i.e. 3–5‰ lower than in the surrounding waters. Waters freshened by the Columbia river discharge in the upwelling area nearby the Oregon coast of North America are carried away by the coastal current far to the south and form the background for the origin of numerous upwelling fronts.

1.3.2. *Coastal upwellings*

The characteristic width of coastal upwelling zones is usually not more than 30–50 miles. The upwelling deep waters are encountered most often within these zones near the surface in the form of isolated cold patches with a salinity equal to that of these waters in the depth and with a characteristic size of 10–30 miles in diameter. Such patches often form along the shelf boundary. A correlation is observed in other cases between the location of the patches and the configuration of the shoreline. Patches of upwelled deep waters and the fronts bounding them appear and are destroyed in a time scale from several days to several tens of days. The stability and sharpness of the fronts confining the patches of upwelling waters depend on the upwelling phase [127] and on the salinity contribution to the density differences across these fronts [103]. Maximum sharpness of the fronts is observed in the upwelling relaxation phase at negative T, S correlation of the space thermohaline variability (the waters rising from the depth have a higher salinity than the surrounding near-surface waters). An important role can be performed by the river discharge into the upwelling zone. Transversal jets, or systems of transversal jets, carrying modified cold coastal waters with less salinity far into the open ocean may arise at upwelling relaxation in the near-surface layer where the upwelling waters initially, or as a result of mixing with the river discharge, experience a certain deficit of salinity (see Section 5.3.2). The actual situations in different regions of coastal upwelling are quite diverse.

The drops in and horizontal gradients of the salinity at the boundaries of the patches are very small in regions where salinity stratification is weak. The drop in salinity, for example, in the region of the Peru upwelling are a maximum of 0.1–0.3‰, and the horizontal gradients are 0.01–0.02‰ per km. The sharpest drops and frontal gradients of salinity are observed in the area of the Oregon upwelling: $\Delta S = 4.5‰$ and $\partial S/\partial x$ reaches 2.0‰ per km. The South African (Cape Town), Brazil (near Cabo Frio Cape), and the Canaries (West African) upwellings occupy an intermediate position. The following drops in and horizontal gradients of salinity are observed in these upwelling regions: in the South African one, $\Delta S = 1.5‰$ and $\partial S/\partial x = 0.1–0.2‰$ per km; in the Brazilian one, $\Delta S = 1.0‰$ and $\partial S/\partial x = 0.05–0.08‰$ per km; and in the Canaries one, $\Delta S = 1.0‰$ and $\partial S/\partial x = 0.01–0.12‰$ per km.

1.3.3. Thermohaline and salinity fronts

Large-scale fronts of climatic origin (e.g. the Oyasio current, subArctic front, etc.) having a typical transversal size of about 100 km and a length comparable to the scales of the ocean per se are characterized, as a rule, by small gradients of salinity, i.e. from 0.01 to 0.025‰ per km. Substantially greater gradients and drops in salinity ($\partial S/\partial x = 0.1–10‰$ per km, $\Delta S = 0.2–1.5‰$) are observed at synoptic-scale fronts which are of unstable character, displace in space, or are of seasonal character. This category includes the above fronts of coastal upwellings and fronts associated with advection of the water on the periphery of synoptic eddies or rings, etc. The typical transversal size of these fronts is 100 m to 10 km, and their length is rarely more than 100 km.

The small number of continuous salinity measurements from a vessel underway makes it impossible to judge the recurrence frequency of purely salinity fronts on the surface of the World Ocean. It is known that salinity fronts do not always coincide with thermal fronts because of the differences in the concrete mechanisms and conditions of frontogenesis. Purely salinity fronts are frequently observed, e.g. in areas of a river discharge. However, they are also found in the open ocean, where they are characterized by unusual sharpness. Salinity fronts are found in the Indian and Atlantic Oceans with horizontal salinity gradients of about 0.4–0.5‰ per 100 m in zones with a width of only 200–250 m. Purely salinity fronts have been found time and time again in the Pacific Ocean with a drop of $\Delta S = 0.5–0.7‰$ and gradients almost equal to 0.07‰ per km which are caused most likely by abundant precipitation. However, on the basis of current data, it may be assumed that the number of salinity fronts in the ocean is much smaller than the number of thermal ones (with and without a salinity contribution).

Thermohaline frontal zones, wherein the space temperature and salinity drops practically coincide, form most often when the sources of space variability of the temperature and salinity act in parallel. A positive and a negative T, S correlation may be observed in these cases at the fronts. The fronts in the area of the Oregon coast of North America, where the upwelling is a source of cold and saltier waters and the Columbia river discharge is a source of fresher and rapidly heating near-surface waters, are examples of fronts with a negative T, S correlation. The negative T, S correlation there gives rise to density fronts with a density drop to 9 units of σ_t.

Fronts with a negative T, S correlation are also encountered in the open ocean. One of these fronts was recorded during the POLYMODE experiment in the northern part of the subtropical convergence ($\simeq 33°$ N) with temperature, salinity, and density drops of 0.5°C, 0.2‰, and 0.45 units of σ_t, respectively. However, according to calculations based on the data obtained during the POLYMODE experiment, these were found in the Sargasso Sea in the summer of 1977 only in 11% of cases of 35 crossings of frontal temperature and/or salinity gradients along a route 1500 km long. Thermohaline fronts were recorded there much more often with a positive T, S correlation (50% of cases) and drops of salinity equal to 0.2–0.4‰ ($\partial S/\partial x = 0.1$–0.6‰ per km) and those of the temperature to 0.2–0.7°C ($\partial T/\partial x = 0.1$–2.5°C/km). Usually, the contributions of the temperature and salinity variations to the variation of the density were not fully compensated, and the fronts were characterized by a small density drop ($\Delta\sigma_t = 0.03$–0.22 units) [103].

As a rule, these fronts were of advective origin and formed at the boundaries of the water 'tongues' that propagated on the peripheries of synoptic eddies. Therefore, the temperature and sanity drops at the fronts were determined by the characteristics of the waters of the afore-mentioned zones of subtropical convergence which was transient between the waters to the north, having in summer a temperature of 25.5–26.5°C and a salinity of 36.2‰, and the Sargasso Sea waters, with mean summer temperatures and salinity of about 28–29°C and 36.5–36.6‰, respectively. The 'background' temperature and salinity gradients observed in this zone were only 4×10^{-3}°C/km and $(1.0$–$1.5) \times 10^{-3}$‰ per km. The same advective mechanism induces the development of considerably sharper fronts in areas where the waters transferred by the eddies differ greatly from the surrounding waters by their T, S characteristics. The frontal zone, which usually forms when the southern periphery of the warm-core ring approaches the Gulf Stream proper, may serve as an example. The sharpness of the fronts in this zone is associated with the convergent inflow of colder and freshened shelf ($S \simeq 32.8$‰) and slope ($S \simeq 34$‰) waters into the Gulf Stream northern boundary area along the eastern periphery of the anticyclonic ring. The 'multifrontality' [103] of frontal zones of this type is associated with the propagation of shelf and slope waters in the form of isolated streams. In this case, the temperature and salinity drops at individual fronts reach 1.5°C and 1.2‰, respectively, and the actual gradients are 10–20°C/km and 5–10‰ per km. The positively correlated contributions of the salinity and temperature do not compensate one another, and the density drops reach 0.7 unit of σ_t (gradients up to 2–3 units of σ_t per km) with a determinative contribution of salinity.

Fronts with a positive T, S correlation are also observed in the cold rings of the Gulf Stream. In this case, the drops in temperature and salinity reach $0.6°C$ and $0.9‰$, respectively, and the maximum gradients $1.0°C/km$ and $0.5‰$ per km. However, such fronts are recorded rather rarely as compared to fronts with a negative T, S correlation and sharp density gradients. The reason for a negative T, S correlation in the near-surface fronts of the Gulf Stream cold ring, which is not associated with the general deformation field of the ring (background large-scale horizontal temperature and salinity gradients in this region are usually correlated positively), is, most likely, the non-uniform wind-driven mixing [103].

1.3.4. *Non-uniform wind-driven mixing*

The intensity of vertical mixing may be different in points withstanding about 10 km from each other. In this case, the temperature in the upper quasi-homogeneous layer may change by tenths per degree and the salinity by tenths p.p.t. Horizontal propagation of the water masses after mixing (overriding the less dense water over the more dense water) may cause either horizontal inhomogeneities of the temperature and salinity in the near-surface layer or the formation of shallow fronts. The correlation sign of the thermohaline inhomogeneities formed depends in this case, on the sign of the product of the gradients $\overline{\partial T/\partial z}$ and $\overline{\partial S/\partial z}$ in the mixing layer. It may be positive or negative in each concrete situation. It is possible that the numerous inhomogeneities of the temperature and salinity with amplitudes of $0.2–0.3°C$ and $0.05–0.1‰$ and more, and with a scale of about 10–30 km, which we recorded in the Sargasso Sea, are of this nature. Wind mixing could also have been the cause of negatively correlated inhomogeneities of kilometre scale ($\Delta T = 0.1°C$, $\Delta S = 0.1‰$,) repeatedly recorded in the vicinity of the frontal zones with a positive T, S correlation of the front proper. They could also be the consequence of water lifts from a small depth in the vicinity of the fronts, compensating the sinking of the waters along the frontal surface. Finally, non-uniform wind-driven mixing could induce the superficial negatively T, S-correlated fronts in the near-surface layer which we observed in the cold ring of the Gulf Stream. The heated near-surface slope waters of the core ($\sigma_t = 23.73$ units) were much lighter than the near-surface waters of the ring shell ($\sigma_t = 23.94$ units) and overrode them, while the situation was quite the contrary at even a small depth.

1.3.5. *Precipitation, ice melting, and evaporation*

Precipitation can exert a rather great influence on the character of the salinity horizontal and vertical distributions in the near-surface layer of the ocean. The depth of the freshened surface layer and the degree of freshening depend, in this case, on the quantity of precipitation, the rain intensity, the intensity of wind (convective) mixing, and the initial salinity stratification on the near-surface layer. Detailed data on the characteristics and space structure of the disturbances in the near-surface layer due to the precipitation effect are given in Section 4.5.3.

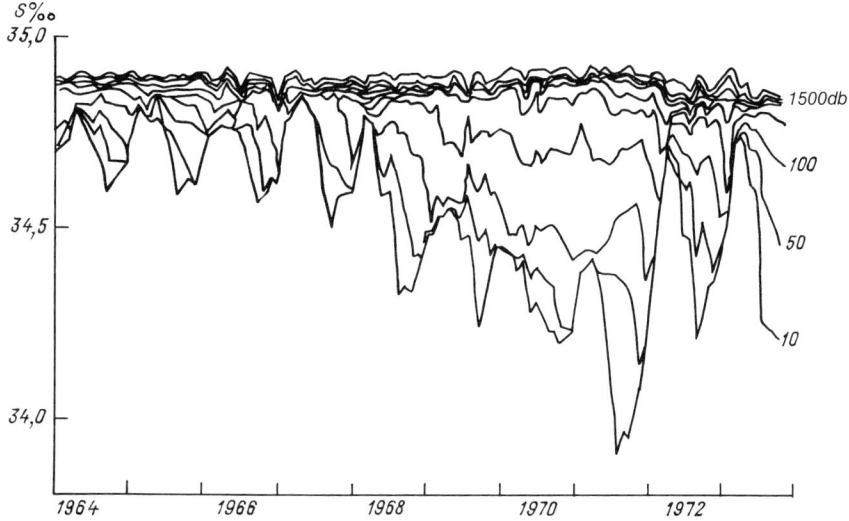

Figure 1.3. Variation of the mean monthly values of salinity S (‰) at 11 horizons at point 'Bravo' in the Labrador Sea in 1964–1973 according to the data in ref. [176].

The decrease in salinity in the near-surface layer of the ocean as a result of polar ice and adjacent land snow cover melting can be quite substantial and can have a long-term effect. According to Lazier [176], who analysed the data of the weather station 'Bravo' (56°30′ N, 51° W) in the Labrador Sea for 1964–1973, the salinity of the near-surface layer in this body of water, which usually fluctuates from 34.6 to 34.8‰, continuously dropped from 1967 and reached a minimum (33.9‰) in the summer and autumn of 1971 (Fig. 1.3). This drop was manifested to a horizon of 100–150 m and caused a decrease in the winter convection penetration depth from 2000 m in normal years to 200 m in 1969–1971. A similar situation was observed in the Norwegian Sea from the weather vessel 'M' in 1956–1957 when a drop in salinity of about 1‰ in the summer of 1957 was detected at the 50 m horizon [157]. It is quite evident that these salinity anomalies in the near-surface layer of the water bodies, where deep waters form, tell later on the T, S characteristics of the deep waters whose thermal and salinity balance is disturbed owing to the long-term lack of annual replenishment.

Evaporation from the ocean surface causes salt enrichment of the near-surface layer which more often propagates to the entire layer of wind-induced (or convective) mixing. Therefore, potential inhomogeneities of the salinity in the near-surface layer during evaporation can form owing to inhomogeneities in the conditions of mass exchange across the ocean–atmosphere boundary (wind-induced inhomogeneities, presence of surfactant films preventing evaporation, etc.) and inhomogeneous depth of mixing propagation. For example, measurements during the 'Polygon-70' programme [110] in a 10 × 10 mile square in an area neighbouring a region of intensive subtropical evaporation in the Atlantic Ocean have allowed us to detect horizontal

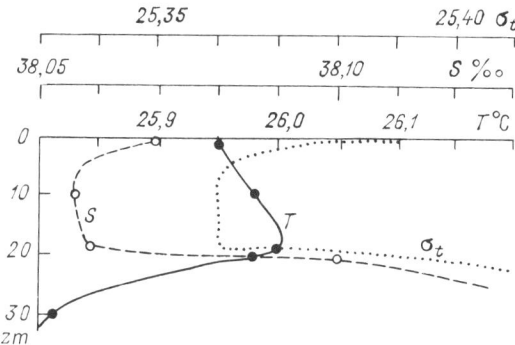

Figure 1.4. Anomalous hydrostatically unstable temperature, salinity, and density distribution in the near-surface layer of the Red Sea; station No. 5249 'Discovery', 1 March 1964, 20°04′ N, 38°27′ E, 2312 GMT.

patchiness of salinity in the UQL with gradients of about 0.01‰ per km. It is also known that the diurnal cycle of evaporation, associated with the diurnal cycle of SST, usually changes the salinity near by the surface with an amplitude of 0.03–0.06‰, while the maximum is at about 1700 local time and the minimum is at night.

When the annual sum total of evaporation exceeds the sum total of precipitation, the salinity of the near-surface layer is usually greater than the mean value for the ocean. The highest values of salinity are observed in the Red and Mediterranean Seas, in the subtropical areas of the Atlantic, and in some coastal, semiclosed water bodies with a specific local regime. In the Red Sea, for example, the total daily evaporation reaches 11 mm [75], which by its absolute value is comparable to the effect of a light steady rainfall. A hydrostatically unstable vertical distribution of salinity, temperature, and density is frequent with such evaporation in a near-surface layer 10–30 m thick (Fig. 1.4). The salinity, which at the sea surface reaches values in excess of 38‰, decreases with the depth by 0.02‰ in a layer several metres thick. Correspondingly, the temperature inversion can reach 0.05°C, and the inversion of density in this layer may exceed 0.03 units of σ_t. Apparently, the existence of this inversion is maintained by the permanent source of salt (evaporation) on the surface even under comparably intensive mixing (wind velocity 10 m/s). Salt enrichment of the near-surface layer near the Red Sea coast to 38.5–39.0‰ owing to continuous anomalous high evaporation induces sliding down of saline and well-heated waters first along the bottom slope and then along the sloping isopycnic lines toward the open sea where quite typical temperature inversions, density compensated by a rise of salinity, form in autumn at the intermediate horizons (40–70 m).

As a rule, individual heavy rains or non-uniform evaporation do not induce the development of profound frontal interfaces. But the seasonal predominance of precipitation over evaporation, or evaporation over precipitation in some regions of the ocean in the presence of convergence currents, or eddy formations near the surface at the boundaries of these regions may induce the formation of sharp salinity fronts.

1.3.6. *Certain general consistencies*

The analysis of sources of salinity variability in the near-surface layer of the ocean, discussed in this section, covers only mechanisms that are currently known, and apparently it is incomplete. It is not known, for example, what kind of 'trace' is left behind in the salinity field by hurricanes and typhoons because this 'trace' is characterized by abundant precipitation, intensive evaporation, upwelling, and wind mixing, and therefore also disturbances in the background salinity stratification. Nevertheless, on the basis of available information, it is possible to form an idea on the range of synoptic variability of the salinity in the near-surface layer in different areas of the ocean. The greatest variability (several per mille) can be anticipated in the coastal regions of the ocean, i.e. in areas of large river discharge, upwellings, and ice melting. Salinity drops of 1–1.5‰ are observed at fronts formed by the advection of alien waters on the peripheries of eddies (rings) which have separated from large currents. The same mechanism of water advection on the peripheries of synoptic eddies induces variations in the salinity within only 0.5‰, mainly due to the meridional gradient of salinity.

Despite the current lack of a complete statistical analysis of the space variability of salinity in the near-surface layer because of the lack of a necessary bank of data, the available data on the repetition frequency of thermal, thermohaline, and purely salinity fronts [103] and the published spectra of salinity make it possible to assume that the main characteristics of the space variability of salinity and temperature in the near-surface layer of the open ocean are similar in the scales of synoptic eddies and fronts. Therefore, it is possible to anticipate the detection of comprehensive contrasts (0.3–0.5‰) and frontal gradients of salinity against a common homogeneous background approximately every 5–10 km in coastal regions, every 50–100 km in areas of climatic frontal zones in the open ocean (zones of western boundary currents and subtropical convergences), and every 500–1000 km in central parts of the open ocean. However, it has become quite evident, currently, that the natural small-scale patchiness of salinity near the ocean surface is manifested appreciably sharper than the small-scale patchiness of the temperature. Sometimes inhomogeneities of salinity with an amplitude of 0.1‰ and a scale less than 1 km are recorded against the background of an entirely homogeneous (within the accuracy of measurements) distribution of the temperature, both horizontal and vertical [110, 223]. In other cases, the characteristic dimensions of the space inhomogeneities in the salinity field are substantially (ten times and more) less than the characteristic dimensions of the space inhomogeneities in the temperature field which is often visible by eye (Fig. 1.5). It is quite obvious that this difference is connected one way or another with the differences in the rates of molecular heat conductivity and diffusion that begin to tell at mechanical fractionation of larger thermohaline inhomogeneities in the near-surface layer into smaller ones.

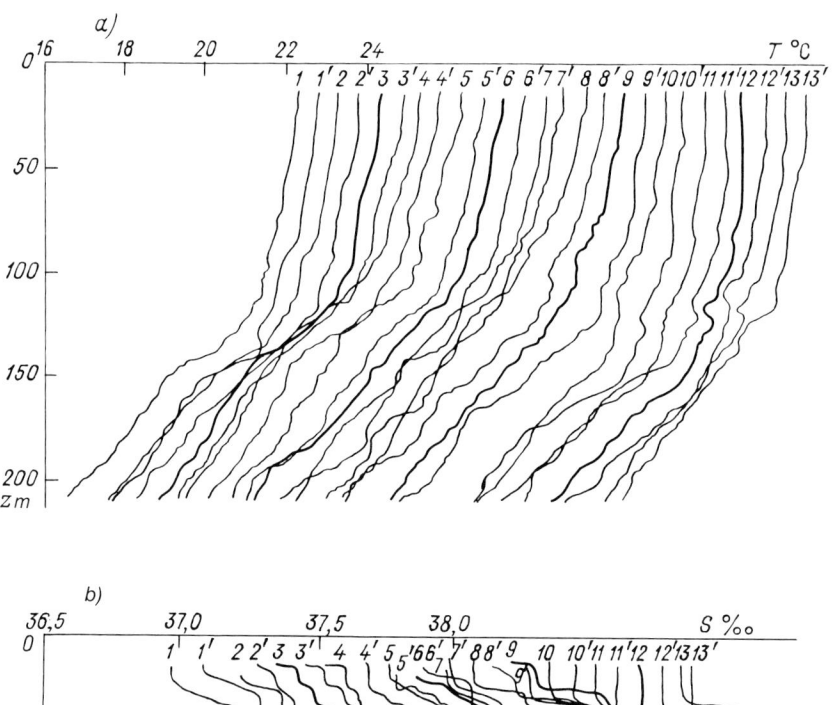

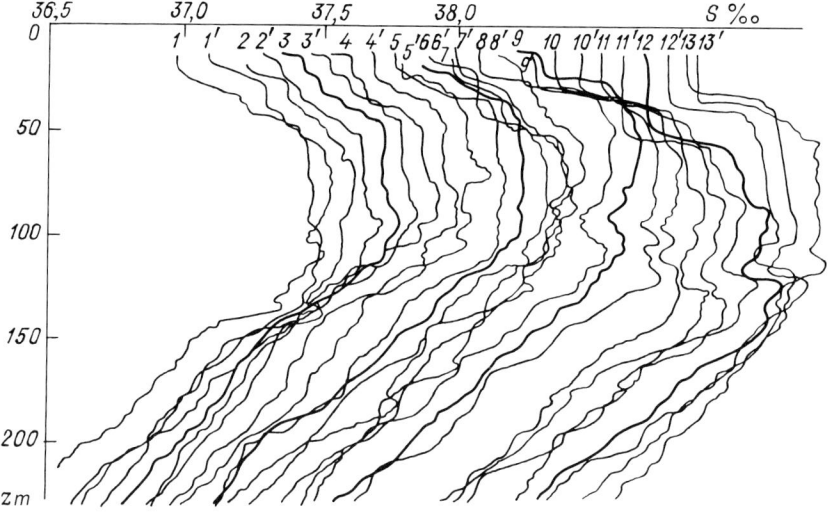

Figure 1.5. Vertical profiles of temperature $T(z)$ (a) and salinity $S(z)$ (b) obtained by the precision probe 'Mark-III' in 'sawlike' mode from a vessel underway. The distance between the adjacent profiles (number without hachure—downward; with hachure—upward) is approximately 2 km (vessel speed 8 knots) at the horizon 150 m. The scales are given for the first profiles. The rest are displaced to the right by 0.5°C and 0.075‰ each. Sharp variations in salinity are visible (up to 0.1–0.2‰ between adjacent profiles) which are not accompanied by commensurable temperature variations. Measurements made by V. T. Paka during the tenth cruise of the research vessel *Mstislav Keldysh* in the Tropical Atlantic in May 1985.

1.4. Methods of studying the near-surface layer of the ocean

Standard devices or instruments practically never existed for hydrophysical measurements in the near-surface layer of the ocean. Researchers interested in the behaviour of the ocean near the free surface usually had to invent new equipment or adapt available equipment, and to elaborate respective methods of measurements. In the majority of cases, the most original devices and methods for the solution of these problems have been developed and applied in the USSR where interest in these problems have not diminished from the very first steps of Soviet oceanology. As far back as 1925 Shuleikin advanced a method and developed a device for measuring evaporation from the seawater surface on board a vessel. Samoilenko performed such measurements by means of this device in 1925 from the vessel *Persius*, and Shuleikin did the same in 1926–1927 from the vessel *Transbalt* [119]. Altberg and Popov, who studied supercooling of the near-surface 1 m layer at freezing on the Neva river, made the first measurements of temperature distribution with a vertical resolution of less than 1 mm by moving a wire sensor of 0.025–0.05 mm diameter from below upwards. The measurements showed an appreciable difference in the surface temperature from the temperature at horizons deeper than 1 mm [2]. In 1949 Arkhipova and Rzheplinsky published original observations of convection in a water reservoir, using chaff as a tracer on the water surface [5]. By means of a string of thermocouples, built-in in the underwater part of light floating masts with floats, Preobrazhensky recorded in 1962 vertical temperature profiles in the near-surface 0–50 cm layer in the Baltic Sea, showing occasionally intermediate temperature maxima at horizons 10–20 cm below the free water surface under conditions of solar heating [87]. Similar measurements were repeated 10 years later by American researchers who used a modified expendable bathythermo-graph [128].

In the mid-1970s, Vershinsky and Solovyev [92] elaborated a special probe, slowly emerging on 'command' from the vessel from a horizon of about 10 m, and used it to study the thermal structure of waters close to the ocean surface. The thin needle-like temperature sensor at the top streamlined end of the profiler did not disturb even very thin layers of water during upward movement, and, upon emerging, pierced the free surface without any disturbances. Later, Solovyev equipped this probe additionally with sensors of electrical conductivity and turbulent fluctuations of velocity [9]. In 1978 Paka developed an original and simple 'Boomerang' probe which he used in the POLYMODE programme from the research vessel *Akademik Kurchatov*. This probe is thrown on a thin cable beyond the zone disturbed by the hull of the vessel and is held close to the water surface to record the temperature in a thin upper layer (5–8 cm), and then it sinks slowly to a depth of 10–20 m. The vertical thermal structure of the near-surface layer under extreme solar heating can be studied only by means of these types of device.

Specific interest in the thermohaline boundary layer near its surface developed alongside general interest in the near-surface layer of the ocean. It is worth mentioning that the first observations of water surface cooling at evaporation and contact heat exchange with the atmosphere and with the formation of what is often called 'a cold film' or 'a skin layer' were made under field conditions [238]* long before it was possible to carry out laboratory experiments correctly. The principal interest of the researchers was directed, in the first place, to studying the value of cooling ΔT_0 (see Fig. 1.1) in the boundary layer under different conditions. High-speed instrumentation was created by Khundzhua *et al.* only at the end of the 1960s and the beginning of the 1970s for the purpose of recording the temperature profile in a thin, 0–30 cm, near-surface layer of the sea by vertical sounding. The first $T(z)$ profiles were obtained with the use of these devices in the cold thermal boundary layer under the natural conditions of the open sea [116]. Nobody could reliably record for many years the near-surface salt enrichment ΔS_0 due to evaporation from the free surface. Only at the end of the 1970s did we, together with Vlasov and Ambrosimov [106], manage to measure the characteristics of the salt boundary layer in the laboratory by means of a laser interferometer (see Section 3.3.4).

Neither of these methods taken separately can produce a full idea of the near-surface layer of the ocean. A complete set of means and devices is necessary for a comprehensive study of the near-surface layer, as well as some common approach combining various contact and remote methods. It is necessary to employ research vessels, planes (or helicopters) and satellites, and the instrumentation should comprise all the available means for measuring the local radiation balance, heat and moisture fluxes at the water–air interface, structures of the thermohaline boundary layer, and of the entire near-surface thickness, including the diurnal thermocline, as well as the vertical distribution of the turbulent fluctuations of the velocity and the optical transparency of the water; the measurements of evaporation and precipitation should be very accurate as well as their mapping; recording of the temperature and salinity at several levels, including those nearby the surface (5–20 cm) when the vessel is underway or while drifting, should accompany all the other measurements. Remote radiometric and radar means for studying the sea surface condition should have a range of resolution in space from 1–10 cm (from aboard the vessel) to 10–1000 m (from a satellite). As far as we know, such a complete set has not been used anywhere; therefore, one may say that systematic study of the physics of the near-surface layer of the ocean is only beginning.

Apparently, platforms, masts, buoys-laboratories, weather vessels, or light vessels, arranged at a substantial distance from the shore, are good as stationary scientific bases for studying the near-surface layer. Their common advantage is the possibility of using any period of time suitable depending on the weather conditions, and of carrying out the necessary measurements in a wide range of local conditions of ocean–atmosphere interaction. Small vessels, launches, and even boats should be used in combination with these bases, being equipped

---

*The stated measurements by Altberg and Popov [2], made by more sophisticated means and a decade earlier, were not measurements of the characteristics of an instable cold film because they were done in fresh water at temperatures below the temperature of maximum density (see Section 3.1).

with devices (e.g. an emergent profiler) that are best applied from small vessels. When drifting or cruising at low speed, these small vessels hardly disturb the near-surface layer under study. Unfortunately, today, at a time of comfortable research superliners and long-range cruises, it is hard to convince the administration and the captain of a vessel of the necessity of equipping and lowering a boat or a launch for oceanologic work even in calm tropical waters. As regards this, one can recall that the hydrophysicists of the research vessel *Akademik Kurchatov* as far back as 1982 for several hours tried in vain to determine by means of a CTD probe the site of emergence to the ocean surface of a sharp thermohaline front several miles NW from the extremity of Isabela Island (Galapagos Islands). The front approached the surface at such a light slope that the vessel, having a draught of 6 m, when reversing the course (to stop for sounding), each time mixed with its screws the entire stratification associated with the front, and made it practically impossible to obtain the necessary result. It was impossible to make measurements near the emergence of the front to the surface.

When discussing a specialized set of means and devices for studying the near-surface layer, it is necessary to stress that its complete composition and methods of utilization should depend on the list of concrete problems being solved in each particular case. Each problem demands its own combination of devices, methods, and sequence of measurements with accurate referencing to local solar time (LST). However, the solution of any problem, or any combination of problems, would require a most complete set of contact measurements of the main hydrophysical characteristics from one or several vessels in the upper layer with a minimum thickness of 100 m. The latter measurements include:

- reconnaisance and background surveys by means of frequent launches of XBT or AXBT;
- frequent CTD-sounding to 50–100 m;
- towing temperature and electrical conductivity sensors or CTD probes on several horizons chosen in conformity with the problem; a towed probe of the 'bat-fish' type or a thermal trawl is an alternative;
- sounding with an emergent profiler;
- a complete set of hourly meteorological and actinometric measurements.

Much of this can be done in parallel with other work. This is always possible even on a vessel of modest size because the required measurements are usually carried out to a depth of not more than 100 m and can be done either underway or at stations with simultaneous deep-water measurements and sampling without any risk of entangling or losing the devices. Our experience demonstrates [24, 102, 108, 109] that it is possible to use in this case also some standard devices having respectively modified the methodology of using them. In particular, the common CTD probe makes it possible to obtain a continuous record of the measured temperatures and salinity of the water at some horizon in the near-surface layer of the ocean in the towing regime from a moving vessel or in a shipboard running-water system [108].

Another modification of the operating mode of a common CTD probe produces reliable and detailed vertical temperature and salinity profiles in the near-surface layer, beginning with the horizon 0.5–0.7 m below the surface and down to 10–20 m with reasonable resolution by the vertical and some estimation of variability at different horizons. Certain conditions should be maintained in this case:

(1)   the measurements should be carried out in waters that have not been disturbed by the hull of the vessel during its movement or drift. Therefore, sounding should be done either on the leeward side (as distinct from common deep-water sounding) or during low speed of the vessel under momentum at a maximum of 1 knot after short-term operation of the marine engines;

(2)   to reduce to a minimum the distortions in the profiles because of the vessel's pitching and rolling, it is necessary to sound with successive lags of the sensor at the work horizons every other 0.5–1 m in the course of time exceeding at least 4- to 6-fold the pitch and roll periods. Subsequent averaging of the values measured at each horizon produces average profiles $\bar{T}(z)$ and $\bar{S}(z)$, and the limits of variability associated greatly with the vertical gradient and amplitude of the pitch and roll. An example of these vertical profiles is given in Figs 4.10 and 4.11 in Section 4.5.3.

Chapter 2

The surface of the ocean and its anomalous conditions

2.1. The role of the free surface of the ocean in the formation of the structure and regimes of the near-surface layer

The role of the free surface of the ocean in the formation of the structure and various regimes of the near-surface layer variability becomes clear if it is imagined for a moment that it can be replaced by a thin, solid heat-conductive cover which is opaque for solar light. The following will occur in this case:

- the kinetic energy of the wind will not be transferred to the upper layers of the water;
- there will be no sea waves, Stokes transport, drift currents, wave–wind turbulence, Langmuir cells, various effects of surfactant films, or any other various surface and near-surface manifestations of the water dynamics;
- there will be no gas exchange through the surface;
- there will be no salinization due to evaporation or freshening due to precipitation;
- the processes of photosynthesis will not function;
- solar heat will be absorbed strictly on the surface, i.e. by the cover proper with subsequent contact heat transfer to deeper layers;
- free convection, whose origin is in the cold boundary layer, will be the sole active factor in the regime of cooling;
- the boundary layer will always be warm and hydrostatically stable in the regime of warming, and heat transfer will occur due to molecular heat conductivity;
- the diurnal thermocline will be essentially thicker and the absolute values of ΔT_D will be much greater.

Summing up this rather fantastic proof by contradiction, it can be stated that the near-surface layer of the ocean, as we observe it, exists only due to the specific properties of the water surface just as a free surface. Certain consequences, resembling those stated above, are actually found under an ice cover. But, in the latter case, an impulse is transmitted to the water by the movement of the ice, salinization can occur when it freezes up, and solar irradiance can penetrate practically into the water even through a substantial thickness of the ice cover. At the same time, the daily cycle of heating and cooling in the water layer under the ice can be sufficiently well manifested only in spring and summer. Therefore, the analogy with a solid cover is incomplete even in this case. At the same time, the above arguments serve as sufficient grounds for anticipating, for example, quite essential specific features in the dynamics and structure of the near-surface layer of the ocean in the boundary zones between ice cover and an ocean surface free of ice (see Section 5.2).

Anyone who observes the sea surface carefully for a long period of time knows well that it is quite variable. Alongside its normal conditions, which are characterized by well-known scales and tables (see Tables 2.16–2.29, pp. 227–243 in ref. [80]), a plurality of unusual (anomalous) conditions are known. It is logical to assume that the near-surface layer, which is closely associated with the ocean surface, also behaves differently and has a varying structure under diverse conditions of the sea surface. As a matter of fact, this was shown in Section 1.2 when indicating the main physical regimes of the near-surface layer. On the other hand, it is natural that certain characteristics of the near-surface layer exert an influence on the condition of the surface. When determining all these feedforward and feedback effects, we are attracted, first of all, to the anomalous variability in the properties of the near-surface layer of the ocean which is manifested best in calm and light-wind weather (i.e. approximately to a sea condition of force 3–4), when anomalous phenomena are more often observed also on the ocean surface. A brief acquaintance with these phenomena is necessary for a better understanding of the contents of the subsequent chapters in this book.

2.2. Anomalous conditions of the ocean surface

Waves on the sea surface are not an anomalous condition by themselves. Rather, a condition of full calmness should be considered to be quite a rare exception to the rule, especially in the open ocean. However, likewise a calm surface of the ocean may not display any specific anomalous features. On the other hand, steep waves with foamy whitecaps in practically calm weather is an obvious anomaly indicating a certain condition of the near-surface layer, and, therefore, it should be considered. Anomalous steep waves associated with the influence of high-speed currents, e.g. in the zone of a strong shift of the speed at the front, also arrests one's attention by their space inhomogeneity and singularity. Long-term observations of the sea surface state has made it possible for navigators and oceanologists to elaborate on the basis of their experience certain ideas on what precisely is the norm under various wind conditions. Seamen use a scale of the sea surface condition (Beaufort scale). Oceanologists, in principle, should be able to state exactly which energy spectra of waves should be considered normal (typical) for given concrete conditions of the wind, its duration, fetch, and sea depth, and which spectra are deviations from the norm or are anomalous states. Any deviations of the sea surface condition from the accepted norm should, in turn, be of interest from the point of view of the probable role of the deep-sea processes in forming various characteristics of the surface.

Anomalous states of the surface, observed in the ocean, are diverse and are most often conditioned by the simultaneous action of several hydrometeorological and hydrophysical factors. The complexity of anomalous surface phenomena, e.g. current rips, is associated with various combinations of the effects arising in this case (see Sections 2.26 and 6.3). Nevertheless, a comparatively small number of well-studied hydrodynamic effects form the basis of all the anomalies of the sea surface. They can be listed as follows:

(1) non-wind generation of wave motions on the free surface of the fluid (e.g. at opposite ends of the spectrum, i.e. lee waves and tsunami);
(2) refraction and diffraction of wind waves (on the specific features of the bottom topography and, probably, on the large-scale specific features of the pycnocline closest to the surface);
(3) addition or interaction of various wave systems, including those subjected to refraction and diffraction;
(4) modulation of the amplitudes and length of the wind and non-wind waves on space inhomogeneous currents, including those generated by internal waves;
(5) phenomenon of blocking wind and non-wind waves on contrary currents as an extreme case of their modulation (item 4);
(6) shear instability and resulting secondary structures in boundary (frontal) zones of strong currents (small vortices, whirlpools);
(7) phenomenon of hydraulic jump or shock in estuaries, channels, or narrow passages with a varying depth and a strong current (turbulent and wave bore and their surface manifestations);
(8) space inhomogeneities in exchange of kinetic energy between waves and turbulence;
(9) interaction of periodic (wave), convective, and progressive motions with formation of cellular circulations (Langmuir circulation, windrows).

Surface phenomena corresponding to these hydrodynamic effects or their combinations will be discussed in Sections 2.2.1–2.2.7. Geographical aspects relating to the frequency and conditions of their detection will be considered in Section 2.3. The modern physico-mathematical interpretation of many of these effects is discussed in refs [73, 85], but interesting ideas are also found in some earlier publications [111, 180, 181, 226, 230]. In particular, a very brief physical interpretation of the varieties of tidal bore can be found in an article by Lighthill [179], and Langmuir circulations are considered not only in ref. [73] and in the references cited there, but also in the works of Ryanzhin (e.g. refs [89, 90]). We would like to discuss here only some of the effects that are considered, in our opinion, insufficiently in the oceanological context.

Lee waves. Lee waves, generated in gaseous and liquid media when flowing round obstacles, represent an effect deserving special consideration. These waves are observed often in the atmosphere, where they are manifested in the form of wavy banks of clouds, and on the lee side of mountain ridges at stable stratification of the air flow crossing over the ridge. A physical description of lee waves is given in ref. [95]. Lee waves appear on the free surface of water when the current flows round an underwater obstacle (banks or shoals). A concrete example of this problem is given by Lamb [62] in the simplest linear statement.

Lee waves are stationary in the classical case, i.e. their crests do not propagate down the currents in relation to the underwater obstacle that generated them. Their phase velocity $c = \omega/k$ is equal in value and opposite in sign to the velocity u of the overflowing flow:

$$c_1 \equiv c + u \equiv \omega/k + u = 0. \tag{2.1}$$

Here, c_1 is the phase velocity of the waves in relation to the obstacle; c is the phase velocity of the same waves in relation to the overflowing flow; ω is the angular frequency; and k is the wave number ($k = 2\pi/\lambda$). Relationship (2.1) also determines all the other relationships of the wave parameters for which $kh_1 = (2\pi/\lambda)h_1 > 1$, where h_1 is the total depth beyond the underwater obstacle, namely:

$$\text{wave length} \quad \lambda = 2\pi u^2/g \tag{2.2}$$

$$\text{wave number} \quad k = g/u^2 \tag{2.3}$$

$$\text{frequency} \quad \omega = (gk)^{1/2} = g/u, \tag{2.4}$$

where g is acceleration due to gravity.

If h is the height of the obstacle and b is its width in the direction of the flow, then the amplitude of the generated waves is proportional to the product hb and is maximum at $kb \simeq 1$. The stated relationships are true if the flow round the obstacle is precritical*, i.e. when the internal Froude number

$$F = u/C_0 = u/(gh_1)^{1/2} \tag{2.5}$$

is less than unity. Here, C_0 is the phase velocity of free long waves in a homogeneous layer of water with depth h_1. It follows from (2.5) that in the case of lee waves $c < C_0$ and λ are always less than λ_0, i.e. the length of free long waves. The group velocity of the lee waves c_g being less than the relative phase velocity c, as in the case of all the gravity waves, conditions the downstream drift of the energy of the short-wave oscillations generated by the obstacle. Only free long waves can propagate upstream in this case, which usually ensures a smooth surface over the obstacle and further upstream. When the flow is in a supercritical condition ($F > 1$), the free long waves are unable to propagate upstream and the influence of the obstacle is sensed only downstream and not necessarily in a wave form, e.g. in the form of turbulization.

As shown by Lamb [62], in the simplest case, the flowing round of a transversal cylindrical obstacle of height h by a homogeneous flow with velocity u over a flat bottom of constant depth h_1 (see the diagram in Fig. 2.1) at $h < h_1$ should generate some local perturbation of the level which is accompanied by a semi-infinite train of stationary periodic waves downstream the obstacle. For example, at $h_1 = 1.5\,\text{m}$, $h = 1\,\text{m}$, and $u = 1.5\,\text{m/s}$ (common velocity of tidal current in shallow waters), we obtain, according to (2.2)–(2.4), the following character-

---

*A flow round an obstacle is always precritical at $h_1 \geqslant 1\,\text{m}$ and $u \leqslant 3\,\text{m/s}$ (i.e. under common oceanic conditions).

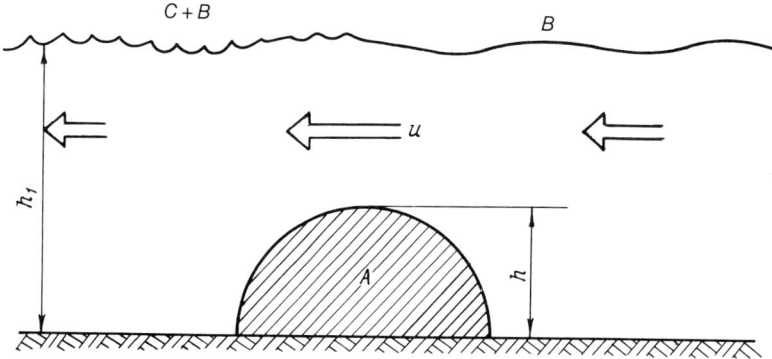

Figure 2.1. Diagram of the flow round an underwater obstacle (A) having a cylindrical section; (B) long waves propagating upstream (see arrow); (C) lee waves.

istics of the lee waves: $\lambda = 1.44$ mm $k = 4.36$ rad/m ($k' = 1/\lambda = 6.9 \times 10^{-1}$ m^{-1}) and $\omega = 6.54$ rad/s ($\omega' = \omega/2\pi = 1.04$ Hz). Although, as follows from observations, the solution of such problems in a linear way underrates the values of the wave amplitudes a, when calculating using the respective formula of Lamb we obtain $a = 4\pi k h^2 e^{-kh_1} \simeq 8$ cm and steepness $2ak' = 0.11$. It is enough to conclude that lee waves should differ little in their characteristics from light or moderate wind waves. Consideration of non-linearity would make lee waves shorter and higher and, consequently, steeper, which at the same h_1 and h would give an utmost steepness ($2ak' \simeq 0.14$), i.e. wave breaking. Apparently, at the chosen values of h_1 and h lee waves are, in general, always unstable and break easily, while at large h_1 (e.g. 3–4 m) their amplitudes (even taking into consideration non-linearity within the Stokes approximation) should be very small due to the factor e^{-kh_1} in the expression for the amplitude.

There are no reliable examples of field measurements of lee waves in the literature. In the case where the presence of these waves was assumed (see, for example, ref. [8]), the actually measured characteristics of the waves did not correspond to the relationships (2.2)–(2.4), namely, the dominant frequencies were several times lower than those prescribed by expression (2.4) at the observed velocities of the tidal current. A qualitative description will be given in Section 2.2.3 of what can be interpreted as patent lee waves. In another case, described in Section 2.2.2, there was, most likely, composition or interaction of lee waves with light wind waves, including the effect of blocking the latter on a contrary current which intensifies over an obstacle. Apparently, a similar situation is observed very often in nature which prompts us to give special attention below also to the effect of blocking.

Effect of blocking at modulation of surface waves by space inhomogeneous currents and internal waves. The blocking effect arises as a limit case of modulating surface waves by a space inhomogeneous current when the latter attains such a velocity that the energy of the shorter surface oscillations cannot propagate in the direction opposed to the vector of water movement velocity, i.e. upstream. Blocking occurs under condition (2.1) in the case of lee waves.

However, in reality, the condition

$$c_g \leqslant |u| \tag{2.6}$$

is sufficient for this effect because the wave energy propagates at group velocity c_g. In principle, the boundary of the blocking area should be determined by the condition

$$c_g = -u, \tag{2.6a}$$

because in the simplest case we deal with the propagation of waves on a contrary current.

The external manifestation of blocking consists in the fact that waves (e.g. wind waves) meeting a contrary current become short and steep, and break without propagating beyond the boundary when approaching the area at whose boundary condition (2.6a) and within which condition (2.6) are carried out. This effect, described apparently for the first time by Unna in 1942 [230], served later as the basis for the working out of a proposal by Taylor [226] to create jet breakwaters for the protection of port buildings. Unna arrived at his conclusions on the basis of a purely kinematic approach, ensuring the greatest simplicity of reasoning (see the discussion in ref. [150]). We will give this argumentation here because it will be useful in several cases discussed below and in Chapter 6.

It follows from kinematic arguments for waves on a deep water surface that

$$c^2 = \lambda g/(2\pi) \tag{2.7}$$

[cf. (2.2) at $c = |u|$]. In its turn, this makes true the relationship

$$\lambda_0/\lambda_1 = (c_0/c_1)^2 \tag{2.8}$$

for any two systems of waves propagating on the surface at different phase velocities. In the general case, the periods of these waves $T_1 = \lambda_1/c_1$ and $T_0 = \lambda_0/c_0$ on still water will be different ($T_1 \neq T_0$). If c_0 is the phase velocity of waves on still water, and c_1 is the phase velocity of the same waves on a current with velocity u, then their period in a fixed coordinate system should be preserved ($T_0 = T_1 = T$); therefore,

$$T = \lambda_0/c_0 = \lambda_1/c_2 = \lambda_1/(c_1 + u) = \text{constant}, \tag{2.9}$$

where $c_2 = c_1 + u$ is the phase velocity of the same waves distorted by the current in a fixed coordinate system.

It follows from (2.9) that

$$\lambda_1/\lambda_0 = (c_1 + u)/c_0, \tag{2.10}$$

i.e. waves on a contrary current ($u < 0$) should become shorter and longer on a following current ($u > 0$). It should be noted that relationship (2.10) holds when the blocking condition [(2.6) or (2.6a)] is not reached, but modulation of the waves by the current takes place. Equating (2.10) with (2.8), we obtain a square equation in relation to (c_1/c_0):

$$(c_1/c_0)^2 - (c_1/c_0) - u/c_0 = 0, \tag{2.11}$$

whose solution is

$$c_1/c_0 = 1/2 \pm \sqrt{1/4 + u/c_0}. \tag{2.12}$$

The phase velocity of waves on a blocking current c_1 at $u/c_0 = -1/4$ is equal to half their phase velocity on still water:

$$c_1 = c_0/2. \tag{2.13}$$

Since the group velocity of the gravity waves on deep water $c_g = c_1/2 = c_0/4$, the root expression in (2.12) is the blocking condition

$$-u = c_0/4 = c_g. \tag{2.14}$$

It follows from (2.13) and (2.14), as well as from (2.10) or (2.8), that the wavelength λ_1 on the blocking current should become four times shorter. It does not occur in reality because the amplitude also changes. Longuet–Higgins and Stewart [180, 181] have shown that the amplitude modulation of the waves on a current is described in the simplest case by the formula

$$(a_1/a_0)^2 = (1 + 2u/c_1)^{-1}, \tag{2.15}$$

where a_1 is the modulated amplitude and a_0 is the amplitude of a harmonic wave undisturbed by the current. Modulation of the phase velocity by a current is determined by expression (2.12), substitution of which in (2.6), with consideration of $c_0 = g/\omega$, results in

$$\left(\frac{a_1}{a_0}\right)^2 = \left[1 + \frac{4\omega u/g}{1 + (1 + 4\omega u/g)^{1/2}}\right]^{-1}, \tag{2.16}$$

whence it follows that different components of the frequency spectrum are transformed differently by the same current. Hence, both the change of the wavelength and the change of the amplitudes contribute to their attaining a limit steepness and breaking near the blocking line. As long as the typical steepness $a/\lambda \sim ak$ of developed wind waves is equal to approximately 0.06, it is enough to increase it only 2–2.5 times to exceed the limit steepness, which is equal to 0.14 in theory and to 0.11–0.12 in observations [65]. Accordingly, to initiate wave breaking it is enough for the wave amplitude a and the wavenumber k to increase only $\sqrt{2}$ times on the contrary current, which explains the commonly observed wide bands of breaking choppy water adjoining the blocking lines.

The mechanism of interaction of the surface waves with the large-scale internal ones in the case where the orbital movement, conditioned by the latter, reaches the sea surface is physically similar to the mechanism discussed above. Formal complications appear because the space inhomogeneous disturbance of the velocity $U = U(x)$ is now transferred in relation to the fixed coordinate system with a phase velocity of the internal waves C. In the general case, it is possible to state here modulation of the sea waves pattern, the degree of modulation being dependent on the relationship c_0, C and U, where c_0 is, as before, the phase velocity of the surface waves on the length $(x = x_0)$ where $U(x_0) = 0$. As shown by Phillips [111], modulation of the surface waves is greatest when the local group velocity of these waves coincides with the phase velocity of the internal waves ('condition of synchronism' or 'resonance condition'). The effect of blocking on the internal waves is reached at points where $c/2 + U - C = 0$ and the phase velocity of the surface waves is $c = c_0^2/[2(c_0 - C)]$, and, in particular, should never be reached at mutually opposed propagation of surface and internal waves. It is known from observations that typical values of C are of the order of several metres per second, while the values of U rarely exceed 0.1 m/s. The degree of modulation of the surface waves by internal ones, which is commonly expressed by the relation of dispersion of the surface slopes $\sigma^2(\varepsilon)/\sigma^2(\varepsilon_0)$, where ε_0 corresponds to areas in which $U = 0$, depends on the relationship $U(x)/C$ so that $\ln(\sigma^2(\varepsilon)/\sigma^2(\varepsilon_0)) \sim U(x)/C$ [73, 85, 162]. This means that the steepness of the waves increases in areas where $U(x)$ and C have the same sign, while smoothing occurs in areas with $U(x)$ and C directed in the opposite directions. When $U(x)/C \geqslant 0.2$, there is an increase of dispersion of the wave slopeness approximately e-times as compared to undisturbed waves. It is often enough for the waves to reach limit steepness and for the appearance of whitecaps which indicate breaking waves. If the phase of the internal waves is judged by the pycnocline elevation, then for internal waves of the lowest mode the maximum values of $U(x)$, coinciding in their direction with C, are observed over the troughs (lowering of the pycnocline), and the opposite extrema of $U(x)$ above the crests (rises of the pycnocline). Accordingly, areas of maximum steepness of the surface waves (bands of choppy water; see Section 2.2.3) are observed most often over troughs or rear slopes of internal wave crests (see article by Basovich, Bakhanov, and Talanov in ref. [85]). However, this connection can be of diverse meaning in the case of multimode composition of the internal waves, which explains why different authors observed different positions of the bands of slicks and choppy water zones in relation to the phases of the internal waves, although in all the cases the distance between the adjacent zones of choppy water or bands of slicks was preserved, being equal to the length λ_m of the main modulating internal wave. The most contrasting patterns of modulation of the surface wind waves by internal waves is observed up to 4–5 m/s and are associated with waves of the decimetre and metre ranges.

The modulating effects of currents and internal waves are complicated by the fact that the surface waves can propagate at an arbitrary angle to them and can reflect from the blocking line. It is indicated in ref. [150] that in the end all the waves meeting a current at an angle less then $\pi/2$ right- or leftward from a strictly contrary direction gradually turn around to meet the current and are blocked by it when the projection of their group velocity to the direction of the current becomes equal and opposite to the vector of the current. It is quite obvious that in this case $|u| < c_0/4$ because $-u = c_0 \cos \theta/4$, where $\theta < \pi/2$, but at the time of blocking it never actually exceeds 35.3°. Table 2.1, adopted from ref. [150], presents the ranges of lengths λ_b of the blocked waves for various values of u_b of the blocking current velocities. The minimum length of the blocked wave in each gradation corresponds precisely to an opposing current,

Table 2.1
Dependence of the length of blocked surface waves on the
velocity of blocking current (data from ref. [150])

Velocity u_b (m/s) of blocking current	Length λ_b (m) of blocked waves
0.25	0.16–0.24
0.50	0.64–0.96
0.75	1.44–2.15
1.0	2.56–3.83
1.25	4.00–5.98
1.50	5.76–8.62

and the maximum length to deviation of the waves from a contrary current by
$\pm 35.3°$.

The minimum wavelengths in each gradation, stated in Table 2.1, correspond
to the phase velocity of the surface waves on the current $c_1 = \sqrt{g\lambda_b/(2\pi)} = 2u_b$
as given in equations (2.13) and (2.14). As far as the reflection of the waves from
the blocking lines is concerned, the new numerical non-linear model of Witting
[237] seems to indicate that this effect is unimportant. At the same time, the
model confirms the full adequacy of the linear kinematic model considered
above [150] from the point of view of predicting quantitative aspects of the
blocking effect.

In regard to the effects discussed, it is necessary to stress the double role of
the bottom topography in forming modulated (anomalous) patterns of the
surface waves. On the one hand, the underwater shoals and banks themselves
can generate steep lee waves with characteristics close to those which, in
conformity with Table 2.1, are blocked by space inhomogeneous currents and
internal waves. On the other hand, when currents flow round underwater
obstacles, the latter cause local intensification of the currents, which leads to
modulation or blocking of weak wind waves with lengths of the metre range.
The generation of modulated conditions of the surface is also possible without
any participation of the bottom topography. As given in ref. [107], anomalous
conditions of the surface are observed often in regions of the ocean with a very
great depth where the influence of the topography is practically excluded. The
role of the bottom topography can be performed in this case by large internal
waves on sharp layers of the jump near the surface, by lifts of frontal interfaces,
etc.

Let us now discuss examples of anomalous conditions of the surface generated
by the effects discussed above and other effects.

2.2.1. *Slicks and smooth patches*

Patches of the surface, smoothed out* as compared to the surrounding areas,
are called slicks. They may be entirely smooth patches against the background

---

*Due to a decrease in high-frequency roughness.

of the surface with light ripples as well as patches with light ripples against the background of the surface with stronger gravity-capillary waves, or with choppy water (see Section 2.2.2). Slicks are usually of irregular patchy shape with a disordered distribution across the sea surface, but they are also observed in the form of ordered structures, including periodic ones. Slicks may be caused by the damping effect of surfactant films, or by factors of the dynamic condition of the near-surface layer such as turbulence, currents, convection, and internal waves, but most often by a combination of all these factors with wind conditions. Near-surface thermal inhomogeneities may be associated with slicks, but the connection is not simple in this case because of the plurality of concrete causes and conditions of slick origin. A large amount of useful visible information on the processes and phenomena occurring in the depth is obtained owing to the configuration of the slicks when the condition of the sea surface is from 0 to force 3–4 according to the Beaufort scale. It is possible to trace, for example, the jets of currents, meanders, and eddies by the configurations of the slicks in the area of a sunglint patch, especially at its edges (see, for example, Fig. 2A in ref. [184] by McAlister and McLeish, which was also reproduced in ref. [73], p. 21). The visibility from a plane, satellite, or even from the deck of a vessel is associated in these cases with the fact that the slicks form where the direction of wave propagation coincides with the direction of water movement on the surface. At the same time, the waves become shorter and steeper (up to breaking) on a contrary current. In these cases, contrasts, which create a visible image corresponding to the structure of the current near the surface, appear in the field of light reflected and scattered by the sea surface.

When surfactant films are the principal cause of slick patches, the slick pattern on the ocean surface corresponds to the distribution of convergence areas in the velocity field to which the films are carried by the surface currents. Narrow zones of downwelling water are detected under the slick bands by towing thermal sensors on horizons close to the surface (3–5 m) during diurnal solar heating. Local temperature rises have the shape of sharp peaks against an even background (Fig. 2.2). Such temperature rises are detected under lines of convergence of Langmuir circulations [102] which are often accompanied by slick bands. However, the distribution of surfactant films on the sea surface may be associated with the field of internal waves. If the character of the latter is stochastic, both the slick patterns on the surface and the associated thermal inhomogeneities [103, 109] are also of irregular character. It is shown in the article by Yermakov, Pelinovsky, and Talipova in ref. [85] that the damping effect of the surfactant films is most evident on surface waves in the length range of 1–10 cm (ripples). The space slick pattern contains information on the bottom topography in shallow waters where the general pattern of the currents is closely associated with the bottom topography.

Smooth patches, which are similar in nature to slicks, are associated with subsurface turbulized regions. For example, smoothing of waves in the turbulent wake of any surface vessel is characteristic. These turbulized regions may appear under natural conditions for many reasons, i.e. when large internal waves break, at shear instability of currents, when strong currents (especially tidal currents)

slick

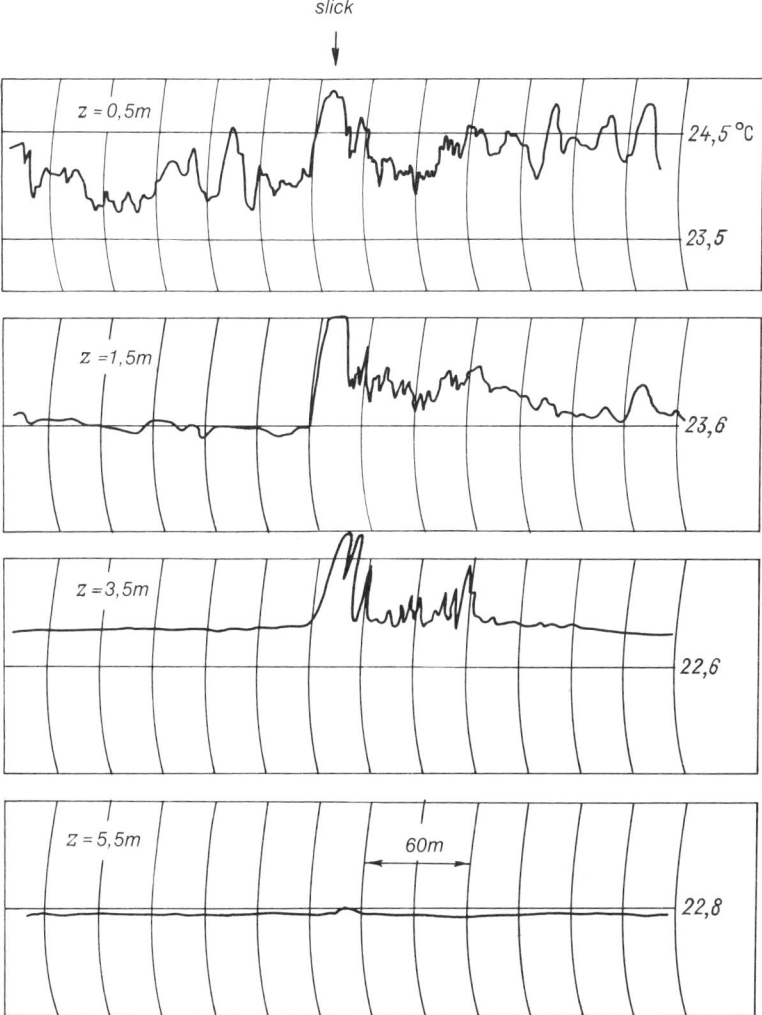

Figure 2.2. Recording downwelling heated water in the convergence zone under the slick by a towed near-surface thermal trawl. The width of the downwelling zone is about 25 m, the temperature contrast between the downwelling and surrounding waters on the horizon 1.5 m exceeds 1°C. Data obtained under the guidance of V. T. Paka during the tenth cruise of the research vessel *Akademik Mstislav Keldysh* in the Tropical Atlantic on 25 May 1985.

flow round shallow areas of the bottom, underwater rocks, etc. The turbulent flux of the impulse 'sucks' the kinetic energy from the surface waves into the depth which results in smoothing of the surface. The latter effect, however, can be of diverse significance, depending on the characteristic lengths of the surface waves and the correlations of the levels of the disturbance and ambient medium turbulent energy. An analysis of this diverse significance is discussed in ref. [85] by Barenblatt and Benilov.

2.2.2. *Choppy water*

'Choppy water' is an unusual state of the sea surface, representing anomalously steep sea waves, sometimes disordered, or having the form of ordered short-standing waves. Choppy water is always three-dimensional and is usually accompanied by foam on the crests of pointed waves. The distinguishing feature of choppy water is the fact that by its external manifestations (sharp breaking crests, foamy whitecaps, roaring or loud hissing) it does not correspond to relatively calm weather with a maximum wind force of 3–5, when it is observed most often. More or less successful descriptions and explanations of choppy water are given in some handbooks and maritime dictionaries. Here is one [63]:

> Choppy water is a variety of interfered waves, is a form of waves when the crests are like isolated hill-like heights while the troughs are like individual depressions. Choppy water forms as a result of wave inter-ference, i.e. the adding of two or sometimes several wave systems. Choppy water is observed when waves are reflected from the coast or when two wave systems meet, e.g. near a cape extending into the sea or in the centre of a cyclone. In the case of choppy water, the waves lose the character of progressive waves and seem to push one another on the same site.
>
> Choppy water is a common case of standing waves, wherein the particles move upward and downward, although the oscillation period is not different from the period of wind-induced progressive waves.

The descriptive part of this definition is remembered by its figurativeness, but the presentation of the causes of choppy water origination is incomplete, as follows from the discussion at the beginning of Section 2.2.

Sometimes irregularly breaking choppy water is present when lee waves combine together with wind waves whose steepness is increased greatly as a result of interaction with a contrary current. An example of this composition was observed many times by one of us during a rising tide in the strait separating Brittany from Île-de-Ba opposite Roscoff. A rock breakwater is situated near the marine biological station and extends about 150–200 m into the strait approximately normal to the isobaths. The height of the breakwater is 3–4 m, and the difference of the high and low water is 8.5 m. Therefore, the breakwater is completely hidden during high water and is fully exposed to the very base arranged on rock flats in the case of low water. A current is observed in the strait during the rising and ebb tide with a velocity up to 1 m/s. Flowing over the breakwater, the tidal current forms a region of steep lee waves from 0.5 to 1 m long directly downstream the breakwater. The predominant westerly winds drive short wind waves in the strait to meet the current. The length of the latter wind waves is also in the 1 m range at winds of 5–10 m/s. Combining with the lee waves, they create irregularly breaking choppy water on an area of several hundred square metres downstream the end of the breakwater where the current is the strongest. At first, the surface turns smooth over the breakwater proper owing to turbulization of the current or to the effect of blocking. When the depth over the breakwater increases during the rising tide, the wind waves start to propagate into this region, while the intensity of the neighbouring choppy water weakens owing to a decrease of the lee waves' amplitude. Anomalies are not observed on the sea surface against the background of wind waves during high water when the tidal current terminates in addition to the effect described above. Observation of a change of situations under a gradual change of the environmental conditions in the simple case of a tidal current flowing round an underwater obstacle (breakwater) turned out to be very instructive.

The strongest modulation of surface wind waves with a length of 10 cm–1 m by internal waves, accompanied by the appearance of very steep choppy water bounding with a smooth surface, occurs as a result of the effect of blocking described earlier. Such a case of the appearance of choppy water on an ebb tide in the westward direction at an opposing westerly wind of about 6 m/s in the Strait of Gibraltar was recorded by aerial photography from an altitude of 100 m (Fig. 2.3) [137]. It is possible that the underwater height in this place (about 6° W) with a maximum depth of 60 m causes essential local intensification of the commonly weak and short-term westward ebb current, whose peak occurred just at the time of observation. The band of choppy water in the case discussed here has a very sharp boundary with a vast slick surface on the front plan of the aerial photograph made in the SW direction. Such sharp boundaries between the bands or patches of choppy water and slick surfaces are frequently observed, including when the effect of blocking occurs as a result of interaction of the surface and internal waves. The patches and bands of choppy water can remain on the site, in particular, when they are associated with the bottom topography

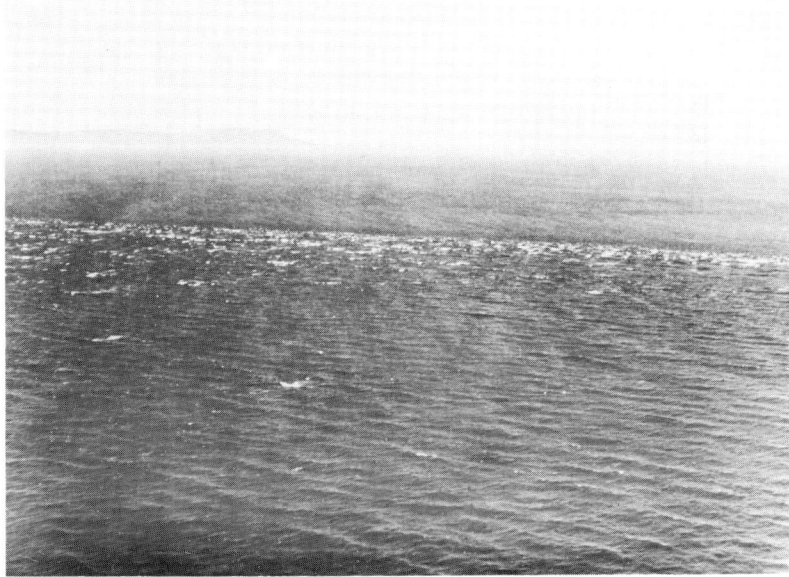

Figure 2.3. Bands of steep choppy water as a result of blocking surface waves on a strong ebb tide current in the Strait of Gibraltar. Aerial photograph from ref. [137].

or with stable oceanic fronts, but they can also move, which is characteristic of the effects of internal waves. In one of the most characteristic examples,* the observers from three vessels reported patches of intensive choppy water with a height up to 2.5 m in the Bay of Bengal in the same place (5°50′ N, 88° E) on different days from February until April 1966. In his comments in the latter article, La Fond reported four observations of choppy water in 1963 in the Andaman Sea. The areas of choppy water moved, and one of them passed through an observation point for 22 min. Apparently, La Fond observed bands of choppy water similar to those that later, from the observations of Osborne and Burch [197] and by space photographs made during the *Soyuz–Apollo* experiment in 1975, were explained by the interaction of packets of large internal waves with surface waves (see also pp. 208–209 and Fig. 3.40 in ref. [103]). In the latter 'Andaman phenomenon', with many subsequently discovered analogues in other parts of the World Ocean, the bands of choppy water alternated with bands of slicks or bands of relatively weak waves. These ordered trains are discussed in the next section.

2.2.3. *Ordered alternation of bands (or patches) of slicks and ripples (or choppy water)*

Alternating bands of slicks and ripples (or choppy water) on the sea surface, representing a periodic or quasi-periodic pattern, are encountered in coastal waters and narrow passages as well as in the open sea. There are no doubts today that in the majority of cases these patterns are associated with the modulating effect of internal waves of various nature. Extreme cases have been previously cited, e.g. gigantic packets observed in the Andaman Sea with 5–6 wide (up to 1–1.5 km) bands of high (up to 1–2 m) and steep choppy water with intervals of calm water, or weak waves (up to 0.6 m) with a width up to 15 km over trains of internal waves of tidal origin moving with phase velocities of about 2 m/s [197]. The pattern of periodic structures with lengths of 50–150 m on the water surface over the rapids in the Night-Inlet fjord (British Columbia, Canada), where trains of very steep internal waves with a height up to 10 m form in the tidal current on the shallow layer of the jump when flowing round the rocky rapids with a peak at a depth of 63 m and at a total depth of 400–500 m in the fjord, is just as impressive by its contrast, though it is inferior by its scales [141] (see Fig. 2.4). In the latter case, the internal waves are displayed on the surface by the contrasts in the steepness of the ripples and in the colour contrasts due to the presence of fine suspended matter of river origin in the layer over the density gradient. Approximately the same scale as in the latter case was in the display of the internal waves during the effect of 'dead water'† following a vessel cruising at a low speed over a shallow layer of the density gradient at a depth of 4 m [162]. The scales of numerous other observations of surface displays

---

* *The Marine Observer*, 1967, **XXXVII**, N215, 10–11.
† The phenomenon of 'dead water' was first described by F. Nansen from his own observations in the Kara Sea.

Figure 2.4. Surface manifestations of internal waves in the Night-Inlet fjord, British Columbia, Canada [141].

of oceanic internal waves can be arranged between the two extremum scales of the examples discussed above (see, for example, Fig. 2.5 and ref. [49]), including observations made from space [100, 122, 213]. Sometimes internal or surface modulating waves propagate in the form of concentric circumferences from a spot source, e.g. from the centre of an underwater shock or explosion. Correspondingly, the pattern of the modulated surface waves takes the shape of expanding concentric circumferences (see Section 2.3).

Ordered periodic alternation of bands with increased and decreased roughness of the sea surface in the centimetre range of wavelengths are also observed as a result of internal waves in the near-water layer of the atmosphere. The latter creates a wind field periodically changing in space near the sea surface. In this case, the wavelength of disturbances (periodic bands) on the sea surface is equal to the length of the atmospheric internal waves, i.e. it is about 10–15 km, while the bands are arranged perpendicular to the direction of the wind. This effect is often observed during the formation of lee waves in the stably stratified near-water layer of the air, e.g. when an air current flows over a mountainous island. This situation was recorded recently (see Fig. 2 on p. 43 in ref. [55]) on an image of the sea surface to the east of Novaya Zemlya obtained by means of a side-looking radar from the *Kosmos-1500* satellite. Another similar effect, when sections of the surface with different roughnesses form an ordered cellular structure, is associated with the cells of atmospheric convection [55]. This effect was also discovered by means of the side-looking radar on the *Kosmos-1500* satellite. In the latter case, the patches of increased roughness (ripples) reached several tens of kilometres in diameter.

The ordered alterations of slick and ripple sections on the water surface can

Figure 2.5. Band of ripples on the calm surface of the ocean, the first in a group of 4–6 bands with spacing of about 1 km, and representing the result of interaction of weak gravity-capillary waves with internal waves. Thirty-fourth cruise of the research vessel *Akademik Kurchatov*, 27 January 1982. Pacific Ocean, East of Galapagos Islands (photograph by K. N. Fedorov).

Figure 2.6. Lee waves on the surface of a tidal current on a shallow bed of sandy littoral in absolutely still weather. Two regions of waves are visible (photograph by K. N. Fedorov).

also be of non-periodic structure. In particular, the bands of ripples on vast areas of a shallow shelf with strong tidal currents can be arranged over the deepest sections of the bottom, repeating quite accurately on the surface the outlines of the underwater topography as a result of interaction with the most intensive jets of the currents following the underwater 'beds'. It is necessary, to this end, that a light wind should oppose the current. In this case, ripples do not form over shallow areas because the currents are weak there and can even be directed to the wind. The band of ripples observed on the water surface over a strong contrary current, owing to the scattering of light by the ripples, looks in the slanting sunbeams like a contrasting (dark) 'river' on the bright (light) calm surface of the rest of the water surface. When the tide rises, the contrast reaches a maximum, then it weakens when the difference in the heights of the bottom topography becomes smaller than the total depth of water over the shelf. One of us has observed many times the generation of the same pattern of ripple bands on a calm surface during the same phase of the tide at a westerly wind of 3–4 m/s in the shallow strait separating the shore of Brittany near Roscoff from Île-de-Ba. Principles similar to those due to which a contrasting image of the bottom bed is obtained on the water surface in the afore-cited case are discussed in ref. [150] in relation to radar images of the ocean surface wherein characteristic features of the bottom topography are clearly visible. The specific features of the bottom topography are manifested on the sea surface also due to lee waves that fall under the definition of choppy water given in Section 2.2.2. One of us photographed clearly outlined areas of immovable waves with a length of about 0.5–1 m on the water surface in the complete absence

Figure 2.7. Accumulations of dynoflagellates on the sea surface in the form of bands in the convergence zone associated with thermal inhomogeneities of calm weather [103] in the Strait of Skagerrak in July 1977 (photograph by K. N. Fedorov).

of wind, but with a strong tidal current (up to 2 m/s) in the sand bed of the littoral at a total depth of water not more than 1–1.5 m (Fig. 2.6). The two areas with a wavy surface on the image are confined to small rises of the bottom in the bed that are well observed in the phase of the ebb, i.e. before the tidal current flows over the bed.

2.2.4. *Bands of foam and debris*

Bands of foam and floating debris and accumulations of plankton appear in areas of surface current convergences. Correspondingly, these bands can mark the lines of convergence (and downwelling) of water in convective circulations, Langmuir cells, internal waves, and at fronts. Examples of similar bands are presented in Figs 2.7 and 2.8. Foam bands may form differently. In upwelling areas and other coastal zones of high bioproductivity, where organic surfactant films cover the whole surface, even light wind waves leave a tremendous number of floating 'islands' of foam and bubbles which are not destroyed for a long time and are carried by the currents to the lines of convergence. As a result, wide (1–5 m) and thick bands of dense foam form along these lines (Fig. 2.8). In other cases, bands of foam form on surfactant films that are carried to the lines of current convergence where the horizontal shear of velocity contributes to the foam formation. A narrow band of foam is then formed and it is often disturbed by whirlpools and eddy-like features. It is possible to observe foam, debris, algae (Fig. 2.9), living and dead plankton cells, pleiston organisms such as physalia, etc., all together in the Langmuir cells and in the surface manifestations of the internal waves on the lines of convergence. The lines of foam and other floating material (e.g. debris) at frontal interfaces (see Section 2.2.7), whose sloping surfaces sink into the depth at a very small angle, may be at some distance from each other, running in parallel and not coinciding with the line of colour contrast that is well observed from a plane [103]. Accumulations of plankton at the time of intensive blooming may serve as an excellent indicator of the velocity field structure of the near-surface currents.

2.2.5. *Eddies, whirlpools, small vortices*

Intensive eddies (whirlpools, also called small vortices), whose diameter is from tens up to several hundred metres and even kilometres, may be generated in straits and narrow passages as a result of the interaction of current jets flowing in different directions, and especially in the case of a favourable configuration of the coast. Most often these whirlpools develop due to tidal currents of very high velocities (up to 4–7 knots). For example, the whirlpools in the Strait of Messina [133], the gigantic Maelstrom whirlpool* in the strait between the Moskenes and the Verey Islands in the group of Lofoten Islands, also whirlpools in the throat of the White Sea, in the Strait of Matochkin Shar, in the Kuril

---

*In the case of NW winds, the Maelstrom whirlpool is also accompanied by very steep choppy water which is dangerous for vessels [135].

Figure 2.8. Band of foam along the convergence line associated with internal waves in the upwelling zone near the Peru coast in February 1982 (photograph by K. N. Fedorov).

straits, and so on are well known. Although eddies and whirlpools sometimes involve a substantial thickness of water, they are also manifested on the surface in the form, for example, of oily (smooth) slick patches creating the impression of very calm waters. Therefore, they are always very dangerous for small vessels, treacherously carrying the latter to rocks and shoals. Eddies sometimes become

Figure 2.9. Sargasso algae along the convergence line of Langmuir circulations. Sargasso Sea, 21 September 1978 (photograph by K. N. Fedorov).

Figure 2.10. Eddy in Night-Inlet fjord visible owing to the confluence of different rivers flowing into the fjord and carrying waters with a substantially different concentration of suspended matter. The eddy was observed from a plane in July 1976 by Professor G. D. Smith (Seattle University, WA, USA), who kindly presented the photograph to us.

visible in colour and brightness contrasts because the waters in the interacting jets of currents are coloured differently or because of the large quantity of fine suspended matter in the upper freshened layer (Fig. 2.10). For example, the latter layer can be broken up in the upwelling region associated with the eddy. Drifting ice, sludge, grease ice, accumulations of phytoplankton, and algae also make eddies and whirlpools readily visible.

2.2.6. *Current rips*

Navigators often encounter long extended bands of extremely agitated water surface in narrow passages, straits, or in large river mouths where the currents are strong. The sea surface of these bands resembles greatly the surface of boiling water even in calm weather. The waves resemble irregular, steep, and sometimes quite high splashes*. Smooth patches with a diameter of 5–10 m are observed along these bands, which are sometimes very narrow, and also accumulations of foam, debris, and small (1–10 m in diameter) whirlpools and small vortices, resembling those commonly encountered in rapid rivers near rocks and piers. Whirlpools and small vortices are best observed against a background of large smooth patches, whose surface structure resembles greatly that in the wake of a common vessel (Fig. 2.11). An upwelling of cold deep waters, accompanied by the formation of fog in the near-water layer, can be observed in these patches

---

*When considered individually, these irregular steep waves are called 'choppy water' (see Section 2.2.2).

(a)

(b)

Figure 2.11. Typical photographs of current rips in the White Sea with patches of smoothing and whirlpools (photographs by A. S. Kazmin).

[8]. The bands of agitated water surface produce a characteristic hissing sound which is audible several kilometres away. Apparently, this sound is associated with the breaking of very steep choppy water. Usually the bands of agitated water, breaking waves, and choppy water have a quite definite arrangement in relation to the islands, the specific features of the shore line, or the topography of the bottom.

Figure 2.12. Parallel bands of current rips in the White Sea off the Solovetski Islands (photograph by V. Kh. Lifshits).

The phenomenon described above in the totality of all or some stated anomalies of the surface condition is usually called current rips, and can be defined briefly as a complicated anomalously agitated state of the water surface in the zone of convergence of strong currents, most often of tidal origin. Sometimes several parallel bands of current rips are observed (Fig. 2.12). For example, the *White Sea Pilot Book* describes on p. 255 'numerous current rips covered with bands of foam' observed along the Karelian coast to the west of the Solovetski Islands. The bands of current rips can shift in space, sometimes slowly and sometimes fast; they weaken or intensify, which depends often on the local tidal phase. At times they have the form of more or less wide sections of irregular shape, more often triangular, rarely rounded. They are well seen on the screen of the vessel's radar, and at night they may be taken for islands. Sometimes one can see by the naked eye bands that glow brightly at night owing to the accumulation of luminescent marine organisms.

Based on an analysis of all the known descriptions of current rips, the conclusion can be drawn that the cause of their origin can be alternating sharp convergences and divergences of streams in combination with a strong vertical and horizontal shear of velocity which develop when the currents (permanent and tidal) flow round Islands, capes, and shoals under conditions of intensive water exchange through narrow passages and straits. The phenomenon of current rips includes the interaction of surface waves with strong space inhomogeneous currents, and the formation of lee waves and eddies with vertical and horizontal axes. Since tidal currents in coastal regions of skerries, near capes, in bays and inlets often reach velocities up to 4–6 knots, which generally exceed the common velocities of permanent currents, the phenomenon of current rips is observed more often in these regions (e.g. in the White Sea or in the Kuril Straits) and mainly in association with tidal currents. Favourable conditions for the origin of current rips exist in the open ocean at sharp near-surface fronts (see Section 2.2.7). The formation of parallel bands of current rips observed in the White Sea demands a special explanation. It may be connected with the jet character of the currents. It is possible that internal waves play an important role in this case alongside tidal currents. The presence of sharp boundaries between the choppy water and the smooth patches (Figs. 2.11 and 2.12) is most likely associated with the local manifestations of blocking (see Section 2.2).

The term *suloy* or *suvoy* (Russian variant of current rips) descends from the coast-dwellers of the White Sea because of the specific characteristic phenomena observed regularly and in many regions of the White Sea. The vocabulary of local coast-dwellers' expressions in the *White Sea Pilot Book* give the following definition of the term *suloy*:

> *Suvoy, suloy*—splashes, whirlpools, and choppy water due to the meeting of two opposing currents, or the wind and a current. (See p. 347 of ref. [67]).

A term of similar meaning, *stromkabbelung* (literally meaning opposition of currents), is found in the German language (see, for example, p. 526 of ref. [172]). Like the Russian term *suloy* (current rips), it is not widespread, nor is it found in many common dictionaries. The equivalents of both terms are also found in other languages [103, 107], since a similar phenomenon has been observed by navigators of many peoples and required its specific designation.

The fact that the manifestations of current rips are extremely diverse in form explains the certain discrepancy between the definitions of the term 'current rips' in different publications. Many publications define current rips as a specific type of sea wave, which, to our mind, narrows greatly the idea of this pheno-menon. As a result, the term 'current rips' is used incorrectly in many recent publications to define bands of choppy water arising at modulation of the surface waves by internal ones. It can be understood in other publications that the characteristic specific feature of current rips is the combination of wave and vortex motion, i.e. choppy water and whirlpools. But mentioning 'vortex motion' or 'whirlpools' without specifying their dimensions leads to the fact that current rips are sometimes erroneously considered a phenomenon of large-scale vortex nature. Evidently, it is necessary to avoid both extremes.

Apparently, it is quite permissible to assess the intensity of current rips by such adjectives as 'strong', 'large', or 'weak'. These assessments are most likely associated with the degree of disorder, the steepness, and height of the choppy water in the bands of current rips. These are possibly the factors that determine in the first place the visibility contrast of the bands of current rips on the ocean surface. The *White Sea Pilot Book* gives the following examples:

The second branch of the tidal current flows through the Morzhovskaya Salma Strait and, skirting Morzhovets Island on the south and south-east, joins with the main branch, forming strong current rips...

The ebb current flows in the reverse direction, forming weaker current rips than during the tide. (See p. 90 of ref. [67].)

Or:

The tidal current flowing from the north along the Tersk coast meets with the quite strong stream of the Ponoi river and forms large current rips to the south of Korabelny Cape. (See p. 108 of ref. [67].)

In the latter case, we deal, apparently, with an interesting and significant effect of stratification in the current rips. The saline and denser waters, carried by the rapid tidal current from the Barents Sea meet with the slowly flowing fresh waters of the river discharge and go under the latter. Since the meeting of waters of various densities occurs by the 'cold front' type, the movement of the thin upper layer near the interface should be extremely unstable, which affects the character of waves and flow on the free surface. All the points stated above can be fully related to the frontal interfaces in the open ocean (see Section 2.2.7). A phenomenon quite similar to current rips has been observed, for example, by Knauss in the equatorial part of the Pacific at the frontal interface propagating by the type of a cold atmospheric front (see the discussion in ref. [103]).

Recently, the phenomenon of current rips has again been subjected to detailed investigations based on direct measurements because of the general increase in interest in the anomalous states of the ocean surface [8, 65] (see also Chapter 6).

(a)

Figure 2.13. Danube discharge front. (a) General view from a distance of several hundred metres; (b) frontal interface from a distance of several tens of metres; a narrow band of foam is visible along the front, the colour and brightness contrasts are also expressed very well; (c) close shot of section of front; vortices at the front can be seen clearly.

2.2.7. *Fronts and their visible manifestations on the ocean surface*

Frontal interfaces (or fronts) can be seen on the surface in the first place because of the difference in the physical characteristics of the water (e.g. optical) on both sides of the front. Figure 2.13a shows a photograph taken by Yemelyanov of the Danube front in the Black Sea. The boundary between the fresh waters of the Danube run-off lens and the Black Sea saline waters proper is distinguished by the brightly coloured contrast ranging between yellow-green and black-blue tints. The yellow-green colour of the Danube front is due to the high content of humus and suspended matter of terrigenous origin. Figure 2.13b shows the same front from a closer distance. The boundary proper between the two waters is clearly designated by a narrow band of foam, indicating the surface currents' convergence at the front. Figure 2.13c, taken from a still closer distance of the same Danube front, shows a trend toward twirling into small whirlpools (vortices) with a diameter of several metres at the very interface. They are visible along the entire interface and testify to the presence of a horizontal shear of velocity at the front. All these points taken together have the character of current rips (see Section 2.2.6).

Current rips manifested in the given case in the form of a narrow band of foam and vortices are, as compared to colour and salinity contrasts, a secondary surface indication of the frontal interface. These secondary visible manifestations of the fronts on the ocean surface are quite diverse in form (see the discussion in ref. [103]). There are recorded observations of ocean surface states near the fronts which have never been mentioned before in the literature. The photograph taken in August 1978 from the research vessel *Ernst Krenkel* (Fig. 2.14) shows the main front of the Gulf Stream on the warm side of which one sees closely located bands of foam, extending (if judging by the crests of the waves) in the direction of the wind and practically perpendicular to the front line. According to data obtained by Yemelyanov, the photographer, the length of these foam bands from the front line to the wind in the direction of the vessel, whence the photograph was taken, reached several cables. The spacing between the bands was about 1–3 m. The wind velocity was not more than 5–7 m/s at the time of observation. Elements of choppy water were observed in the area of the bands.

Figure 2.14. Bands of foam on the warm side of the Gulf Stream (photograph taken in August 1978 by M. V. Yemelyanov).

They are visible in the photograph. The area of cold waters beyond the front was practically calm. Apparently, in this case we were dealing with a peculiar form of current rips, wherein the foam which accumulated in the convergence zone at the front drifts to the wind toward the warm waters, and serves as a tracer of convergent lines in Langmuir circulations that develop on the warm side of the front where the thickness of the upper quasi-homogeneous layer is very small, i.e. only about 1–3 m. It is interesting to note that similar phenomena have also been observed in the Baltic Sea (personal report by Sustavov, Leningrad State Marine Institute).

There is other evidence supporting the fact that current rips appearing at the emergence of the frontal interface to the surface are extremely widespread in the open ocean. There is every reason to assume that the numerous cases of current rips (*stromkabbelung*) analysed in the works by Remer and Schumakher (see the discussion in ref. [103]), especially those relating to the region to the south of Cape Verde Islands, are associated with sharp oceanic fronts encountered often in that region. The fact that the frequency of the observed cases of current rips is of seasonal character (see Fig. 1.1 in ref. [103]) gives additional weight to the latter point of view.

The presence of a frontal interface does not mean by itself the automatic appearance of such phenomena on the surface as current rips. The practice of our field observations demonstrates that the simultaneous observance of certain conditions is necessary for the origin of a complex of surface phenomena of the frontal current rips type. Some of the conditions that are known today are listed below:

(1)   the state of the sea surface should not be higher than force 3–4;
(2)   the presence of a sharp density gradient across the front;
(3)   strong convergence of the surface currents at the front;
(4)   the presence of a horizontal velocity shear at the front.

High bioproductivity near the front is an additional factor contributing to the appearance of foam and debris in the bands of convergence.

2.3. Anomalous phenomena on the ocean surface through observations of seamen

In the conclusion to this chapter it is necessary to give the reader an idea of where and in what parts of the World Ocean anomalous states of the ocean surface are observed most often. A lot of information useful for this purpose on such anomalous phenomena on the ocean surface as current rips, choppy water, slicks, whirlpools, and 'small vortices', manifestations of internal waves, lines of current demarcation, lines of foam and debris accumulations, discoloured patches, phosphorescence of the sea, etc. are contained in oceanographic literature of the first half of the twentieth century, as well as in ship-log books of military and merchant ships, and in pilot books that are continuously published. The quarterly journal of the Marine Department of the Meteorological Service of Great Britain, the *Marine Observer*, published since 1924, has been invaluable in this respect. We analysed more than 350 reports on various surface phenomena published in this journal in 37 years from 1935 to 1939 and from 1952 to 1983.* 283 observations of particular interest have been selected from these reports. The results of the analysis are given below.

Surface phenomena that have drawn our special attention can be divided into four categories:

(1) isolated bands and patches of anomalously restless water, stronger waves, whitecaps, or choppy water, or, vice versa, isolated smooth sections of the surface (most often with an indication of space orientation), frequently accompanied by changes of water temperature, ship 'yawing', etc.;

(2) isolated lines of separation or demarcation (with an indication of space orientation) expressed by a change of colour, sharp change of water temperature, debris or foam accumulation;

(3) isolated bands, areas of various shape, and demarcation lines (with an indication of certain space orientation), combining different elements of items (1) and (2) and, furthermore, whirlpools, 'small vortices', acoustical manifestations, and accumulations of plankton (sometimes luminescent ones), also marine animals and birds; observed often at complete calm against the background of a calm sea;

(4) multiple alternating parallel bands of ripples, choppy water, or strong waves and more calm water (or slicks) accompanied by yawing of the ship when crossing the band boundaries, with an indication of space orientation of the bands.

While items (1)–(3) contain information relating more to the surface manifestations of frontal interfaces of various origin, the information in item (4) is unequivocally interpreted as visible manifestations of internal wave trains on the ocean surface, arising sometimes also because of fronts. Item (3) deals with a phenomenon commonly called 'current rip' (see Section 2.2.6). It is necessary to point out certain terminological difficulties encountered when reading the *Marine Observer*. The same English terms, i.e. 'rip(s)', 'current rip(s)', and 'tide rip(s)', also 'disturbed water', 'rough water', 'choppy water', etc., are used in the titles and in the texts written by the observers, discussing without any discrimination at least three categories of information [items (1), (3), and (4)]. The first three terms are exact English equivalents of the Russian term for *suloy* (current rips), and the other three relate to the Russian term for *tolcheya* (choppy water). The information was classified into the four stated categories only on the basis of detailed descriptions, but not because of the titles or terms used by the observers. The scientific comments following some of the reports were not always of help because the authors (workers of the National Oceanological Institute of Great Britain) most often explained the observed phenomena

---

*The issues of the journal from 1924 to 1934 and from 1940 to 1951 are not available in libraries.

by the effects of opposing current meetings or by the interaction of tidal and permanent currents. It was only in 1964 that the *Marine Observer* (1964, Vol. 34, No. 205) published comments of the well-known British oceanologist Lee, who explained the alternating bands of choppy water and slicks by the effect of internal waves. The report and comments deserve to be given here, word for word:

> Barents Sea
> s.t. St. Loman. Skipper J. E. Dobson.
> 11 August 1963. When the vessel was steaming on a 300° course at 14 knots, I suddenly noticed alternate lanes of violently rippled and glassy smooth water. Each lane was about 100 feet wide and lying in a straight line in a N–S direction as far as the eye could see. In the rippled lanes the small wavelets were moving in all directions and there were numerous 'white horses', giving the appearance of a confused sea. The color was darker in the disturbed water than in the glassy lanes.
> Our position on entering these rips was 76°15′ N, 16°55′ E; we left them in 76°21′ N, 16°10′ E. Outside these positions the sea was normal for a force 2 wind. The radar was distinctly showing blank areas and lanes of clutter on the three-mile range.
> Position of ship: South of Spitsbergen.

Note. Dr. A. J. Lee of the Fisheries Laboratory at Lowestoft comments:

> In the Spitsbergen area the water column in summer is often stratified with a shallow layer of cold water of low salinity resting on warm water of much higher salinity. The water temperature measured by Skipper Dobson shows that he was in cold surface water. At the boundary between such water masses of different densities undulating swells known as 'internal waves' occur. These 'waves' create zones with a convergent water circulation at the sea surface and hence a concentration of the surface film in bands parallel to the 'waves'. Under certain conditions of wind and lighting and provided that sufficient organic matter is present in the water, these bands show up as glassy streaks of calm surface water alternating with bands of ruffled water with wavelets.

In regard to these observations it is necessary to note that it is the most northern recorded case, known to us, of manifestations of band-like structures on the sea surface associated with internal waves. The report does not contain sufficient grounds to make an unequivocal conclusion in favour of the film mechanism of surface wave modulation by internal waves, as was done by A. J. Lee. It should be noted that alternating bands of slicks and ripples were interpreted as surface manifestations of internal waves for the first time in 1950 [139].

The information taken by us from the *Marine Observer* (1935–1983) and its classification into several categories make it possible to determine at least

Table 2.2
Distribution of observations over different categories

Category of phenomenon (i)	n_i	$n_i = k_i + m_i$		n_i/N (%)
		k_i ($H < 500$ m)	m_i ($H > 500$ m)	
1	97	18	79	34
2	78	24	54	28
3	32	8	24	11
4	76	30	46	27

Note. $N = \sum_{i-1}^{4} n_i = 283$.

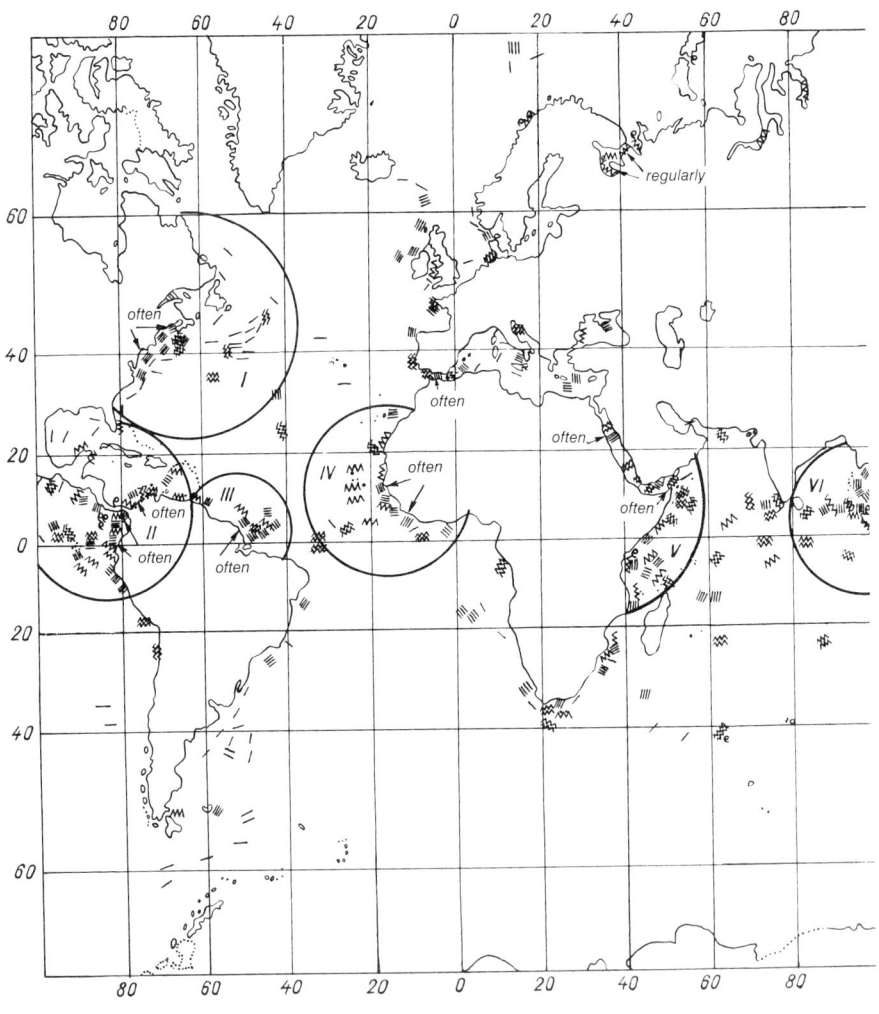

approximately the predominant types of anomalous phenomena on the ocean surface, and to realize the degree of influence of the shores, narrow passages, islands, bottom topography, and general ocean circulation elements on the various surface phenomena. Table 2.2 presents the total number of observations n_i, relating to each category ($i = 1$–4), and the distribution of the observations over 'shallow water' ($H < 500$ m) and 'deep water' ($H > 500$ m) areas.

It follows from the table that the greatest number of observations (73%) relates to various frontal phenomena [items (1)–(3)], including current rips. The fronts relating to category 2 are encountered more often (42%) in association with the main currents (Gulf Stream, Kuroshio, Falkland, etc.). The quantity

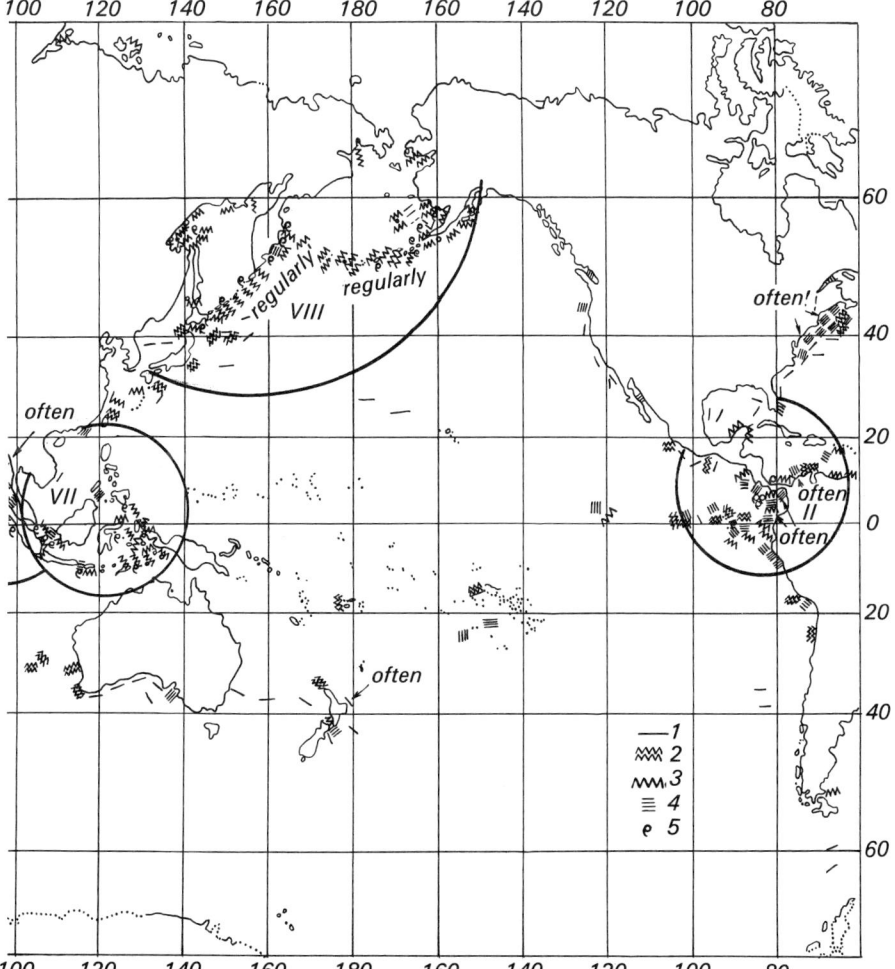

Figure 2.15. Phenomena on ocean surface according to data of the *Marine Observer*, pilot books, and known field and remote observations [107]. (1) Frontal interfaces manifested by temperature and colour; (2) frontal interfaces accompanied by current rips, choppy water, etc. (3) current rips; (4) traces of internal waves; (5) small eddies and whirlpools.

of frontal phenomena in upwelling regions reaches 23%, while in areas near river mouths it is 8%, and the part of fronts in category 2 relating to 'deep water' areas reaches 69%. Frontal phenomena relating to category 1 also occur more often (82%) in 'deep waters', and the most often in upwelling regions, in areas of increased variability, meandering, and convergence of currents. About 40% of observations of frontal phenomena of this type relate to the Red Sea, Aden-Somali region, and the Indian Ocean as a whole.

A comparatively small number (32 cases) of indisputable observations of current rips makes up only 11% of the total quantity of reports, including eight cases (i.e. 25%) relating to 'shallow waters' (five of the cases in straits). The greater number of current rip cases was recorded in areas where currents meet (converge).

It should be noted that the very small number of reports on current rip observations is associated with the fact that seamen usually recorded uncommon (anomalous) phenomena only when they were encountered unexpectedly. Current rips, well described in the pilot books and regularly appearing in the White Sea, Kuril Straits, or in the region of the chain of Aleutian Islands, as a rule were not recorded by seamen in the ship's log books. Their frequency is so high that the reports would number in the thousands, if it were not for the precise and clear warning of the pilot books. The per cent predominance of observations of current rips and fronts in deep waters is associated with the same circumstance. Even phenomena that seem quite common (and are not recorded) near the shores, in straits, and over banks and shoals might look anomalous in the open sea.

Surface manifestations of internal waves are observed approximately with the same frequency in 'shallow water' regions (40%) and where the total depth is more than 500 m (60%). They were recorded in 30% of the cases near by islands, banks, and underwater ridges (22 observations); in 8% of cases in straits; and 4% of the observations relate to areas nearby river mouths, while the rest to upwelling regions, the shelf boundary, and the area of currents' meetings.

The geographical distribution of all the observations of the surface phenomena given above is shown in the map drawn by us (Fig. 2.15); the symbols to depict the four types of phenomena are chosen so as to present an idea of the orientation of the observed bands in relation to the cardinal points. In addition, some published field data and results of remote observations of surface manifestation of internal waves and ocean fronts known in the literature were used when drawing up the map.

The general conclusion that can be drawn when looking at the map (Fig. 2.15) is very similar to the one made on the frequency of the thermal fronts' repetition in the ocean according to the data of surface temperature measurements from aboard ships [103], namely, that the surface phenomena of all four categories are observed more often near the coast than in the open ocean. Surface phenomena are most rare in the open part of the Pacific, while in the Indian Ocean anomalous surface phenomena are observed comparatively often by seamen even in its open part.

The following regions may be distinguished in the map (Fig. 2.15) where the most frequent observations of the ocean surface anomalous states are concentrated (without division into categories):

I   NW Atlantic;
II   Panama–Galapagos and Caribbean regions;
III   region of influence of Amazon and Orinoco rivers;
IV   West African region;
V   Aden–Somali region (including the Red Sea);
VI   Andaman Sea and Bay of Bengal regions;
VII   region of Sunda Archipelago;
VIII   NW Pacific (including the Sea of Okhotsk, Bering Sea, Kuril and Aleutian Island arcs).

Parts of regions II, III, IV, and VI, which are most interesting from the point of view of the phenomena considered here, are given in a larger scale in Figs 2.16a, b, c, and d, respectively.

The Panama–Galapagos region (II, Figs 2.15 and 2.16a) is abundant in phenomena of frontal character and in surface manifestations of internal waves with lengths of $\lambda \simeq 0.2$–$1\,\mathrm{km}$, which are linked with a complicated system of currents in this region and with the specific features of the bottom topography (islands, banks). Internal waves with a length of several kilometres observed to the west of the Galapagos Islands ($\simeq 94°30'\,\mathrm{W}$, Fig. 2.16a; $\simeq 103°\,\mathrm{W}$, Fig. 2.15) are conditioned, most likely, by the Cromwell current, whose upper boundary in this region is at a depth of not more than 15–$20\,\mathrm{m}$ and is characterized by a large velocity shear [49].

The fronts in the Caribbean Sea are connected with the Magdalena river discharge, with the interaction of the Caribbean counter-current with the South trade current, and upwelling in the region $\simeq 72°\,\mathrm{W}$. According to the *Marine Observer*, the upwelling frontal zone in this area is distinguished by its increased variability. Internal waves with a length of $\bar{\lambda} \simeq 2.2\,\mathrm{km}$ were observed in the vicinity of the Magdalena river discharge front [100].

Frontal phenomena and surface manifestations of internal waves are also observed often in the vicinity of the area of propagation of the Amazon river discharge (III) and in the adjacent region over the Demerara upland which is several hundred miles away from the continent in the NE direction (Figs 2.15 and 2.16b). A series of measurements of internal waves with lengths from several hundred metres and up to 3 km, making it possible to determine their association with tidal currents, was carried out during the 23rd cruise of the research vessel *Akademik Vernadsky* [76] slightly more to the north (Fig. 2.16b) at the boundary between the inter-trade counter-current and the North trade current.

The West African region (IV) near $10°\,\mathrm{N}$ (Figs 2.15 and 2.16c) is extremely rich in surface manifestations of internal waves and current rips. On the average, the position of the observation points in Fig. 2.16c is very similar to the average monthly position of points of recorded choppy water and current rips in this region on the map drawn by Remer (see Fig. 3.27 in ref. [103]). The observation

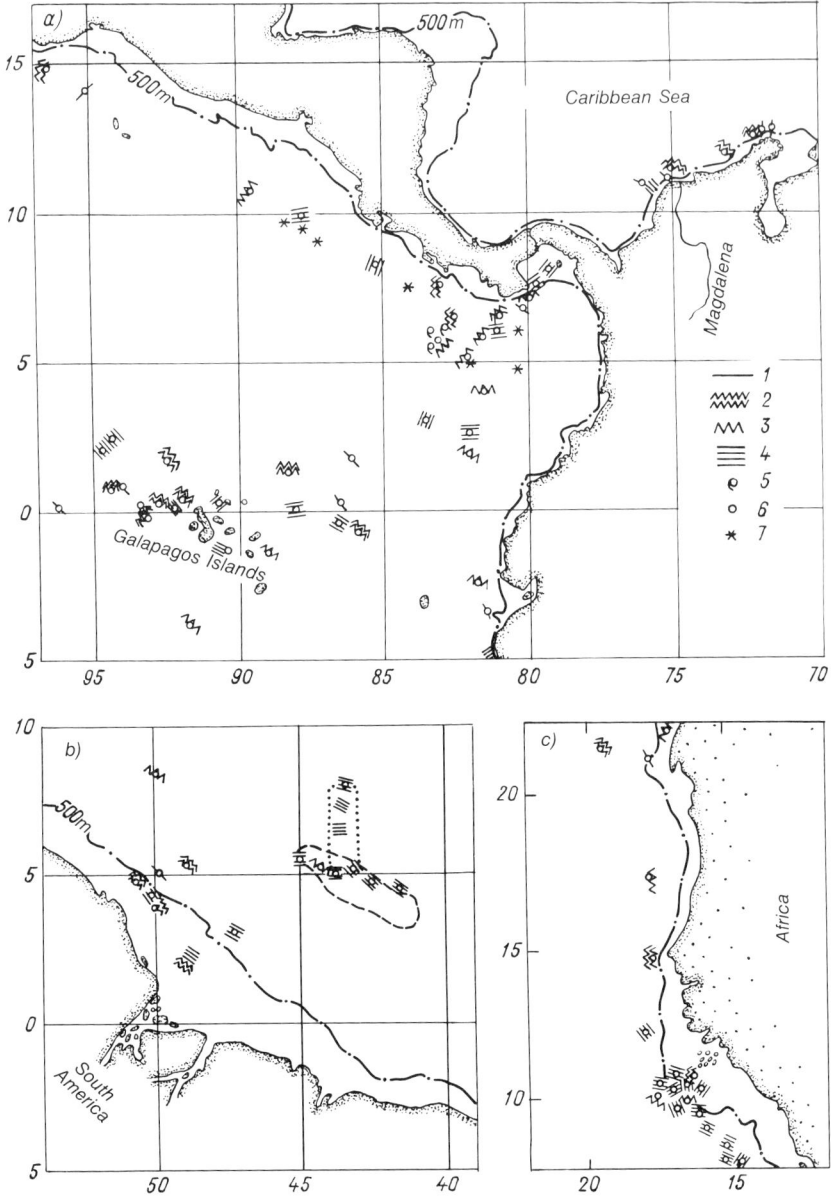

points are concentrated in both cases in the region of transition from the shelf to the slope; therefore, the assumption made in ref. [103] on the connection of the observed phenomena with convergence in the upwelling circulation at the break of the shelf is quite probable. It is also interesting that the frequency of occurrence of choppy water and current rips in this region and somewhat to the south is associated with the seasonal cycle intensity of upwelling (the

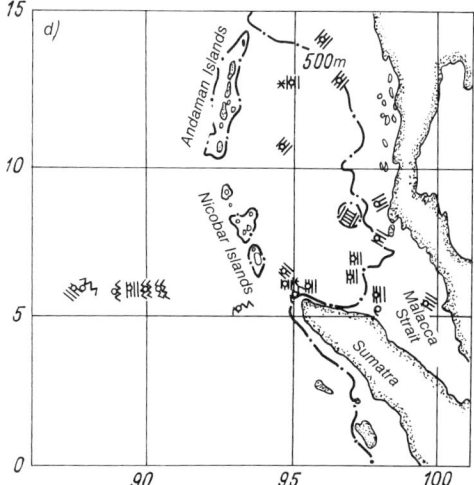

Figure 2.16. Enlarged fragments of map in Fig. 2.15 of regions II(a), III(b), IV(c), and VI(d). Symbols as in Fig. 2.15. (6) Exact site of observation; (7) closeness of bank or underwater upland. The dashed line indicates the Demerara upland in (b), the dotted line the area of investigations of the research vessel *Akademik Vernadsky* in (b), and the area of internal solitons survey during the USSR–US *Soyuz–Apollo* experiment [197] in (d).

maximum is in October [103]). A similar dependence of the frequency of occurrence of current rips on the season in the region westward of the African coast (in the vicinity of Cape Verde Islands, Fig. 2.15) is obvious in the maps of Schumakher (see Fig. 1.1 in ref. [103]), and is associated, apparently, with the seasonal variability of the directions and intensity of the currents in this region.

The Andaman Sea is another region of regular surface manifestations of internal waves with lengths from several hundred metres to 15 km (VI, Figs 2.15 and 2.16d; see also Sections 2.2.2 and 2.2.3). Osborne and Burch [197] think that the most probable regions of their formation are the vast banks and underwater ridges to the north of Sumatra Island and in the vicinity of the Nicobar Islands where the tidal wave is destroyed. The Andaman Islands probably play a similar role for the region which is situated more to the north. The commentators of the *Marine Observer* consider the meeting of the current, flowing out of the Strait of Malacca, with the equatorial counter-current to be the cause of surface manifestations on the internal waves, fronts, and current rips in the southern part of the Bay of Bengal. As the equatorial counter-current in the Indian Ocean has a seasonal cycle due to monsoon circulation in the atmosphere, it would be interesting to study the dependence of the frequency of occurrence of current rips and traces of internal waves in this region on the season. It is impossible to determine this dependence by the limited number of observations analysed by us. It is necessary only to state that practically all the cases of observations relate to the period February–April. According to La

Fond evidence in the *Marine Observer* (1967, No. 215), large internal waves in the south of the Bay of Bengal do not always occur, unlike the situation in the northern part of the bay and in the Andaman Sea, which are ideal for studying them. Large internal waves of the soliton types were also found recently in the Sulu Sea by means of satellite and ship measurements [123]. Although the 'visibility' of the internal waves on the ocean surface is associated most often with the modulation of surface wind waves in alternating bands, it can also be conditioned by modulation of the thickness of the upper freshened and turbid layer in the region of river discharge [100]. The effects of internal waves are manifested quite often in accumulations (in the area of convergence) of blooming plankton (e.g. in the Strait of Skagerrak, in the Brazil current region). Sometimes bands of modulated waves are observed at night due to phosphorescence, e.g. in the Bay of Aden. Seamen noted in some observations variations in the water temperature in areas where internal waves were recorded, but the temperature variations were not linked with the character of the bands being crossed.*

The observations analysed give an idea of the conditions under which surface manifestations of internal waves arise. These manifestations are observed most often in the vicinity of fronts, in places with a rugged bottom, in the vicinity of islands and banks, in straits, in regions with run-off lenses, over the shelf, and at the shelf boundary, in places of interaction or shear instability of strong currents, when a sharply expressed sloping pycnocline is situated close to the ocean surface.

In the majority of cases mentioned above, it is possible to assume the significant role of tides in the generation of internal waves whose traces are observed on the ocean surface. Tidal waves and currents could interact with the permanent currents and sloping fronts, or deform under the influence of the bottom topography, shallow areas and configuration of the coast. It has been possible to prove the patent roles of the tides only in several cases, e.g. the internal waves over the Atlantic shelf of North America [122, 213], in the Andaman Sea [197] and Tyrrhenian seas [120]. In some cases, a link was observed between the intensity of manifestation of the internal waves on the ocean surface and the magnitude of the tide. For example, internal waves over the Mascarene ridge were clearly observed visually for several days after the spring tide and were practically absent in the neap tide [17]. The most noticeable manifestation of large internal wave traces on the surface of the Andaman Sea are also confined to the syzygy [197].

Tidal currents often induce current rips in coastal zones. The regions of regular and most intensive formation of current rips are the White Sea and the areas of the Okhotsk and Bering Seas, and the seas of the Sunda Archipelago marked in Fig. 2.15 (regions VII–VIII). Our analysis of the pilot books of several regions of the World Ocean (see also Section 2.2.6) shows that current rips form often in places with a rugged bottom (over banks, reefs, shoals), near capes, in straits, in the zone of interaction of the river discharge with the tidal current

---

*See Section 4.5.4 on modulation of the SST field by internal waves.

(e.g. the mouths of the Ponoy and Mezen rivers flowing into the White Sea, the mouth of the Lena river flowing into the Laptev Sea), at areas of convergence of different branches of tidal currents (e.g. the White Sea), or at the boundaries of the currents (e.g. the boundary between the coastal tidal and monsoon current to the south of Cape Guardafui is marked by a line of current rips and breakers). Current rips show up in different parts of straits depending on the direction of the tidal current (tide or ebb tide), as it is, for example, in the Unalga Strait in the Bering Sea. However, current rips propagate most often along the entire length of the strait during spring tides. And, quite the contrary, current rips are not observed at all in the neap tide in areas with weak tidal currents.

The influence of the wind on the formation of current rips is quite variable. Sometimes current rips appear either when the wind opposes the tidal current (e.g. the Strait of Akun in the Bering Sea), or reaches in this case the maximum intensity (in the straits between the islands to the north of Sumatra Island, or in the case of the Maelstrom, Lofoten Islands). However, current rips arise often on an absolutely calm sea and at light winds (at the Shantarskie Islands and at Iona Island in the Okhotsk Sea; at Urup Island of the Kuril group of islands; in the Unalga strait in the Bering Sea where current rips are dangerous for small ships even in calm weather and in the absence of sea waves). In this case, the wind only induces choppy water.

In many places (e.g. in the NE strait of the Shantarskie Islands) there are current rips which are accompanied by whirlpools and smooth patches. Whirlpools with a smooth water surface followed by bands of breakers are observed often at the northern coast of Sumatra; in the latter case, silt surfaces from the bottom and creates sharply outlined areas of turbid water.

Smooth round patches, surrounded by zones of high waves against the background of a relatively calm sea, are also found in the open ocean over great depths. Such a patch with a diameter of $\simeq 270$ m was observed in the East China Sea ($\simeq 30°$ N) to the west of a chain of islands and banks (*Marine Observer*, 1966, No. 214, in Fig. 2.15 shown as current rips). A large smooth patch with a diameter of $\simeq 3.6$ km surrounded by waves $\simeq 1$ m high that were breaking in the direction of the patch's centre was observed to the north of Sumatra in the vicinity of the region of recorded internal waves. The cause of its formation has not been determined, and the probable association of the patch with the bank ($6°15'$ N, $95°00'$ E, depth 100 m against the background of the surrounding 1500 m depth) and with the destruction of the internal waves in its vicinity has not been confirmed (*Marine Observer*, 1968, No. 222).

In the bands and smooth patches that are encountered in connection with current rips, fronts, and internal waves it is quite often possible to observe small eddies and ripples propagating in a direction opposite to the background wind waves beyond the bands or patches. Sometimes a change in the velocity or direction of the wind is observed in the vicinity of these bands and patches where a vessel frequently yaws several degrees off course. For example, such a band with a width of about 270 m, extending from one horizon to another, was observed at the northern boundary of the Gulf Stream ($41°24'$ N, $64°33'$ W). Upon approaching it, the wind force increased from 4 to 6, dropped to 0 over it, then again intensified to 6, finally dropping once more to force 4 (*Marine Observer*, 1961, No. 193).

We did not practically deal in this analysis of reports published in the *Marine Observer* with numerous observations of sharp variations of the water colour in the ocean or with numerous cases of water phosphorescence at night. From the character of arrangement of the turbid, decoloured, or phosphorescent bands it was possible in many cases to draw a conclusion about the modulating influence of the internal waves. A series of observations offer evidence that the phosphorescent bands of water coincided with bands of substantially smoother waves. It is possible that this effect is associated with the fact that phosphorescent plankton release certain organic surfactants which dampen the high-frequency components of the surface waves. Other examples (see, for example, *Marine Observer*, 1953, **XXIII**, N161, p. 134) show that the origin of current rips is sometimes connected with underwater eruptions and shocks. A most interesting observation of concentric waves was reported recently by Marinin, chief mate of the vessel *Professor Pavlenko*. The waves were observed on 8 June 1984, from 1140 to 1220 Moscow time, in the Bay of Neretva in the Adriatic Sea about 6–7 miles off the coast at a depth $H = 50$–60 m, $43°01.6'$ N, $17°21.5'$ E. It started with a bright whitish patch 1–2 miles from the vessel. Then waves began to propagate from the patch in the form of concentric circumferences of light-green colour against a background of darker water (Fig. 2.17). The wave front propagated toward the vessel at a velocity of about 100 m/min (1.67 m/s) with an average wavelength $\bar{\lambda}$ (judging by the photographs) of 6–7 m. The visibility of the waves was associated with modulation of the suspended matter

Figure 2.17. Concentric waves generated by an underwater shock in the Bay of Neretva (Adriatic Sea) which are visible due to modulation of the suspended matter concentration in the thin near-surface layer (photograph by L. K. Marinin).

concentration in a thin freshened near-surface layer with a thickness of not more than 0.3–0.5 m, and, probably, with modulation of the thickness of this layer. The latter layer is created by the Neretva river discharge (average discharge 280 m^3/s), carrying to the bay not only fresh water, but also a considerable quantity of suspended matter of lime composition due to the karst character of the locality in which the river originated. The suspended matter was uniformly distributed in the entire thin near-surface layer before the formation of waves. The forming waves produced alternating bands of divergence and convergence in the near-surface movements of the water that formed progressively further away from the source alongside the propagation of the waves. The initial white patch apparently formed as a result of the convergence of currents over the site of an underwater shock caused, most likely, by an abrupt karst hole in the bottom. The observed velocity of the wave front propagation corresponds to the group wave velocity c_g, which, as stated before (see Section 2.2), is equal to half the phase velocity. Accordingly, the phase velocity c of the concentric waves should have been 3.3 m/s. This high phase velocity at a relatively small wavelength $\bar{\lambda}$ demonstrates that in the given case we are dealing not with internal waves, but with surface ones. As the given case deals with 'deep' water ($\lambda \ll H$), the correctness of the latter conclusion is confirmed by the relationship $c = (g\bar{\lambda}/2\pi)^{1/2}$, whose performance is easily checked in this case. The internal waves have a considerably lower phase velocity (0.3–0.4 m/s) in the given situation. According to some photographs taken by L. K. Marinin, the light wind waves and ripples also smoothed out slightly over the light bands of concentric waves.

All the information discussed above gives an idea of the règions with the most frequent manifestations of fronts, current rips, and internal waves on the open surface, and on the qualitative variety of these manifestations. Although it is impossible to consider the map (Fig. 2.15) and other data contained in this section to be complete and final, we think it is necessary to acquaint specialists with the information analysed here in the interests of developing current investigations of phenomena on the ocean surface. Replenishment of such a map may be based on archival and new field data, including satellite information purposefully collected for individual regions, first of all regions with the greatest repetition of phenomena in which researchers are interested.

Chapter 3

The thermal boundary layer of the ocean and the primary scales of convection

A hierarchy of thermal, salt, and viscous boundary layers can be distinguished owing to the specific conditions and processes of heat, mass, and impulse exchange between the ocean and the atmosphere near the interface [58]. The primary boundary layers in the water in this hierarchy, which we intend to consider here, are very close to the surface and are the thinnest ones. Molecular processes (heat conductivity, diffusion, viscosity) play a dominant role in their dynamics, as a result of which they concentrate the maximum vertical gradients of temperature, salinity, velocity, and many other characteristics. In principle, the entire sphere discussed below is more often called a viscous molecular sublayer. As our attention will be concentrated mainly on its thermal structure, we will consider the part of the sublayer where the maximum vertical temperature gradients are concentrated as a 'thermal boundary layer'. The expression 'salt boundary layer' will be used in the same sense but as applied to salinity. This approach is justified the more so as the stated boundary layers should have a different thickness inside a viscous molecular sublayer owing to the different coefficients of heat conductivity and diffusion. Comprehension of the regularities of the behaviour and formation of the thermal boundary layer structure is of paramount importance when investigating the characteristics of the near-surface layer of the ocean.

Numerous measurements made under field (in the ocean and freshwater bodies) and laboratory conditions (see refs [2, 25, 26, 28, 29, 69, 82, 116, 140, 151, 167, 169, 200, 210, 219, 236, 238]) have shown that the water surface temperature T_0 can differ greatly from the temperature of the underlying layer T_w (or T_h depending on the profile $T(z)$, see Fig. 1.1). The difference $\Delta T_0 = T_0 - T_w$ (or $T_0 - T_h$), its sign and the character of the temperature distribution in the depth $T(z)$ in the thermal boundary layer and its thickness are determined by the conditions of local interaction of the ocean and atmosphere and, first of all, by the value and direction of the heat flux q_Σ per unit of surface area, which is equal to the sum of the components (with respective signs) due to evaporation q_E, contact heat exchange with the atmosphere q_T, long-wave effective radiation q_R (difference of fluxes due to self-radiation of the ocean and atmospheric counter-radiation), and the flux of penetrating solar radiation q_s:

$$q_\Sigma = q_E + q_T + q_R + q_s = q_0 + q_s, \qquad (3.1)$$

where the positive is the direction of the heat flux from the air into the water. For the sake of simplicity and a better understanding of the physics of the thermal boundary layer, we will neglect the component q_S,* assuming that $q = q_0$. Methods of determining the components of the flux q_E, q_T, and q_R are given in numerous works (e.g. refs [69, 82]) and are not discussed here. We will note only that in the majority of cases, as applied to the ocean, the principal contribution to q_0 is introduced by the component of evaporation q_E. The difference $\Delta T_0 > 0$ at $q_0 > 0$ (surface heating), $\Delta T_0 = 0$ at $q_0 = 0$, and $\Delta T_0 < 0$ at $q_0 < 0$ (surface cooling). The latter case is most typical for the ocean [69, 116]. The presence or absence of sources of turbulence and convective motions in the water, the wind velocity, the character of the waves, and the presence of surfactant films on the water surface, etc. are additional factors which determine ΔT_0 and the structure of the layer.

3.1. On 'warm' and 'cold' boundary layers

The character of the temperature distribution in water near the interface has attracted the attention of scientists for at least 50 years. Measurements made by Altberg and Popov on the Neva river [2] were a pioneer study in this direction. They demonstrated for the first time that the maximum temperature gradient in water was observed near the surface in a layer 1 mm thick. A similar result was obtained later in works carried out by different authors, who also showed that the greater part of the temperature drop ΔT_0 was often concentrated in the top 1 mm layer. Discussions began on the top 1 mm phenomenon which resulted in, in our opinion, unsatisfactory terms, e.g. 'cold film' ($\Delta T_0 < 0$), 'warm film' ($\Delta T_0 > 0$), 'skin layer', etc. The drawback of these terms is associated mainly with the fact that the thickness of both the 'cold' and 'warm' layers may be quite variable, from several millimetres to tens of centimetres and more, depending on the character of the density stratification near the surface. In water with a salinity $S > 24.7\%$ (freezing temperature above maximum density temperature), positive values of ΔT_0 are always accompanied by a stable density distribution, and negative values by an unstable distribution which contributes to the development of convection. At $S < 24.7\%$, the same value of ΔT_0 can be associated with a stable and an unstable distribution of the density; therefore, the sign of ΔT_0 is unable to characterize the condition of the boundary layer. The latter should be considered when analysing the results of field measurements, because stable and unstable vertical distributions of the density near the water surface are accompanied by an essentially different structure of the boundary layer (see below). In particular, the measurements of Altberg and Popov [2] ($\Delta T_0 < 0$) made in fresh water in a temperature range below $+4°C$ cannot be considered from the same positions as other measurements in the cold boundary layer at $T_w > 4°C$, as is done in some works (e.g. ref. [69]). Despite the negative

---

*The influence of the volume absorption of the total solar radiation on the thermal structure of the near-surface layer is discussed in Chapter 4, and the accounting q_s, when estimating the parameters of the cold boundary layer, is considered in Section 3.3.7.

sign of ΔT_0, the thermal structure in this case corresponds to the hydrostatic stability of the boundary layer.

Measurements near or below the maximum density temperature are not considered in the text here; therefore, the terms 'warm' and 'cold' will be associated only with a quite definite structure and condition of the boundary layer.

3.2. The structure of the warm boundary layer

The interface remains continuous and heat exchange through it is molecular under normal interaction of the ocean with the atmosphere (water and air). Therefore, it can be assumed that the variation of the vertical temperature profiles with time should be similar to their variation in a solid body when a water layer with a temperature above the maximum density temperature and in the absence of permanent heat sources and turbulence inside this layer is heated from the surface ($q_0 > 0$). The heat flux through the surface can be assumed constant during the short periods of time considered in this chapter. As follows for the boundary condition of the second type ($q_0 = $ constant), which is closer to the conditions of heat exchange through a free water surface [25, 26], the temperature profiles at $T(z,0) = T(\infty, t) = T_w$ are

$$T(z,t) = T_w + \Delta T(z,t) = T_w + \frac{2q_0}{c_p \rho k_T} \sqrt{k_T t} \left[\frac{1}{\sqrt{\pi}} e^{-\xi^2} - \xi \operatorname{erfc} \xi \right], \qquad (3.2)$$

where $\xi = z/2\sqrt{k_T t}$, ρ is the water density, c_p is the specific heat capacity of the water at constant pressure, and t is the time. A comparison of the theoretical profiles (3.2) in the dimensionless form $\Delta T(z,t)/\Delta T_0$ [here $\Delta T_0 = \Delta T(0,t)$] for various values of t with the experimental values, obtained by us (Fig. 3.1a) in the laboratory for fresh water, indicates that the stated similarity is actually true at all $T_w > 4°C$ [25]].

According to (3.2), the temperature drop ΔT_0 in a warm layer and the thickness of the layer are determined in the absence of wind or a velocity shear in the water by the value q_0, which is associated with contact molecular heat exchange, and the time of heating. However, the absence of various disturbing factors is hardly typical of the ocean or natural water bodies. A wind can cause the development of shear instability of the Kelvin–Helmholtz type and the origin of eddies with a horizontal axis due to which inversion and quasi-homogeneous layers arise on the temperature profiles of the warm layer. This has been demonstrated by the field observations of Thorpe in Loch Ness [227] and the recent laboratory experiment of Dykhno [45] (Fig. 3.1b). Molecular heat transfer is preserved in this case only near the very surface, while a common turbulent boundary layer appears below the warm boundary layer. The wind can even change the initial direction of the heat flux due to the increasing evaporation from the water surface which induces the development of convection and conversion of the warm boundary layer into a cold one. These transitions of a thermal boundary layer from a stable condition to an unstable one and back have been observed in the laboratory experiments of Ewing and McAlister

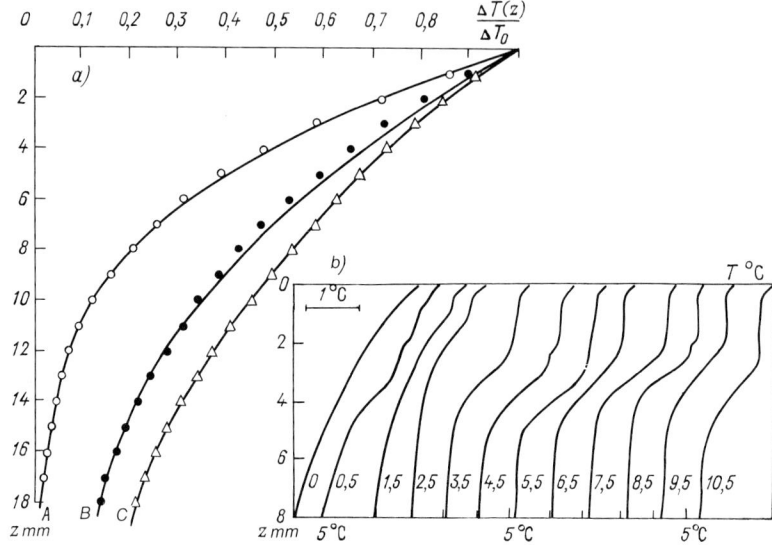

Figure 3.1. Vertical temperature profiles in a warm thermal boundary layer. (a) In the absence of air blowing. A, B, C (solid lines): profiles calculated by (3.2) at $t = 120\,\text{s}$, $t = 720\,\text{s}$, and $t = 1020\,\text{s}$, respectively; circles and triangles: experimental points [25]. (b) In the presence of air blowing ($u_0 = 2.2\,\text{m/s}$); the numbers indicate the time of measuring profiles in minutes (0 corresponds to air blowing being switched on) [45].

[140] with saline water, and in laboratory and field (freshwater body) measurements made by Panin [82].

The presence of other factors under natural conditions which are not associated directly with the wind, e.g. volume absorption of solar radiation, subsurface shearing current, salt convection due to evaporation, can also have an appreciable influence on the stability and structure of the warm layer; therefore, the real profile $T(z)$ near the surface at $q_0 > 0$ may, apparently, differ greatly from the theoretical one (3.2). Nevertheless, it is obvious that the thickness of the warm boundary layer can reach several metres under favourable conditions.

3.3. The cold boundary layer

The structure of the cold boundary layer is conditioned by the simultaneous effect of two mechanisms, i.e. the spread of cooling into the depth of the layer in conformity with the law of molecular heat conductivity according to (3.2), and convective instability. It is convection that limits the thickness of the cold boundary layer to several millimetres, which is its principal distinction from the warm boundary layer.

The development of convective movement associated with the cooling of water from the surface was noted in the very first works devoted to the cold boundary layer (the field visual observations of Woodcock [238] and the laboratory experiments of Ewing and McAlister [140] and Spangenberg and Rowland (see ref. [30]). However, the opinion of the stability of this layer under natural conditions lasted for a long time (see ref. [69]), while convection was entirely ignored as a mechanism limiting the cooling of the boundary layer as compared to the main thickness of the water mass. This point of view descended in many respects from the deceptive simplicity of using the Rayleigh number as the criterion of stability:

$$Ra = g\alpha\Delta T_0\delta^3/k_T v, \tag{3.3}$$

where g is the acceleration due to gravity, α is the coefficient of volume thermal expansion of water, v is the kinematic viscosity, and ΔT_0 and δ are the temperature drop and layer thickness, respectively. The point is that the concrete values of Rayleigh critical numbers, Ra_{cr}, are different under various conditions of origin of convective instability and configurations of the experiment. They depend on the number and character of the boundaries and boundary layers in the area, and on some other factors. It is also important to know definitely by which thickness of the layer δ it is necessary to calculate Ra_{cr} in each concrete case. Therefore, formal application of Ra_{cr} values, taken from experiments or models that are not in conformity with the situation under study, may be the cause of erroneous conclusions. For example, according to the estimates of Ball (see the discussion in ref. [30]), in the case of unsubstantiated adoption of $Ra_{cr} = 657$, the maximum possible value of $|\Delta T_0|$ which does not cause convective instability in a boundary layer of thickness $\delta = 1.5$ mm is 13°C. The latter estimate was used in several works to draw the conclusion that convective instability never arises in the thermal boundary layer because the value of $|\Delta T_0|$ never reaches 13°C under real conditions of ocean–atmosphere interaction. But $Ra_{cr} = 657$ relates to a thin layer of fluid with *two* free boundaries with no friction on them and is not applicable to conditions near one free boundary in a real viscous fluid. In the case of a cold boundary layer at the free surface of water, as is shown in ref. [30], the value of the critical boundary number Ra_{cr} is 64 (see Section 3.3.3). When determining Ra_{cr}, the thickness of the boundary layer is used in the capacity of δ instead of the full thickness of the fluid layer involved in convection. We will use the designation Ra in cases where it is necessary to calculate the Rayleigh number by the entire thickness of the fluid layer.

Detailed investigations of the thin structure of the cold boundary layer and the consistencies of its cooling made in the laboratory by us [25, 26, 28, 29] and Katsaros *et al.* [169] have shown that the boundary layer is not stationary during heat transfer from the surface, and its behaviour is inseparably linked with free (in the absence of wind, waves, and velocity shear in the water) or forced convection. Although the results obtained in refs. [25, 26, 28, 29, 169] relate to freshwater, in the majority of cases they can be applied, as is shown in refs. [31, 106], to water of oceanic salinity. The contribution of salinization at evaporation from the ocean surface to the formation of the cold boundary layer is discussed in Section 3.3.4.

3.3.1. *The nature of convective motion and the primary scales of convection developing near the free surface of fresh- and salt water*

The development of convection in water with a negative resulting heat flux from the surface ($q_0 < 0$) is well observed by contact measurements when a thermal sensor is placed near the surface on several horizons [12, 25, 26, 169], and by various visualization techniques [23, 169], or by obtaining an image of the surface temperature by means of an IR radiometer [3].

The beginning of cooling (in the laboratory experiment it is the moment of removing the cover from the basin, $t = 0$ in Fig. 3.2a) is accompanied by a sharp drop in the water temperature on the surface and near by it as against the initially homogeneous by depth value of T_w. When the difference $\Delta T_0 = T_0 - T_w$ reaches a certain critical value (point B in Fig. 3.2a), instability sets in* and convection commences. From this time the surface temperature T_0 and the values of ΔT_0 gain a fluctuating character (Fig. 3.2).

Analysis of the histograms $\Delta T_0 / \overline{\Delta T_0}$, where the overbar indicates averaging by the chosen time period or the period of measurements, has shown [30] that the maximum value of the temperature drop in the boundary layer under free convection is not more than $1.5|\overline{\Delta T_0}|$ and the minimum is practically never less than $0.5|\overline{\Delta T_0}|$. The law of distribution of the $\Delta T_0 / \overline{\Delta T_0}$ values is close to normal, and the average value of $\overline{\Delta T_0}$ is close to the most probable one. This is the value which is used below to characterize the cooling of the boundary layer (Fig. 3.2a).

The character of the temperature fluctuations changes with the depth. If at the surface they are of more or less stochastic nature with a relatively high frequency (Figs 3.2a and 3.2b), then beginning with the level of approximately 1 mm cold quasi-periodic 'outbursts' from almost constant temperature are observed (Fig. 3.2b). The mean time interval between these 'outbursts' t_c is approximately constant by depth in the upper layer with a thickness of several millimetres. The amplitude of the cold 'outbursts' of temperature decreases with increasing distance from the surface, while t_c increases due to heat exchange with the surrounding water. Beginning with a depth of about 12–14 mm, the 'outbursts' are practically not sensed at a sensitivity of the measuring method of $\simeq 0.01°C$. The character of the temperature change with depth in the regime of forced convection is qualitatively similar, but the fluctuations of the surface temperature are of higher frequency and with a better expressed random character (Fig. 3.2c).

The different character of the temperature fluctuations at various horizons is associated with the specific features of the three-dimensional structure of the horizontal and vertical convective motions near the surface. Examples of the space pattern of convection in fresh- and salt waters obtained by using shadow visualization [23] are given in Figs 3.3a–3.3d. The light lines in the photographs correspond to the surfaces, along which the water cooled in the boundary layer

---

*The time of instability initiation depends on the value of q_0.

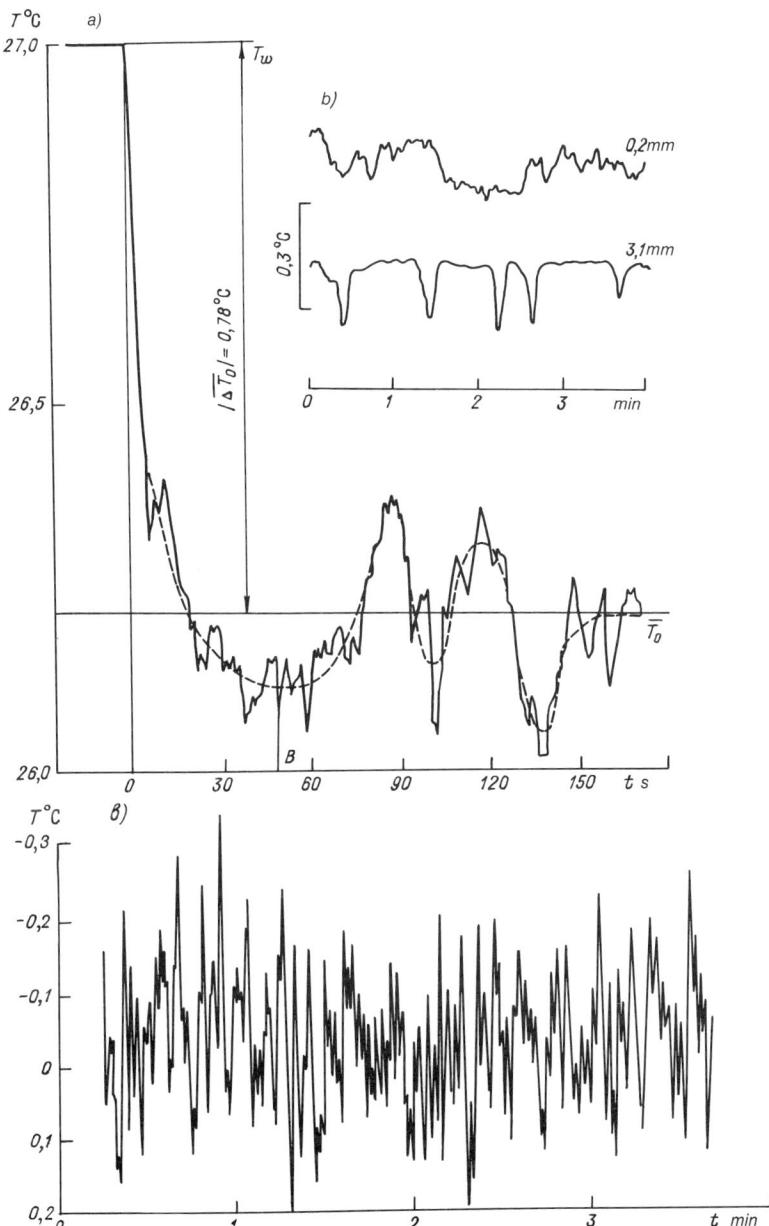

Figure 3.2. Character of the temperature fluctuations close to the free surface in laboratory experiments. (a) At free convection development ($z = 0.2$ mm, $q_0 = -320$ W/m^2, $t = 0$ corresponds to the time of removing the cover from the basin); (b) at developed free convection on two horizons ($z = 0.2$ mm and $z = 3.1$ mm, $q_0 = -120$ W/m^2); (c) at forced convection ($u_0 = 2.4$ m/s, $q_0 = -637$ W/m^2).

(a)

(b)

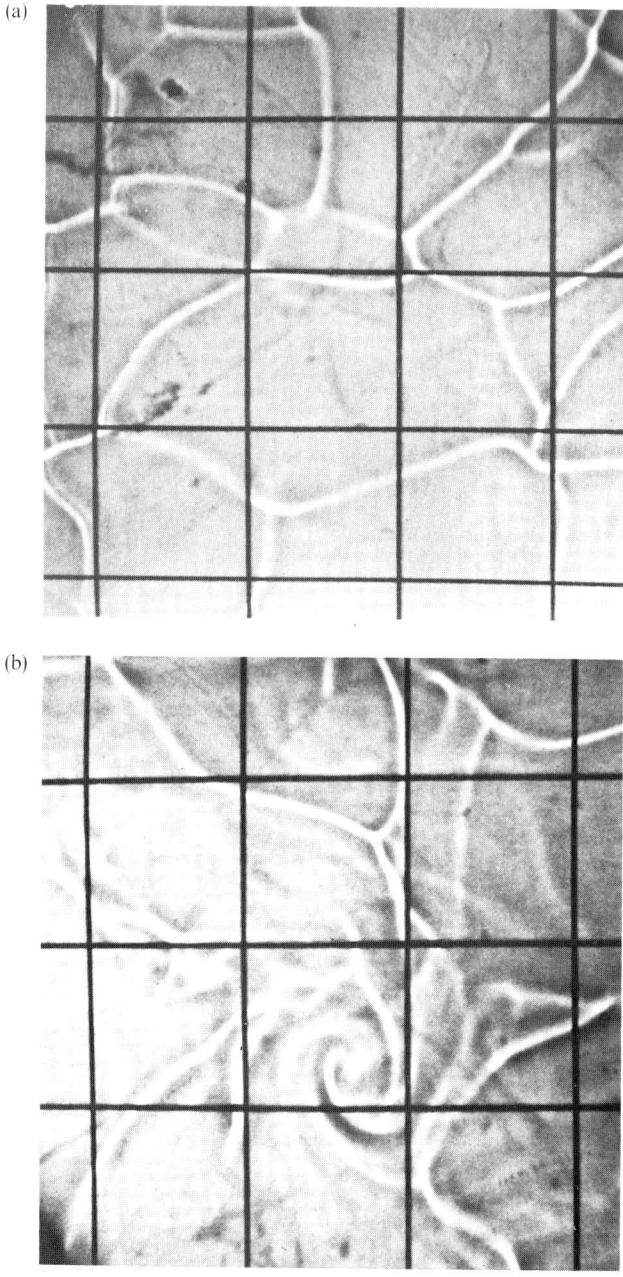

goes down, and to boundaries of the sinking cold convective elements. The instantaneous space distribution of the freshwater surface temperature T_0, obtained in ref. [3] by an IR radiometer, repeats the specific features of Figs 3.3a and 3.3b.

In the case of turbulent convection $(Ra > 10^7)$, the lines of sinking fluid

(c)

(d)

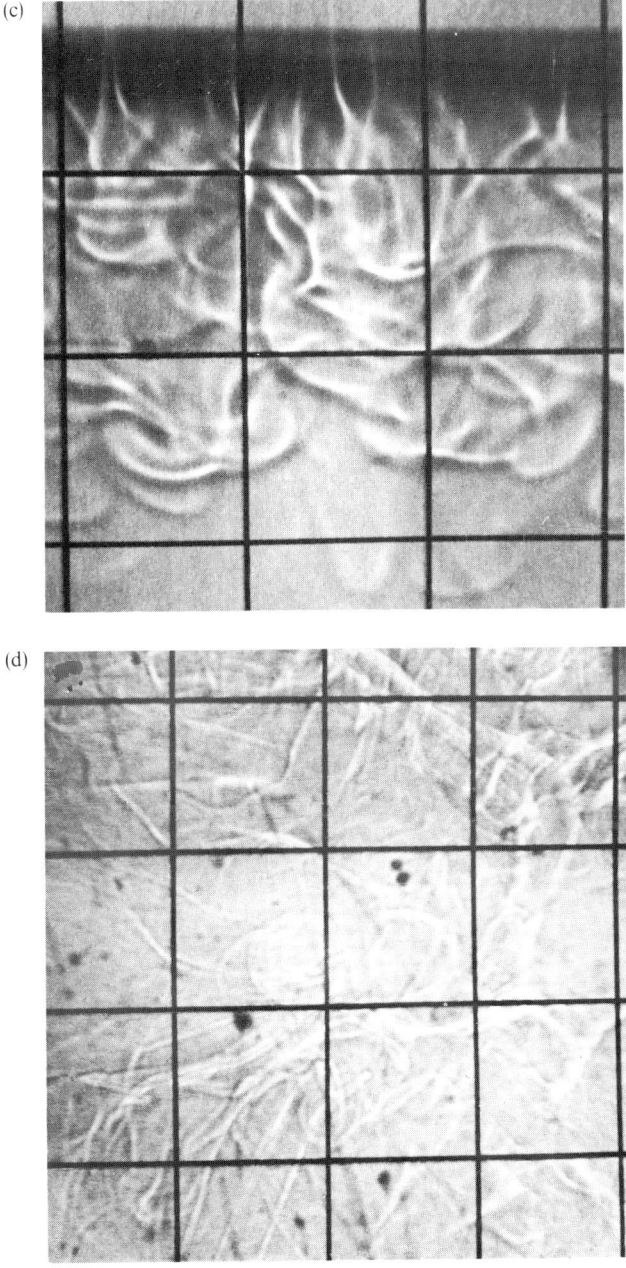

Figure 3.3. Images of convection obtained by the shadow visualization method in layers of fresh- (a, b, c) and salt (d) waters. (a) Top view, $H = 1$ cm, $Ra \simeq 2 \times 10^4$; (b) top view, $H = 10$ cm, $Ra \simeq 2 \times 10^7$; (c) side view, $Ra \simeq 10^7$; (d) top view, $S = 150\%_0$, $H = 10$ cm, $Ra \simeq 10^7$. Scale of graticule 2×2 cm^2.

(Fig. 3.3b) displace chaotically in the horizontal plane at an average speed of about 1 mm/s. Some of them disappear in time, while others form again. The water cooled near the surface moves convergently in the direction of these lines (surfaces of sinking) and along them. Vortical water motion is observed often at the crossing of two or several lines (vortex in Fig. 3.3b), and water sinking either along the vortex tubes or in the form of isolated vertical convective elements, thermals (Fig. 3.3c). The sinking speed of the convective elements is several millimetres per second.

The continuous surface cooling of the water is due to molecular thermal conductivity, its horizontal movement in the direction of the continuously displacing lines of convergence in the very upper layer with a thickness less than 1 mm (see Fig. 3 in ref. [169]), and the upwelling of the warmer water in the space between the regions of sinking creates a complicated non-stationary temperature field close to the surface which is recorded by a point thermal sensor in the form of fluctuations (Figs 3.2a and 3.2b).

When moving away from the surface, mainly the horizontal movements are gradually replaced by more ordered vertical ones, i.e. by comparatively rapid sinking of the convergence zones, and slow lifting in the spacings between them. The share of sinking regions in the total surface area, as given in Figs 3.3a and 3.3b, is rather small. When any element of convection passes through the measuring point, the thermal sensor, which is arranged at a depth of several millimetres, records a sharp drop in the temperature ('outburst'), whose duration, when recorded on a strip chart (Fig. 3.2b for $z = 3.1$ mm), is proportional to the time of displacement of the convective element through the measuring point. The average interval between the 'outbursts' ($\bar{t}_c$), which is the time scale and the characteristic of turbulent convection intensity [23], is determined by the value of the heat flux q_0 through the interface.

The character of temperature fluctuations on horizons below 1 mm (Fig. 3.2b) may give rise to the assumption of a cyclic (with a period of $\bar{t}_c$) character of the convection process with the layer growth in conformity with the molecular thermal conductivity laws during a greater part of the cycle (in the interval between the 'outbursts'), and destruction of the layer on reaching instability (during the 'outburst'). This is the scheme that was laid on the basis of the one-dimensional convection model of Howard [160] and in the models of Liu and Businger and Foster elaborated on the basis of the latter (see the discussion in ref. [26]). In reality, there is no regular cyclic growth and destruction of the boundary layer near by the free surface, and withdrawal of the cooled water from the free surface into the depth is continuous along the vertical and slanting sinking surfaces displacing in space. Nevertheless, the models of Howard and Foster are convenient for receiving averaged quantitative characteristics of the convection process and produce a result which is in accordance with the experimental data (see Sections 3.3.2 and 3.3.3). In particular, the theoretical dependence $\bar{t}_c = f(q_0)$, obtained by Foster [146], has been confirmed in our laboratory experiments [31], which gave the following relationship:

$$\bar{t}_c = 12.1 [v\rho c_p/(\alpha g q_0)]^{1/2} \tag{3.4}$$

for freshwater and

$$\bar{t}_c = 12.9[v\rho c_p/(\alpha g q_0)]^{1/2} \qquad (3.5)$$

for salt water (the discrepancy with the model is in the value of the constant factor, i.e. 12.1 and 12.9* instead of 14). The experimentally discovered centimetre scale of convection (Figs 3.3a and 3.3b) is also well in accordance with the predictions of Foster [146].

Convection in salt water ($S = 35‰$) is on the whole similar to the situation described above. However, the presence of frequent and thin light lines (Fig. 3.3d), existing simultaneously with still brighter lines, observed also in freshwater, is characteristic of the convective pattern of the salt water surface. A thermal sensor at a depth of 3 mm does not 'sense' thin lines, but in the upper millimetre layer thin secondary lines are sometimes accompanied by a slight drop in temperature. As the consistencies of heat exchange and the time scales of convection in salt and fresh-waters are practically the same (except for the low-temperature zone) [31, 106], these lines are most likely the result of purely mechanical crushing of the salt boundary layer in the process of convergent movement of the water in the thermal boundary layer toward the lines of convergence and along the latter. Then the forming 'folds' of the salt boundary layer are observed as secondary light lines on the shady image of turbulent convection in salt water. Their wavenumber (Fig. 3.3d) is approximately ten times greater than that of the convergence lines that characterize thermal gravitational convection, which corresponds to the actual correlation of the thicknesses of the two boundary layers.

As the surfaces of fresh- and salt waters are free ones, a natural question arises on the role of surface tension in the development of the observed convection. As is generally known, in fluids with a free surface, in the case of evaporation and cooling gravitational convection can develop due to buoyancy forces (when the Rayleigh number exceeds the critical value), as well as thermocapillary convection due to surface tension forces (when the critical value of the Marangoni number is exceeded). In water layers with a thickness more than 2.5 cm, both types of convection could act together, in principle, in the same direction. However, a comparison of the consistencies of convection in water (fresh and salt) and in fluid, where the thermocapillary mechanism deliberately operates simultaneously with the thermogravitational one (ethanol), has shown [23] that:

(1) a small-scale (millimetre) mobile pattern, which is characteristic of thermocapillary convection, is not observed in water regardless of the thickness of the layers;

(2) convection does not develop in water in general (at given heat fluxes) at a thickness of the layer less than 1 cm, while thermocapillary convection

---

*The value of the factor 12.9 for salt water was obtained by a considerably smaller number of measurement than for freshwater [31].

is observed in layers of other fluids (e.g. alcohol) of thickness even less than 1 mm;

(3)   the fine structure of the thermal boundary layer near the free water surface is similar to that observed near the solid boundary;

(4)   the characteristics of heat exchange through a free water surface do not practically differ from the characteristics of heat exchange through a solid boundary, while thermocapillary convection in alcohol greatly intensifies heat exchange through a free surface at a smaller drop of ΔT_0 in the boundary layer;

(5)   concavity of the surface, which is characteristic of gravitational convection, is observed where the cooled water sinks into the depth;

(6)   artificial decrease of the surface tension of water by scattering powdered soap on the surface does not affect the convective pattern of the boundary layer or on the characteristics of heat exchange. The convective pattern disappears only when the powder settles on the surface, and in this area there is upwelling of water from the depth which is easily recorded by a thermal sensor. The pattern of convection recovers in several seconds.

Hence, thermocapillary convection does not develop in water in general no matter what the thickness of the layers. Apparently, this is explained by the fact that water always adsorbs a great amount of surfactants and dust on its surface which protects it well from the development of thermocapillary convective effects. The behavioural consistencies of the boundary layer, its structure, and characteristics of heat exchange in water are determined at free convection only by the thermogravitational mechanism. Therefore, the term 'microconvection' proper and the explanation of the water cooled at the surface sinking into the depth by its gathering into drops by the forces of surface tension, as stated in some works (see ref. [69]), are illegitimate.

The observed centimetre scale of convection is, evidently, the primary scale of convection both in the laboratory and in the ocean.* It is confirmed by the visual observations of Woodcock [238] of the surface state in small bays, lakes, and in the open sea (on the surface of large waves) in the presence of a recorded cold boundary layer and wind. According to Woodcock, small-scale streaming is well distinguished under these conditions with continuously varying distances between the lines of flows (within several centimetres) lining up to the wind. The surface water draws into lines and lowers along the latter. 'Individual lines grow and die away within a few seconds time; sometimes converging upon other lines, sometimes diverging and disappearing' [238]. Comparing this pattern with the observations of free convection under laboratory conditions discussed above, it is easy to notice an almost complete analogy of the convective pattern behaviour and the convection scale in both cases with the exception of the fact that the lines of convergence are usually extended in the direction of the wind under natural conditions, as it should be in the case of convection under shear conditions [238].

---

*The hierarchy of convection scales in the ocean is discussed in Section 4.5.5.

operating a mixer in the water. In this case, the profiles become more linear, approaching in their upper part a straight line $Nu = 1$, and the transient layer between the linear part and the underlying layer with a zero vertical temperature gradient decreases sharply in thickness.

Analysis of the profiles, recorded under various regimes, makes it possible to determine the trend in the changes of the thickness of the boundary layer structural parts and their correlations depending on the degree of turbulization created in the water due to external (wind) and internal (convection and mixer) sources. The thickness of the layer with a quasi-linear part of the profile δ_l wherein the value $\Delta T(z/\delta_0)/|\Delta T_0|$ differs from an analogous value for a straight $Nu = 1$ by not more than by 3% is the least a free convection (0.4–1 mm). When shifting to forced convection at light winds, δ_l increases to 1–1.5 mm. At the same time, the proportion of δ_l in the total thickness of the boundary layer δ, whose lower boundary is conditionally adopted at level z, where $\Delta T_0(z/\delta_0) = 0.97\Delta T_0$, continuously increases with increasing water turbulization. In the limit $\delta_l/\delta \rightarrow 1$, although the total thickness of the cooled layer δ decreases. In an experimental case of creating intensive small-scale turbulence in the liquid by means of a mixer $\delta_l = \delta \simeq 0.5$–1 mm (Fig. 3.4c) and the temperature profile gains a bend, while the transient layer practically disappears in it.

Thus, all the experimental profiles under various conditions of the boundary layer are confined to the range between two theoretical dependences, i.e. on the one hand, profile (3.8), and on the other, a linear one

$$\overline{\Delta T(z)}/\overline{\Delta T_0} = 1 - z/\delta_0 \qquad (3.11)$$

at $z \leqslant \delta_0$ corresponding to $Nu = 1$.

3.3.3. On the critical boundary Rayleigh number for cooling water from the free surface

As stated above, formal utilization of inadequate values of the critical Rayleigh number (3.3) may lead to incorrect conclusions on the possibility of developing convective instability. Therefore, it seems important to estimate the critical value of the boundary Rayleigh number which is the criterion of most probable cooling of the water surface ΔT_0 under given heat exchange conditions as compared with the underlying layers.

The case of a cold boundary layer with a free surface, constant heat flux through the interface, penetration of convection outside the boundary layer downward, into the water thickness, and non-linearity of the temperature profile in the lower part of the layer is not in strict conformity with any known theoretical patterns of convection. Determination of Ra_{cr}^* directly from equation (3.3) is difficult because of the absence of a clear lower border of the boundary layer and, therefore, uncertainty in selecting δ. An experimental method of determining Ra_{cr}^* is suggested in ref. [30], wherein the boundary layer thickness proper is not used and which rests exclusively on easily measured values. The latter method of determining Ra_{cr}^* is based on the following considerations, resting on the known physical consistencies of free convection and the phenomenological theory of Howard [160] (with consideration of the reservations made in Section 3.3.1 and the results discussed in Section 3.3.2).

(1) The following correlation between the Nusselt (3.10) and Rayleigh (3.3) numbers is correct for free convection:

$$Nu = A\, Ra^{1/3}, \tag{3.12}$$

which does not depend on the thickness of the layer considered (A is an experimentally determined constant).

(2) The growth and destruction of the boundary layer are a cyclic process. Heat transfer through the boundary layer is molecular up to the occurrence of instability; therefore, the $Nu = 1$ equality is still true when the surface reaches maximum cooling. Then from (3.12)

$$Ra^*_{cr} = A^{-3}. \tag{3.13}$$

(3) As ΔT_0 is a continuously fluctuating value (Fig. 3.2), the critical boundary Rayleigh number should be found for the most probable value of ΔT_0, which is $\overline{\Delta T_0}$ (Fig. 2 in ref. [30]). The constant A, found in ref. [28] for $\overline{\Delta T_0}$ and constant q_0, is equal to 0.25. Hence, $Ra^*_{cr} = 64$ according to (3.13).

The value of Ra^*_{cr} obtained corresponds to conditions of developed convection at a mean drop of $\overline{\Delta T_0}$. It is natural that application of the maximum value of $|\Delta T_0|$, corresponding to the primary development of convective instability from an initial undisturbed condition of the water layer (the time soon after removing the cover, see Fig. 3.2a), would lead to smaller values of A and larger values of Ra^*_{cr}, respectively [30]. In particular, for $\Delta T_0 \simeq 1.2\overline{\Delta T_0}$ as in Fig. 3.2a, $A = A' \simeq 0.198$ and $Ra^*_{cr} = Ra'_{cr} \simeq 128$.

3.3.4. On the role of salinization of evaporation in the thermal boundary layer structure and dynamics. Salt boundary layer

The formation of a cold boundary layer in the ocean is accompanied by its salinization due to evaporation, which is an additional contribution to its instability. Therefore, it can be expected that a smaller space scale of convection and smaller values of cooling $|\overline{\Delta T_0}|$ will be observed in sea water than in fresh-water under the same heat exchange conditions. However, the pattern of convection under laboratory [23] and field [16, 238] conditions and the time scales of convection for fresh- and salt waters are approximately the same in practice [31] (see Section 3.3.1). The values of $|\Delta T_0|$ measured in salt waters (in the laboratory) are only 10% smaller than those calculated by us in accordance with the dependence in ref. [28] and coordinate well with the results of field measurements in the sea (see Section 3.3.5), while the thickness and structure of the thermal boundary layer in fresh- and salt waters are practically the same (within the measurement error) [106, 167], and are close to those obtained by Khundzhua et al. under field conditions in summer in the Black Sea [116]. This gives rise to the assumption that at least at a water temperature of 20–30°C the influence of the salinity on $\overline{\Delta T_0}$, the structure of the thermal boundary layer, and the character of convection is insignificant. We explained this fact in ref. [31].

Let us compare the speed of growth with time of the purely thermal and purely salt boundary Rayleigh numbers ($Ra^*_{(T)} = g\alpha\Delta\delta^3_T/\nu k_T$) and $\beta = (\partial\rho/\partial S)/\rho_0$ $Ra^*_{(S)} = g\beta\Delta S\delta^3_s/\nu D$), which are criteria of stability of the respective boundary layers. Here, D is the coefficient of salt diffusion; δ_T and δ_s are the thickness of the thermal and salt boundary layers, respectively; ΔS is the value of salinization of the boundary layer; $\alpha = -(\partial\rho/\partial T)/\rho_0$; and $\beta = (\partial\rho/\partial S)/\rho_0$, where ρ_0 is the initial density of water.

On the basis of formal analogy between thermal and salt convection, we adopt the varying in time t values $\delta_T = \sqrt{\pi k_T t}/2$ and $\delta_s = \sqrt{\pi D t}/2$ as measures of the thickness of the purely thermal and purely salt boundary layers, and we use the known laws of molecular diffusion of heat (3.2) and salt to calculate ΔT and ΔS:

$$\Delta S(z,t) = \frac{2F\sqrt{Dt}}{\rho_0\beta D}\left(\frac{1}{\sqrt{\pi}}e^{-\xi^2} - \xi\,\mathrm{erfc}\,\xi\right), \tag{3.14}$$

where $\xi = z/(2\sqrt{k_T t})$ in the case of heat and $\xi = z/(2\sqrt{Dt})$ in the case of salt (when calculating $Ra^*_{(T)}$ and $Ra^*_{(S)}$ we assume z equal to δ_T or δ_S, respectively); and F is the salt flow per unit of surface.

The results of calculating $Ra^*_{(T)}$ and $Ra^*_{(S)}$ in water with a salinity of $S = 35‰$ at water temperatures of 30 and 0°C are presented in Fig. 3.6a. In the calculations it was assumed that the heat flux was determined in both cases totally by the component due to evaporation ($q_0 = L_T E$), whose rate E was predetermined in each case so that it was in conformity with real oceanic conditions at the given values of the temperature (L_T is the latent heat of vaporization). At 30°C: $E/\rho_0 = 5.6 \times 10^{-6}$ cm/s, $F = 2 \times 10^{-7}$ g/(cm^2 s) and $|q_0| = 136.5$ W/m^2. At 0°C: $E/\rho_0 = 3.3 \times 10^{-7}$ cm/s, $F = 1.2 \times 10^{-8}$ g/(cm^2 s) and $|q_0| = 8.4$ W/m^2.

It is shown in Fig. 3.6a that at $T_w = 30°$C the value of $Ra^*_{(T)}$ reaches the critical value of $Ra^*_{cr} = 64$ much faster than $Ra^*_{(S)}$ does. As in salt waters salinization and cooling effects are summarized in layers of considerably different thickness, the conclusions of linear analysis of stability are not applied in this case. Actual instability of the thermohaline boundary layer should occur in the interval between the moments designated by the letters A ($Ra^*_{(T)} = 64$) and C ($Ra^*_{(T)} + Ra^*_{(S)} = 64$). In the given case, the interval is equal to 2 s only, and it can be considered in all the practical cases that the occurrence of instability in the boundary layer is determined fully by cooling.

The situation is quite different at $T_w = 0°$C. In this case, the salt boundary Rayleigh number $Ra^*_{(S)}$ slightly outstrips $Ra^*_{(T)}$ in its growth, and salt instability ($Ra^*_{(S)} = 64$) occurs 10 s before the thermal one. Here, the uncertainty in estimating the occurrence of thermohaline boundary layer instability reaches 100 s, which indicates a considerable contribution of salinity to convective instability.

The different contribution of temperature and salinity to the boundary layer convective instability at various water temperatures is also shown in Fig. 3.6b,

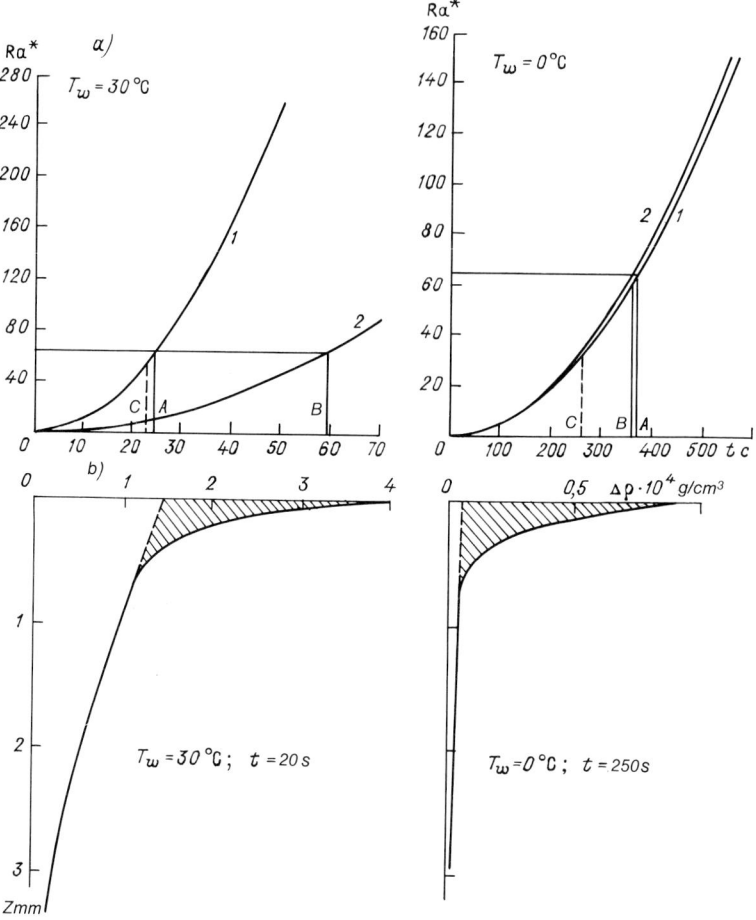

Figure 3.6. Difference in salinization contributions to convective instability of the thermal boundary layer at $T_w = 30°C$ and $T_w = 0°C$ in sea-water ($S = 35‰$). (a) Growth with time of thermal (1) and salt (2) boundary Rayleigh numbers at $T_w = 30°C$ and $T_w = 0°C$; (b) profile $\Delta\rho(z)$ at $T_w = 30°C$ and $t = 20$ s, and at $T_w = 0°C$ and $t = 250$ s. Area of integral salinization contribution is shaded.

where the calculated total density inversions $\Delta\rho(z) = \Delta\rho_{(T)}(z, t) + \Delta\rho_{(S)}(z, t)$ in the upper 3 mm of the boundary layer at $T_w = 30°C$ and $0°C$ are given. The contributions of cooling $\Delta\rho_{(T)}(z, t) = -\rho_0 \alpha \Delta T(z, t)$ and salinization $\Delta\rho_{(S)}(z, t) = \rho_0 \beta \Delta S(z, t)$ are calculated in conformity with (3.2) and (3.14), and the time moments t to which the profiles relate are chosen so that they are somewhat to the left of point C ($Ra^*_{(T)} + Ra^*_{(S)} = 64$) on the X-axis in Fig. 3.6a, i.e. in a deliberately stable situation: for $T_w = 30°C$, $t = 20$ s, and at $T_w = 0°C$, $t = 250$ s. The salt addition $\Delta\rho_{(S)}(z)$ is shaded in both cases. It can be seen clearly that the integral increment of density due to cooling (unshaded area) is considerably greater than the integral increment due to salinization (shaded area) in the case of warm water. The situation is just the opposite at $T_w = 0°C$.

A similar conclusion can be drawn when comparing time scales of convection for thermal $\bar{t}_{c(T)}$ and salt $\bar{t}_{c(S)}$ boundary layers [31]. Applying the relationship

$$\bar{t}_{c(T)}/\bar{t}_{c(S)} = (c_p F/\alpha q_0)^{1/2} \tag{3.15}$$

and assuming for simplicity that $q_0 = q_E$, we obtain at $q_E = L_T E$ and $F = ES$:

$$\bar{t}_{c(T)}/\bar{t}_{c(S)} = (c_p S/\alpha L_T)^{1/2}. \tag{3.16}$$

It follows from (3.16) that $\bar{t}_{c(T)}/\bar{t}_{c(S)}$ depends mainly on the temperature at a fixed salinity of the sea-water, as c_p and L_T remain practically constant in a wide range of natural temperatures and only α demonstrates a strong temperature dependence. Relationship (3.16) equals unity at $T_w \simeq 0.3°C$ for water with a salinity of 35‰. At all temperatures above 0.3°C, $\bar{t}_{c(T)} < \bar{t}_{c(S)}$ and at $T_w < 0.3°C$, $\bar{t}_{c(T)} > \bar{t}_{c(S)}$. Hence, instability of the thermal boundary layer occurs faster at high temperatures than instability of the salt boundary layer, and the situation is just the opposite at low temperatures. Then, in the first case the elements of thermal convection carry an excess of salt into the water thickness from a salt boundary layer which is not fully developed. In the second case, the elements of salt convection act in a similar way on the thermal boundary layer which is not fully developed.

It follows from the aforesaid that convection conditioned by evaporation is mainly thermal in character in a wide temperature range characteristic of the sea surface in temperate and low latitudes, and not salty although it is accompanied by the transport of excessive salt from the surface into the water thickness. The latter explains the satisfactory coincidence of the field measurements of $\Delta \bar{T}_0, \delta$, and the structure of the thermal layer in seas and oceans with the measurements in freshwater bodies and estimations obtained by formulae for freshwater. Only when the water temperature approaches freezing point in high latitudes do the contribution of salinity to the development of instability and its influence on the dependence $\Delta \bar{T}_0 = f(q_0)$ become more sensible and in some cases even determinative.

However, it should be taken into consideration that the estimations and constructions made above assumed equality of the heat flux q_0 to the component due to evaporation (q_E). In real conditions, $|q_0|$ is always greater than $|q_E|$, especially in conditions close to water freezing when the water–air temperature difference may be great and the flux component increases due to contact heat exchange, respectively. Therefore, thermal instability most likely will be determinative in the development of convection, even close to the freezing point. The presence of considerable drops of $\Delta \bar{T}_0$ (T_0 is below the freezing temperature) before the sea-water freezes (see ref. [167]) testifies in favour of the latter statement. Only in cases where the buoyancy flux is clearly determined by the salt component (e.g. when freezing-on ice) should salt convection develop, which is confirmed by the laboratory experiments of Foster (see the discussion in ref. [31]) wherein thermal convection in water was replaced by salt convection with considerably smaller space scales immediately after the formation of ice on the water surface (water cooling through a solid plate).

To estimate the potential salinization $\overline{\Delta S_0}$ of the boundary layer and its thickness δ_{OS}, it is possible to apply the known consistencies of convection development in freshwater assuming complete similarity between the heat conductivity and diffusion of salt [31]. Then the dependence determined in ref. [28] for $\Delta \overline{T_0}$ (see Section 3.3.5) is transformed into the expression

$$\overline{\Delta S_0} = 2.83 F^{3/4} \beta^{-1} (g \rho_0^3 D^2 / v)^{-1/4}. \tag{3.17}$$

Substituting (3.17) into the formula $Ra^*_{(S)} = g \beta \Delta S \delta_S^3 / vD$, at $Ra^*_{cr} = 64$ we obtain

$$\delta_{OS} = 2.83 D^{1/2} [v \rho_0 / (gF)]^{1/4}. \tag{3.18}$$

Estimation in conformity with (3.17) and (3.18) gives $\overline{\Delta S} = 0.32‰$ and $\delta_{OS} = 0.33$ mm for water of salinity 36‰ at $T_w = 28°C$ and $F = 1.2 \times 10^{-7}$ g/(cm^2 s) (evaporation rate 2.85 mm/day, an average value for oceanic conditions). From the relationship $\delta_{OS} = \sqrt{\pi D \bar{t}_{c(S)}}/2$ we receive a time scale of salt convection equal to 89 s. In real conditions, when thermal convection dominates, the values $\overline{\Delta S_0}$ and δ_{OS} at the same T_w and F should be less than the estimates 0.32‰ and 0.33 mm because $\bar{t}_{c(T)}$ is not in excess of 30 s and the salt boundary layer does not manage to reach a state of instability. The obtained estimation of δ_{OS} and Fig. 3.6b demonstrate that the thickness of the salt boundary layer is considerably smaller than the thickness of the thermal one, which agrees well with the assumption of Saunders [210] on the basis of Prandtl's theory, i.e. $\delta_{OS}/\delta_0 \sim (D/k_T)^{1/3}$. It is the thinness of this layer that makes it difficult to measure its thickness and $\overline{\Delta S_0}$. The first and, as far as we know, the only measurements made up to the present time giving an idea of the structure of the salt boundary layer were carried out by us together with Vlasov and Ambrosimov under laboratory conditions with the application of optical interferometry [106], making it possible to obtain space resolution of fractions of a millimetre without introducing disturbances into the structure of the thin layer under study. Variations in the salinity could be determined according to this method by the variations in the density, with an accuracy not worse than $\pm 0.01‰$.

The salinity profile was determined in the boundary layer of the sea-water by subtracting the profiles of the refraction index variations in salt and distilled water measured under the same conditions (at the same air humidity and the same difference in water and air temperatures). An example is given in Fig. 3.7 (profile A) of the refraction index $\Delta n_s(z)$ or density $\Delta \rho_s(z)$ vertical profile variation recorded in one experiment (see respective scales) in sea-water at $q_0 = -105$ W/m^2 and $T_w = 27.2°C$. The vertical profile (B) of the temperature variation $\Delta T(z)$ is distilled water at the same temperature is given in the same figure. The difference in profiles A and B gives the salinity profile (C). When recalculated to density in conformity with

$$\Delta \rho = \rho_0 (\beta \Delta S - \alpha \Delta T) \tag{3.19}$$

the contribution of ΔS_0 to $\Delta \rho$ on the water surface is approximately 1/3 and two times less than the contribution of cooling ΔT_0, respectively. Profile B for distilled water practically coincides with profile $\Delta n_s(z)$ for sea-water (A) at a

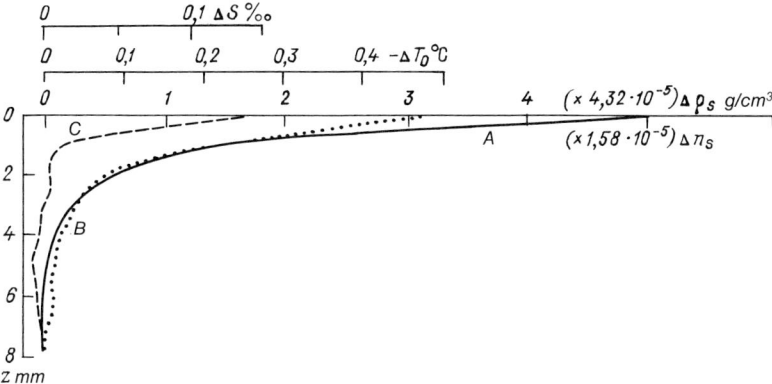

Figure 3.7. Vertical profiles obtained by the optico-interferometric method [106]. (A) $\Delta n_{\mathrm{s}}(z)$ or $\Delta\rho_{\mathrm{s}}(z)$; (B) $\Delta T_0(z)$ in distilled water; (C) salinization $\Delta S(z)$ in boundary layer.

depth below 1 mm, which confirms the identity of the temperature profiles in fresh- and salt waters under similar heat exchange conditions. Salinization on the surface ΔS_0 (profile C) reaches 0.15–0.16‰, which is twice as small as the maximal value of $\overline{\Delta S_0}$ obtained above for purely salt convection at $T_{\mathrm{w}} = 28°\mathrm{C}$. Note also that all the measured values of $\Delta\rho_{\mathrm{s}}(0)$ on the sea-water surface were systematically greater than the values of $\Delta\rho_0(0)$ in distilled water under other similar conditions. This means that salinization was present in all cases in the thin boundary layer of the sea-water. The thickness of the salinized layer was not more than 1 mm in all the measurements, while the thickness of the thermal boundary layer reached 5–7 mm, which coordinates well with contact measurements at free convection.

To estimate $\overline{\Delta S_0}$ in conditions of forced convection, Saunders [210] suggested a formula based on the similarity of the processes of heat and salt transfer through the interface:

$$\overline{\Delta S_0}/S = \lambda(k_{\mathrm{T}}/D)^{2/3}(\nu c_{\mathrm{p}}/\chi_{\mathrm{T}})E/(\tau/\rho)^{1/2}, \qquad (3.20)$$

where λ is a constant equal to 5–10 as in the case of temperature (see Section 3.3.5) and τ is the tangential wind stress. As E and $\tau^{1/2}$ are proportional to the wind velocity, $\overline{\Delta S_0}$ should depend only on the moisture contrast between the ocean surface and the atmosphere. Estimation according to (3.20) indicates that $\overline{\Delta S_0}/S$ is not in excess of 2% in real conditions [210] and, therefore, $\overline{\Delta S_0}$ is always less than 0.7‰. Measurements made by one of us during the 22nd cruise of the research vessel *Akademik Kurchatov* in the tropical zone of the Indian Ocean (April–May 1976) gave $\overline{\Delta S_0}$ equal to 0.477‰ in one case and to 0.243‰ in another (ΔS_0 was determined as the difference between the surface film salinity, gathered by means of a special device with a kapron gauze of the Garrett type, and the salinity at a depth of several tens of centimetres).

It should be borne in mind that the dependences (3.17) and (3.20) can give an estimate of the limit value of $\overline{\Delta S_0}$ only due to evaporation. When measurements are made in the ocean, the latter value can be much greater owing to the contribution of suspended matter to S_0 whereon sorption of soluble salts and organic substances of anthropogenic origin occurs. According to Mikhailov [71], the salinity of water samples collected from the ocean surface by the Garrett net was different from their salinity after filtration (removal of suspended matter and petroleum products) by 0.08‰ in the open ocean and 0.65‰ in coastal zones. The method of measuring is an additional factor influencing the values of ΔS_0 obtained in field conditions (see ref. [22]).

3.3.5. On the quantitative relationship $\overline{\Delta T_0} = f(q_0)$ during free and forced convection. Results of laboratory and field measurements

One of the first attempts to obtain the dependence $\overline{\Delta T_0} = f(q_0)$ by the dimensional argument was made by Saunders [210]. Saunders postulated the existence of a thin quasi-laminar viscous layer in the water close to the surface whose thickness δ was determined by the viscous shear stress τ' transmitted by the wind to the water, and also by the kinematic viscosity v and density ρ:

$$\delta_v \sim v/(\tau'/\rho)^{1/2}. \tag{3.21}$$

From the assumption that the greater part of ΔT_0 is concentrated in the layer of molecular heat conductivity within the viscous layer of thickness δ_v, it follows that

$$q_0 \sim \chi_T \Delta T_0/\delta_v. \tag{3.22}$$

The following expression is obtained from (3.21) and (3.22):

$$\Delta T_0 = \frac{\lambda q_0 v}{\chi_T (\tau/\rho)^{1/2}}. \tag{3.23}$$

Here, λ is the coefficient of proportionality, considering also the difference between the viscous shear stress τ' and the full wind stress τ applied to the water surface:

$$\tau = \rho_a c_D u_a^2 = \rho_a u_*^2 \tag{3.24}$$

(ρ_a is the air density, c_D is the coefficient of water surface resistance, u_a is the wind velocity measured at height a, and u_* is the friction velocity). According to Saunders, on the basis of a few field measurements, $\lambda = 5$–10.

Various authors have also suggested other expressions to determine ΔT_0 (see ref. [28]), but they almost all demonstrate the same dependence $\Delta T_0 \sim q_0/u_a$ and should give comparable results. A variant of formula (3.23), considering the thickness relationship of a layer with molecular heat conductivity and a viscous layer, was suggested by Wesely [236]:

$$\Delta T_0 = \frac{\lambda' q_0 v}{\chi_T (\tau/\rho)^{1/2}}, \tag{3.25}$$

where $\lambda' = \lambda Pr^{1/3}$ ($Pr = v/k_T$ the Prandtl number).

Formula (3.23) has been derived by Saunders under the assumption of similarity in the processes of heat exchange in the ocean boundary layer and close to a smooth rigid boundary. This similarity is actually observed in the case of free convection. Nevertheless, its existence is problematic in the case of forced convection. Therefore, the satisfactory quantitative correspondence of the measured values of ΔT_0 and those calculated by formula (3.23) could be random, all the more so as the spread of the λ values was great (5–10). However, there is still no other dependence than $\Delta T_0 = f(q_0)$ for forced convection conditions, and it is formula (3.23) that was experimentally verified under laboratory [28, 29] and field [151, 200, 210, 219, 236] conditions.

To estimate ΔT_0 under conditions of free convection, Saunders [210] suggested using the dependence following from the well-known relationship in the theory of convection between the Nusselt and Rayleigh numbers:

$$Nu = A\,Ra^n. \tag{3.26}$$

At $n = 1/3$ and taking into consideration (3.3) and (3.10), an expression is obtained which is independent of the layer thickness:

$$\Delta T_0 = A^{-3/4}(g\alpha c_p \rho \chi_T^2 / \nu)^{-1/4} q_0^{3/4}. \tag{3.27}$$

This dependence is verified most easily in laboratory conditions.

Results of laboratory measurements. Our laboratory experiments [28] and those of Katsaros *et al.* [169], made in freshwater in heat-insulated tanks of approximately the same dimensions, confirmed the equality $n = 1/3$ and gave slightly different values for constant A: $A = 0.22–0.25$ [28] and $A = 0.156$ in ref. [169]. This difference is most likely a consequence of the various methods of measuring $\overline{\Delta T_0}$. In ref. [169], the mean value of two averaged values $\overline{T}_{01}$ and $\overline{T}_{02}$, obtained as a result of uniform horizontal movement of the sensor (film resistance, diameter 15 μm, length 1.2 mm) on two levels, was adopted for $\overline{\Delta T_0}$: $\overline{T}_{01}$ corresponded to contact of the sensor with the surface (from above), $\overline{T}_{02}$ to a depth of 25 μm. In this case, the values of $|\overline{T}_{01}|$ could be overstated due to evaporation from the sensor surface as a result of its partial sinking into the water, which in turn could cause an understated value of A.

On the other hand, the value of constant $A = 0.25$ for freshwater obtained in ref. [28] could be slightly too high because it was not actually the surface temperature that was measured, but the temperature at some depth $z = 0.1$ mm due to the final size of the sensor (the diameter of the bead was 0.2 mm). The relative error of determining $\overline{\Delta T_0}$ was not more than 7%. However, taking into account that the measured values of $\overline{\Delta T_0}$ are approximately 10% smaller in salt water than in freshwater (with the same q_0) (see Section 3.3.4), it is possible to consider the value of the constant $A = 0.25$ to be the most appropriate for seawater. It also turns out to be the closest to the theoretical estimate on convection models (recalculated to one boundary layer) [28]. Thus, the dependence $\overline{\Delta T_0} = f(q_0)$ at free convection is qualitatively and quantitatively similar to the corresponding dependence for cooling through a rigid plate (3.27).

As we have shown in ref. [29], expression (3.27) is also applicable in a certain range of wind velocities over the water surface. This is clearly seen in the dependence of $m = \overline{\Delta T_0}/\overline{\Delta T_{00}}$ on u_0 in Fig. 3.8a, where $\overline{\Delta T_{00}}$ is the temperature drop in the boundary layer in the regime of free convection calculated by (3.27) at $A = 0.25$ with the same q_0 with which the drop of $\overline{\Delta T_0}$ is determined when the surface is blown over by the wind, and u_0 is the wind velocity determined at a height of 2 cm above the water surface. The relationship obtained in the laboratory investigations of Wu (see the discussion in refs [28, 29]) was used for approximate estimation of the respective values of the dynamic velocity u_*:

$$\frac{u_a}{u_*} = 5.75 \log \frac{a\rho u_*^2}{\sigma} + 4.29, \tag{3.28}$$

where σ is the surface water tension equal approximately to $7.3 \times 10^{-2}\,\mathrm{H/m}$ at $20°C$. The calculated values of u_* for various u_0 ($a = 2$ cm) and respective values of u_{10} [wind velocity at a height of 10 m estimated with consideration of (3.24)] are presented in Table 3.1.

The value m is actually the index of deviation of the boundary layer cooling regularity from the regime of free convection. In a wind velocity range from zero to some critical value $u_{ocr} = 2\text{--}2.5\,\mathrm{m/s}, m \simeq 1$* and, consequently, the dependence (3.27) is true. At u_{ocr}, the relation m reduces stepwise from 1 to 0.36 and does not change essentially at subsequent increase of the wind velocity.

The stepwise change $\overline{\Delta T_0}/q_0$ does not follow in itself from dependence (3.23). Comparison of this dependence with our experiment (Fig. 3.8b) demonstrates that λ is not a constant at all. With an increase of the wind velocity u_0 from 0.3 to $\simeq 2\,\mathrm{m/s}$, λ increases from 5 to 13, and with a subsequent increase of u_0, it again reduces to 5. Considering the spread of the experimental points, it can be assumed that $\lambda \simeq 6.5$ at $u_0 > u_{ocr}$. Different values of λ from 4 to 15 with a trend of a reducing λ with an increase of the wind velocity were also obtained in the laboratory experiments of other authors (see Table 3.3). It is obvious that

Table 3.1
Correspondence between u_0, u_*, and u_{10}

u_0 (m/s)	u_* (cm/s)	u_{10} (m/s)	u_0 (m/s)	u_* (cm/s)	(u_{10} m/s)
0.2	5.0	1.4	2.2	21.0	5.8
0.4	7.5	2.1	2.4	22.0	6.1
0.6	9.0	2.5	2.6	23.5	6.5
0.8	11.0	3.0	2.8	25.0	6.9
1.0	12.5	3.5	3.0	26.0	7.2
1.2	14.0	3.9	3.2	27.0	7.5
1.4	15.5	4.3	3.4	28.0	7.7
1.6	17.0	4.7	3.6	29.5	8.1
1.8	18.0	5.0	3.8	30.5	8.4
2.0	19.5	5.4	4.0	32.0	8.8

*Some difference of m from 1 in Fig. 3.8a is apparently related to periodic sensor emergence from the water when air is blowing over the surface.

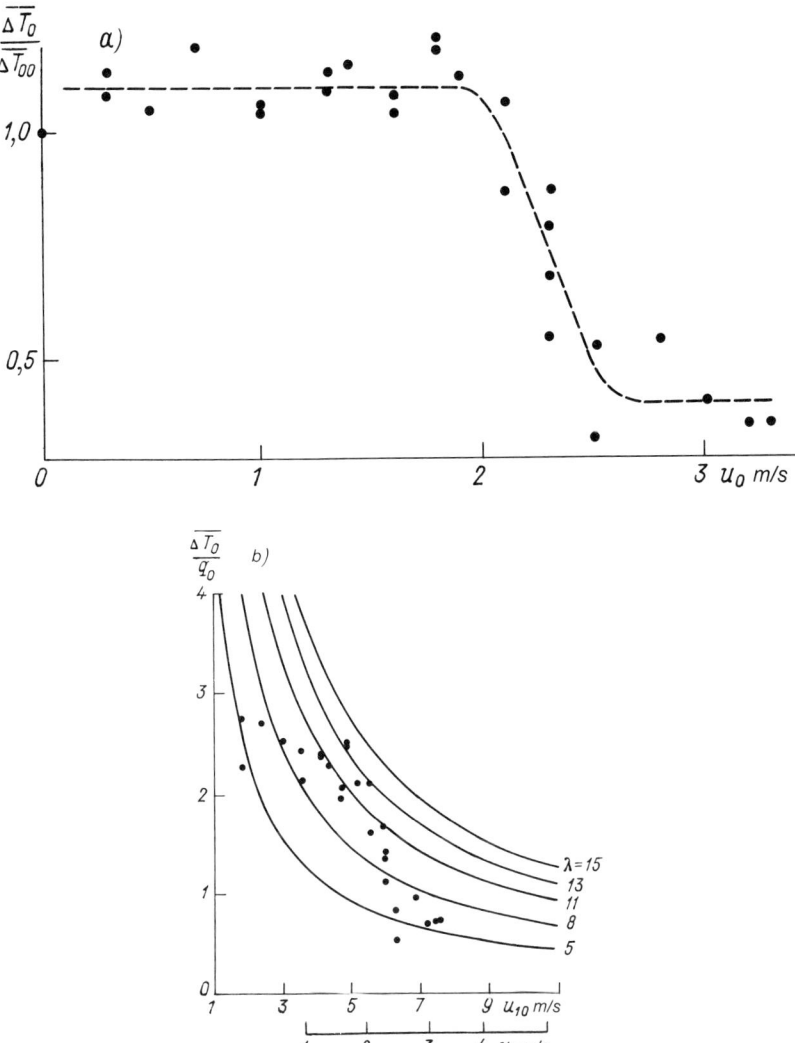

Figure 3.8. Change of boundary layer cooling regularity in a wind velocity range of $u_0 = 0-3\,\text{m/s}$. (a) $\overline{\Delta T_0}/\overline{\Delta T_{00}} = f(u_0)$. Dots on the graph correspond to mean values for each measuring cycle at $u_0 = $ constant and $T_w \simeq 18-25°\text{C}$; (b) $\overline{\Delta T_0}/q_0$ at $T_w = 20°\text{C}$. Experimental dots are the result of single measurements of ΔT_0 and q_0 at $u_0 = $ constant; hyperbolas correspond to dependence (3.23) at various values of λ, $T_w = 20°\text{C}$, and $c_D = 1.32 \times 10^{-3}$.

the dependence $\overline{\Delta T_0} = f(q_0)$ in a forced convection regime is considerably more complicated than the relationship (3.23), which can be a good approximation only at $u_0 > u_{\text{ocr}}$ ($\lambda = 6.5$).

It is an important problem to learn what conditions a change of the boundary layer cooling regularity at an excess of the critical wind velocity which is observed in the laboratory experiment. We have reported [29] that a sharp decrease in

$|\overline{\Delta T_0}|$ in the range $2 < u_0 < 2.5$ m/s, which is accompanied by an increase of the capillary waves' steepness in the tank, is associated with an increase in turbulence in the water due to shear stress under the effect of wind. This fully agrees also with the fact of the boundary layer thinning (see Section 3.3.2). Other sources of turbulence in water, i.e. mixing the water inside with a mixer (in the absence and presence of air blowing), artificial creation of shear in the water (by passing a sheet of paper over the surface when the latter is blown over with air) [29], or creation of waves on its surface by means of a mechanical source [167], render a similar influence in regard to $\overline{\Delta T_0}/q_0$ and the layer thickness. The experimental value of the critical velocity of air blowing, $u_{ocr} = 2$–2.5 m/s, obtained by us coincides practically with the values of wind velocity recorded by other researchers (see the discussion in ref. [29]) when an increase in the turbulence intensity is observed in the water, as well as the appearance of steep capillary waves, and a sharp increase in the gas exchange intensity across the air–water interface. Approximately the same air blowing velocity occurs at the formation of eddies of shear instability close to the surface at deliberately stable stratification in the water (laboratory experiment of Dykhno [45]). Consequently, the wind velocity $u_0 = 2$–2.5 m/s (or $u_* = 20$–23 cm/s) in the laboratory experiment is the velocity at which a sharp change apparently occurs in the behaviour of the boundary layer, both cold and warm.

A stepwise change of $\overline{\Delta T_0}/q_0$ may be caused by two factors in the case of a cold layer, i.e. by the presence of steep capillary waves on the surface and by dynamic instability of the boundary layer. It follows from the theoretical analysis of Witting (see the discussion in ref. [29]) that a change in the geometry of the boundary layer at the appearance of capillary waves (increase of the water surface area and thinning of the boundary layer, respectively, various thicknesses of the layer over the wave crest and troughs) without consideration of convective instability or turbulence in the water may cause a 7- to 8.5-fold decrease in $\overline{\Delta T_0}$ at $q_0 =$ constant as compared to a smooth surface, depending on the amplitude, wavenumber, and phase velocity of the waves. On the other hand, when the free surface is air-blown, dynamic instability of the boundary layer may arise alongside convective instability as it is close to a smooth rigid wall. The investigations of Klein et al. (see the discussions in refs [60] and [167]) have shown that periodic outbursts of water from the boundary layer into the depth occur in this case irrespective of the temperature gradient sign near the boundary on account of entrainment by turbulent eddies forming in the underlying layer. The mean period of these outbursts, which because of the similarity to convective processes can be called a time scale of dynamic instability t_d, is determined by the dependence [60]

$$\bar{t}_d = c_v v u_{*w}^{-2}, \tag{3.29}$$

where u_{*w} is the dynamic velocity in water ($u_{*w} = \sqrt{\rho_a/\rho}u_*$) and c_v is a constant. In the publications of Kudryavtsev and Luchnik (see ref. [60]) and the later publications of Kudryavtsev and Solovyev [60], application of the relationship of time scales of convective $\bar{t}_c$ and dynamic $\bar{t}_d$ instability [(3.4) and (3.29), respectively] is suggested as a criterion of the change of the boundary layer regime, i.e. regime of free convection at $\bar{t}_c/t_d < 1$, regime of forced convection at $\bar{t}_c/t_d > 1$.

An excess of the critical value by the flow Richardson number within the limits of a viscous underlayer is another form of the criterion of transition from free to forced convection [60]:

$$Rf_0 = \alpha g q_0 v/u_{*w}^4 > Rf_{ocr}. \tag{3.30}$$

According to estimates made in ref. [60] on the basis of laboratory [29] and field [151] experimental data, $Rf_{ocr} = -(1.4-1.5) \times 10^{-4}$.

The proportion of capillary waves ('geometrical factor') and turbulence per se in water in the total recorded decrease of $\overline{\Delta T_0}/q_0$ at $u_0 > u_{ocr}$ is unknown. But it is obvious that the increase of the capillary wave steepness is a result of their interaction with the velocity shear in the water which increases due to strengthening of the wind, which is confirmed by our laboratory experiment, i.e. artificial shear in the water created by passing a sheet of paper over the surface under light air blowing, which does not usually cause a deviation from the free convection regime, induces a sharp increase of the capillary wave steepness and a decrease of $|\overline{\Delta T_0}|$ (Fig. 3 in ref. [29]). Therefore, it is the critical value of the velocity shear and, consequently, a certain value of the wind velocity which cause a change in the layer behaviour and in the cooling regularity $\overline{\Delta T_0} = f(q_0)$.

Comparison of results of laboratory and field measurements. The values of constants A and λ in the dependences (3.27) and (3.23), obtained in laboratory and field experiments, are presented in Tables 3.2 and 3.3, respectively (the values of λ for refs [69] and [117] were calculated by us on the basis of the results of measurements discussed in the latter works). The data of Table 1 in ref. [28] have been used in Table 3.3 (Nos. 1–3, 6).

It follows from Tables 3.2 and 3.3 that the values $A = 0.25$ and $\lambda = 6.5$ (at $u_0 > u_{ocr}$) obtained in our laboratory experiment agree well with the results of field measurements. It has also been noted in refs [151] and [200] that λ depends on the wind velocity at $u_{10} < 3-4$ m/s, i.e. λ decreases upon abatement of the wind velocity. At $u_{10} > 3-4$ m/s, λ is practically unchangeable. The values of

Table 3.2
Values of constant A according to data of laboratory (l) and field (f) measurements

No.	Type of experiment	Author	Experimental conditions and measuring method	A
1	l	Katsaros *et al.*, 1977 [169]	Thermal sensor, $u_0 = 0$	0.156
2	l	Ginsburg and Fedorov, 1978 [28]	Thermal sensor, $u_0 = 0$	0.22–0.25
3	l	Kazmin, 1983 [53][a]	Thermal sensor, $u_0 = 0$	0.26
4	f	Shigayev *et al.*, 1982 [117]	Sea of Japan, vertical probing with thermocouple, $u_{10} < 0.5$ m/s	0.25

[a]The measuring method in ref. [53] is similar to that used by us [28].

Table 3.3
Values of λ according to data of laboratory (l) and field (f) measurements

No.	Type of experiment	Author	Experimental conditions and measuring method	λ
1	l	McAlister and McLeish, 1969 (see the discussion in ref. [28])	IR radiometer, fetch 70–160 cm, $u_\infty = 4.5$ m/s	4.5
2	l	Paulson and Parker, 1972 (see the discussion in ref. [28])	IR radiometer, small fetch (wave formation practically excluded), $u_0 = 1.39$ m/s $u_0 = 3.64$ m/s	15 15
3	l	Hill, 1972 (see the discussion in ref. [28])	IR radiometer, fetch 44.7 cm, $u_\infty > 7$ m/s ($u_* < 35$ cm/s) ($u_\infty > 7$ m/s) ($u_* > 35$ cm/s)	11 4
4	l	Ginsburg and Fedorov 1978 [29]	Thermal sensor, fetch 37 cm, $u_0 < 2.0$ m/s $u_0 > 2.5$ m/s	5–13 (increase in line with u_0) 6.5
5	f	Saunders, 1967 [210]	IR radiometer, except calm weather and light wind	5–10
6	f	Hasse, 1971 (see the discussion in ref. [28]	T_0 is assumed equal to T_a at neutral stratification, $u_{10} = 1.45$–11.35 m/s	8
7	f	Grassl, 1976 [151]	IR radiometer, $u_{10} = 1$–4 m/s	2.5–6 (increases in line with u_{10}) 5.5–6.5
8	f	Wesely, 1979 [236]	$u_{10} = 4$–10 m/s	
9	f	Malevsky-Malevich, 1974 [69]	IR radiometer, $u_{10} = 3$–8.5 m/s	6–7
10	f	Simpson and Paulson, 1980 [219]	IR radiometer, $u_{10} = 4$ m/s	6
11	f	Paulson and Simpson, 1981 [200]	IR radiometer, $u_{10} = 5.5$–9.2 m/s	7
12	f	Shigayev et al., 1982 [117]	IR radiometer, $u_{10} < 3$ m/s $u_{10} = 3$–11 m/s	<6.5 (decreases in line with u_{10}) 6.5
			Vertical probing with thermocouple, $u_{10} = 2$ m/s $u_{10} = 3$–4.5 m/s	5.4 6.3

u_∞ is the velocity of air flow in an aerodynamic tunnel outside the boundary layer.

u_{10} at which a deviation from $\lambda = $ constant is observed under field conditions are slightly different (3 and 4 m/s in refs [200] and [151], respectively) and differ from $u_{10} = 5$–6 m/s, which corresponds (see Table 3.1) to the wind velocity $u_{ocr} = 2$–2.5 m/s in our laboratory experiment. This variance of the velocities may be the result of applying different coefficients of c_D in refs [151] and [200] when shifting from u_* to u_{10} (3.24). Recalculation of the data in refs [151] and [200], done by Wu at the same c_D and supplemented with measurements in [219], gave [245]

$$\lambda = 2 + (5/7)u_{10} \quad \text{at } u_{10} < 7 \, \text{m/s};$$
$$\lambda = 7 \qquad \text{at } 7 < u_{10} < 12 \, \text{m/s}. \tag{3.31}$$

Apparently, the values of u_{10cr} should be subsequently defined more precisely by using more data of field measurements. It is also of interest to verify whether the dependence of this velocity on q_0 is preserved as follows from (3.30). Before these measurements are made with consideration of everything stated above, it is possible to recommend application of the dependences (3.23) and (3.27) in the case of respective wind velocity ranges at $A = 0.25$ and $\lambda = 6.5$ in the form

$$\overline{\Delta T_0} = 2.83(g\alpha c_p \rho \chi_T^2/\nu)^{-1/4} q_0^{3/4} \quad \text{at } u_{10} < 5.5 \, \text{m/s} \tag{3.32}$$

and

$$\overline{\Delta T_0} = \frac{6.5 q_0 \nu}{\chi_T(\tau/\rho)^{1/2}} = \frac{6.5 q_0 \nu}{\chi_T u_{10}(c_D \rho_a/\rho)^{1/2}} \quad \text{at } u_{10} > 5.5 \, \text{m/s} \tag{3.33}$$

Table 3.4
Values of $-\overline{\Delta T_0}$ for water of oceanic salinity ($S = 35$‰) calculated by (3.32)

$-q_0$		T_w (°C)					
(W/m²)	(cal/cm² min)	5	10	15	20	25	30
35	0.05	0.23	0.20	0.18	0.17	0.16	0.15
70	0.10	0.39	0.34	0.31	0.28	0.26	0.25
105	0.15	0.53	0.65	0.41	0.38	0.36	0.34
140	0.20	0.65	0.57	0.51	0.47	0.44	0.42
175	0.25	0.77	0.67	0.61	0.56	0.52	0.49
209	0.30	0.88	0.77	0.70	0.64	0.60	0.56
244	0.35	0.99	0.86	0.78	0.72	0.67	0.63
279	0.40	1.09	0.96	0.86	0.79	0.74	0.70
314	0.45	1.20	1.04	0.94	0.87	0.81	0.77
349	0.50	1.30	1.13	1.02	0.94	0.88	0.83
384	0.55	1.39	1.21	1.10	1.01	0.94	0.89
420	0.60	1.49	1.29	1.17	1.08	1.01	0.95
454	0.65	1.58	1.37	1.24	1.14	1.07	1.01
487	0.70	1.67	1.45	1.31	1.21	1.13	1.06
523	0.75	1.76	1.53	1.38	1.27	1.19	1.12
558	0.80	1.84	1.61	1.45	1.33	1.25	1.18
593	0.85	1.93	1.68	1.52	1.40	1.31	1.23
628	0.90	2.01	1.76	1.58	1.46	1.36	1.29
663	0.95	2.10	1.83	1.65	1.52	1.42	1.34
698	1.00	2.18	1.90	1.72	1.58	1.47	1.39

to estimate $\overline{\Delta T_0}$ by the measured values of q_0, or the dependence (3.23) in the entire range of wind velocities (except light wind) with consideration of (3.31). Table 3.4 presents values of $\overline{\Delta T_0}$ for water of oceanic salinity calculated by (3.32).

Evidently, the temperature drop in the cold boundary layer is not more than 0.5–0.6°C in the most typical ocean–atmosphere heat-exchange conditions, which is confirmed by field observations [69]. The trend of increasing $|\Delta T_0|$ (at $q_0 = $ constant) with decreasing water temperature conditioned by the temperature dependence of the kinematic viscosity v and coefficient of thermal expansion of the water α, which is absent in formula (3.33), is well manifested. Apparently, the real values of $|\Delta T_0|$ at low temperatures will be smaller than those stated in the table because of the increasing contribution of salinity to the total instability of the boundary layer at a decrease of the temperature (see Section 3.3.4). Experimental data for estimating the latter decrease of $|\overline{\Delta T_0}|$ are currently not available.

3.3.6. *Estimation of the thermal boundary layer thickness*

As shown in Section 3.3.2, it is difficult to determine the thickness δ because of the lack of a clear lower border of the boundary layer. The measure or scale of the layer thickness, as mentioned before, is the value δ_0, which is determined by formula (3.7). Proceeding from (3.7) and applying relationships (3.32) and (3.33), it is possible to derive formulae to estimate δ_0 in various wind velocity ranges:

$$\delta_0 = 2.83\chi_T^{1/2}(g\alpha c_p\rho/v)^{-1/4}q_0^{-1/4} \quad \text{at } u_{10} < 5.5 \text{ m/s} \tag{3.34}$$

and

$$\delta_0 = \frac{6.5v}{(\tau/\rho)^{1/2}} = \frac{6.5v}{u_{10}(\rho_a c_D/\rho)^{1/2}} \quad \text{at } u_{10} > 5.5 \text{ m/s.} \tag{3.35}$$

In the case of free convection and light winds, the value of δ_0 decreases with increasing heat flux, and it is inversely proportional to the wind velocity at $u_{10} > 5.5$ m/s. The same is true for the total thickness of the layer δ which equals $\simeq (1.8–2.3)\delta_0$ under free convection [25, 26] and approaches δ_0 with an increase of turbulence and shear in the water (see Section 3.3.2). The estimate by (3.34) gives $\delta_0 = 2.4$ mm at $q_0 = -60$ W/m^2 and $T_w = 24$°C, and from (3.35) we obtain $\delta_0 \simeq 0.7$ mm at the same T_w and wind $u_{10} = 6$ m/s, which agrees well with the results of laboratory (Fig. 3.4) and field [116] measurements.

As mentioned before, the thicknesses of the thermal boundary layer and the viscous underlayer practically coincide; therefore, the thickness of the former can be estimated by means of various formulae for the thickness of the viscous underlayer given in the work by Kraus [58] or Panin [82].

3.3.7. *Effects of various factors on the value of $\overline{\Delta T_0}$ under field conditions*

When applying the dependences (3.32) and (3.33) and interpreting the results of measuring $\overline{\Delta T_0}$ and T_0, it is necessary to consider the potential effects of certain factors on the boundary layer and the local conditions of heat exchange, i.e. solar radiation, cloud cover, presence of surfactant films, gravity and internal waves, etc. Let us consider individually the influence of each of these factors on the ΔT_0 drop.

Solar irradiance. Volume absorption of solar radiation close to the water surface does not usually cause termination of convection, with the exception of several hours close to midday during small heat transfer from the surface (see Section 4.3). However, radiation q_{sa}, which is absorbed in the boundary layer, can be commensurable with q_0; therefore, it is necessary to use $q = q_0 + q_{sa}$ instead of q_0 when calculating ΔT_0 by formulae (3.32) and (3.33). The component q_{sa} can be estimated as follows:

$$q_{sa} = q_s(0, t) - q_s(\delta_0, t) \tag{3.36}$$

[$q_s(0, t)$ ans $q_s(\delta_0, t)$ are fluxes of solar radiation at the surface and in the depth δ_0 at the time moment t, respectively]. As δ_0 is not more than 3 mm even under free convection, q_{sa} is not more than 8% of $q_s(0, t)$ according to Table 4.2. For example, $q_{sa} \simeq 45 \text{ W/m}^2$ at $q_s(0) = 560 \text{ W/m}^2$ (intensive solar heating). This causes a decrease of $|\Delta T_0|$ from 0.42 °C, corresponding to $q_0 \simeq -140 \text{ W/m}^2$ and $T_w \simeq 30°C$ in the absence of insolation, to $\simeq 0.31°C$ (see Table 3.4). The contribution of solar radiation to the change of ΔT_0 decreases with increasing wind velocity and decreasing δ_0. Obviously, the value q_{sa} is characterized by a diurnal cycle and reaches a maximum at local midday.

Cloud cover. Observations are known (see ref. [61]) which demonstrate a rise of the ocean surface temperature when a cloud cover appears over the point of measurements. The rise of T_0 is associated with a decrease (and probably a change of sign) of a component (q_R) of the total heat flux and, consequently, a change of ΔT_0. Usually the change of $|\Delta T_0|$ (decrease) under the influence of clould cover is several tenths of a degree. However, it is possible that the change of q_R may cause a change in the sign of q_0 and result in a transition from the cold to the warm boundary layer when the effective radiation q_R is the basic component of q_0. But even in this case it is hardly possible to anticipate a rise of T_0 by 3°C at the appearance of cloud cover as reported by Schooley (see the discussion in ref. [61]). The changes in T_0 and ΔT_0 may also be caused by circulations in the atmospheric convective cells which are accompanied by characteristic cloud structures (see Section 4.5.2) and also by local wind intensification when powerful cumulo-nimbus systems pass by.

Precipitation. A change (more often a decrease) of the ocean surface temperature under the influence of precipitation, as observed by several researchers when making measurements by means of an IR radiometer (e.g. [19, 200]), is the result of the effect of rain on the near-surface layer temperature (see Section 4.5.3) and on ΔT_0. Depending on the rain character, its influence on ΔT_0 may be diverse. Large raindrops create intensive mixing near the surface and completely destroy the cold boundary layer [167]. Precipitation, freshening the water, creates density stratification in the near-surface layer and may cause a rise of $|\Delta T_0|$ directly [12] and through a stabilizing influence on the waves and turbulence [167]. The quantitative aspect of the problem and the behaviour of the boundary layer in the case of precipitation has not been studied up to now.

Surface waves. The influence of capillary waves on $\overline{\Delta T_0}$ is discussed in Section 3.3.5. According to the theoretical analysis of Witting (see the discussion in ref. [29]), gravity waves can decrease $|\Delta T_0|$ almost 1.5 times, owing to modulation of the boundary layer thickness (thickening at the crest and thinning in the wave trough). The dependence of T_0 and $\overline{\Delta T_0}$ on the wave phase has been demonstrated by the laboratory experiments of Miller and Street and the field measurements of Simpson and Paulson from the research platform 'FLIP' [219] with the use of an IR radiometer [the difference between the mean and maximum values of T_0 was adopted for $\overline{\Delta T_0}$ in the assumption that at least once T_{max} is equal to T_w during the observation period (2 min) due to maximum thinning of the boundary layer]. It was found that the phase shear between T_0 and the wave is $-30°$ at the peak of the wave spectrum ($\simeq 0.06$ Hz). In this case, the maximum of T_0 (minimum of $|\overline{\Delta T_0}|$) is on the windward side of the gravity wave and is associated with an increase of the wind tangential stress and a decrease of δ either due to growth of τ or due to the appearance of capillary waves. The phase shear increases to $+100°$ at high frequencies (0.4 Hz). In the latter case, the maximum of T_0 corresponds to the steep parts of the wave on its lee side near the crest and is associated with instability (breaking) of the wave, or with an increase of the capillary wave amplitude [219].

Wave breaking causes local destruction of the cold boundary layer, which is recovered in about 12 s [140]. Therefore, the cold boundary layer exists at high wind velocities in the form of patches, and the mean value of $|\overline{\Delta T_0}|$, measured by a radiometer with an averaging scale greater than the wavelength, should be smaller than the value calculated by (3.33).

Internal waves. The influence of the internal waves on $\overline{\Delta T_0}$ has not been studied. It can only be supposed that the internal waves exert an influence on $\overline{\Delta T_0}$ through surface wave modulation, e.g. in the areas of surface smoothing (slicks). In this case, the character of the change of $\overline{\Delta T_0}$ can be diverse depending on the slick nature. If the slick on the surface is conditioned by a change in the turbulence intensity in the near-surface layer due to local breaking of the internal waves, it is possible to anticipate thinning of the boundary layer and a decrease of $\overline{\Delta T_0}$. But the appearance of a slick due to the accumulation of surfactant films in the convergence zones will result in capillary wave damping and an increase of $|\overline{\Delta T_0}|$ if the surfactants do not cause a decrease of the heat flux component as a result of evaporation.

Experimental evidence of the influence of internal waves on the boundary layer cooling regularity is not available. It is only known that the surface temperature T_0 can change under the influence of the internal wave field due to the redistribution of heat and space modulation of the temperature T_h in the near-surface layer mainly during the time of solar heating (see Section 4.5.4). Experimental research carried out by Kudryavtsev and Kuftarkov in the International programme JASIN-78 during the 18th cruise of *Akademik*

Vernadsky (see pp. 147–157 in ref. [85]) also demonstrated the relation of the ocean surface radiative temperature to the internal wave field under conditions of practically complete absence of temperature gradients near by the ocean surface (fluctuations of T_w within the UQL were in the limits of 0.02°C). The experiment was carried out under conditions of a relatively light wind ($u_{10} < 4$ m/s) during the evening and night, where in the daytime bands of slicks were observed from a distance of about 1 mile. Simultaneous measurements of the radiative temperature (by an IR radiometer) and the temperature at some horizons in the layers of sharpest temperature gradients in the seasonal thermocline (38–48, 50–60, 61–71 m) have shown that the fluctuations of the radiative temperature are correlated and are in phase with the vertical displacement of the thermocline, and are out of phase with the respective temperature fluctuations on the fixed levels in the thermocline. Radiative temperature deviations of 0.2 K correspond to a 5 m displacement of the thermocline. According to the authors [85], the latter deviations are due to a change in the radiating capacity of the ocean surface as a result of redistribution of the surfactant films by the internal waves.

Slicks, oily films, surfactants. Laboratory and field measurements carried out by means of an IR radiometer made it possible to determine that the radiative temperature of the water surface, T_{rf}, covered with various films (surfactants, oil, traces of biological activity), differs from the radiative temperature of pure water, T_r [59, 167, 200, 210]. The temperature difference $\Delta T_r = T_{rf} - T_r$ can be positive or negative, and its value apparently depends on the film thickness and the conditions of local ocean–atmosphere interaction. According to Kropotkin *et al.* [59], at oily film thicknesses less than $\simeq 10\,\mu m$, $\Delta T_r < 0$ and changes from $\simeq -0.1$ K in calm weather and under intensive radiation $[q_s(0) = 550$ W/m$^2]$ to $\simeq 1$ K at force 3 waves and the same $q_s(0)$. At film thicknesses more than $10\,\mu m$, ΔT_r and its sign in calm weather depend on $q_s(0)$. A 1 mm thick film (aqueous–oil emulsion) gives $\Delta T_r = -0.6$ K in calm weather and $q_s(0) = 0$, and $\Delta T_r = 5.5$ K in calm weather and $q_s(0) = 550$ W/m^2.

As none of the known radiometric measurements were accompanied by contact ones, it is impossible to state whether the recorded contrasts of ΔT_r appear only due to changes in the radiating capacity of the water in the presence of a film, or whether they reflect, at least partially, the real changes in the ocean surface thermodynamic temperature T_0. It seems physically verisimilar that the presence of a film can cause a change of T_0 (and $\overline{\Delta T_0}$), and, therefore, the contrast of ΔT_r, at least in some cases, indicates a change of the ocean surface thermodynamic temperature. Besides the mechanisms of film influence mentioned above (decrease of evaporation, damping of waves), there probably exist also others that are associated with their influence on the hydrodynamics of the boundary layer and its thickness [167, 210]: the film makes the surface substantially less mobile and increases the thickness of the viscous heat-conductive layer, which may cause growth of $|\Delta T_0|$. The measurements made by Kazmin [53] by vertical displacement of a thermal sensor near the surface covered with an oily film (0.5–10 μm) under conditions of free convection have shown that the structures of a boundary layer with and without an oily film are practically similar although the temperature fluctuations at the surface are of different character ($\sigma_T/|\overline{\Delta T_0}| \simeq 7\%$ unlike 15–18% at a clean surface).

It is obvious that the problem of the influence of films on the regularities of cooling (heating) the boundary layer demands subsequent studies as applied to each concrete type of film. Apparently, it is necessary to use the dependences (3.32) and (3.33) with caution in the presence of films.

3.3.8. On the space–time variability of $\overline{\Delta T_0}$ and the temperature of the ocean surface T_0

As follows from Fig. 3.2a, the response of the thermal boundary layer to a change in the local heat-exchange conditions is very quick, i.e. from several seconds until the ΔT_0 drop reaches the value $0.9\overline{\Delta T_0}$, to several tens of seconds until a steady character of T_0 fluctuations is reached. Approximately in the same time (15–20 s) ΔT_0 adapts to a new wind velocity or to some other turbulence level in the water (see Figs 2–4 in ref. [29]), which is confirmed by the rapid recovery (in 12 s [140]) of the cold layer after wave breaking. Therefore, the space and time variability of $\overline{\Delta T_0}$ is determined by the respective variability of the factors conditioning this difference (see above). They are seconds and tens of metres for waves, minutes and tens of minutes (and more) and kilometres for cloud cover, tens of metres to tens of kilometres for different types of films on the water surface, and so on. As the heat flux and the wind velocity are the main determinative factors, the sharpest change in $\overline{\Delta T_0}$ can be anticipated in the region of atmospheric fronts. A daily and seasonal variability of $\overline{\Delta T_0}$ should also be observed. Mapping of $\overline{\Delta T_0}$ on the basis of climatic data on q_0 or on data of direct measurements of $\overline{\Delta T_0}$ is, most likely, of no real physical sense for the reasons mentioned above.

The ocean surface temperature is determined by the sum of the components $\overline{T}_0 = T_w + \overline{\Delta T_0}$ (or $\overline{T}_0 = T_h + \overline{\Delta T_0}$); therefore, its variability is a result of variability of the near-surface layer temperature and of $\overline{\Delta T_0}$. With this is associated the greatly more 'high-frequency' character of fluctuations of the radiation temperature T_r, measured by an IR radiometer, as compared with the simultaneously measured temperature at a depth of 1 m (see Fig. 3 in ref. [200]) at coincidence of the common character of T_r and T_w. The maxima in the spectra T_r, T_w, and ΔT_0 in the experiment [200] (measurements from the buoy-vessel 'FLIP') were at the frequencies 1×10^{-5} Hz (1 day) and 2.5×10^{-6} Hz (5 days) and corresponded to the diurnal and large-scale atmospheric effects.

3.4. Methodological questions on measuring the temperature difference in the thermal boundary layer of the ocean

Measurement of ΔT_0 in the boundary layer presupposes determination of the ocean surface temperature T_0 and the temperature T_δ on the lower boundary of the layer δ within which this difference is concentrated. In the case of warm boundary layers, δ can reach tens of centimetres and more (see Section 3.2), and it is limited to several millimetres in the case of cold ones.* Therefore, the determination of ΔT_0 in the cold boundary layer requires a special method of measurement under field conditions. Three methods of measuring ΔT_0 are used most widely, i.e. (1) determination of ΔT_0 by vertical probing in a 0–30 cm water layer with the use of a low-inertial thermal sensor (thermocouple) [116, 117]; (2) measurement of T_0 by an IR radiometer, and T_δ by a contact thermal sensor. In this case, the temperature at a depth of several centimetres–tens of centimetres [69, 82, 151, 236] or at the depth of sampling by a bathometer is adopted as T_δ; (3) measurement of both T_0 and T_δ by an IR radiometer. The temperature of an artificially mixed layer of water with a thickness of several metres [140] (see also the discussion in ref. [22]), or the surface temperature at destruction of the boundary layer by the breaking wave [219] is adopted as T_δ.

Obviously, the value of ΔT_0 at correct determination of T_0 will be determined just by measuring T_δ. If at the time of measurement the water layer is isothermal within several metres below the boundary layer, then it makes no difference at what depth z_δ the reading of T_δ is done. The latter situation is possible either under intensive wind mixing or in the case of developed convection in the absence of solar heating. If there are deviations from isothermy in the upper several metres, which is quite characteristic in the case of intensive solar heating under conditions close to calm, then ΔT_0 is determined by the depth of measuring T_δ or by the thickness of the layer artificially mixed before measuring T_δ. Hence, the value of ΔT_0 measured by different methods depends greatly on the vertical distribution of $T(z)$ in the several upper metres of the ocean [22]. Let us demonstrate this by examples of profiles (Fig. 3.9) obtained by several researchers in approximately the same region of the Atlantic in calm weather and a light wind. As there is no information on the temperature difference near the surface at the time of measurement, the profiles in Fig. 3.9 are completed in their upper part with a boundary layer with a difference of $\Delta T_0 = -0.3°C$ [for $q_0 = -140\ \mathrm{W/m^2}$ and $q_s(0) = 560\ \mathrm{W/m^2}$, see p. 95] for approximate determination of T_0. The value of ΔT_0 in this case is of no importance because the errors in determining T_0 are much less than the methodological errors, which will be discussed below.

It follows from Fig. 3.9 that, depending on the character of $T(z)$ and the method of measuring T_δ, it is possible with the same actual $\Delta T_0 = -0.3°C$ to obtain high negative as well as high positive fictitious values of ΔT_0; $-1.3°C$ for profile b in Fig. 3.9 at $z_\delta = 0.3\ \mathrm{m}$; $2.0°C$ for profile c at $z_\delta = 1.7\ \mathrm{m}$, or at mixing up to horizon 7 m. Thus, the absence of information on the temperature variability in the upper 10 m (especially in the 1 m) layer of the ocean at the point of measurement may cause substantial errors in determining ΔT_0, which

---

*T_δ is equivalent to T_w or T_h depending on the circumstances (see Fig. 1.1).

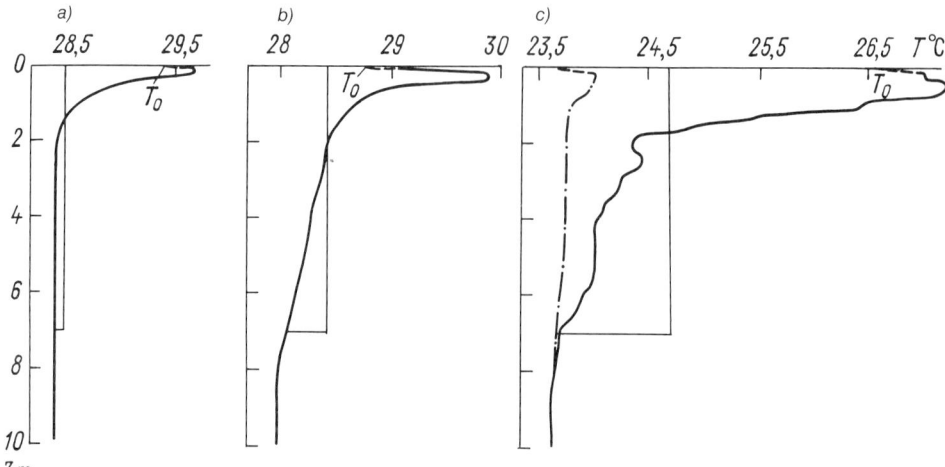

Figure 3.9. Vertical temperature profiles obtained in the near-surface layer of the ocean by A. V. Solovyev on 1 October 1977 at 1447 LST with an emergent profiler [92] (a); by V. T. Paka and K. N. Fedorov on 25 August 1978 at 1400 LST with a 'Boomerang' probe and thermal sensors suspended on floats (b); by Bruce and Firing on 28 May 1973, with a modified expendable bathythermograph at 1840 LST (solid line) and at 2125 LST (dash-dotted line) [128] (c). The thin line shows the water temperature profile after mixing the 7 m layer.

should be taken into consideration when performing field measurements, especially in tropical latitudes.

The possibility of such errors is eliminated when using the method of vertical probing [116, 117]. However, there are certain difficulties associated with the determination of T_0 when the thermal sensor passes from the air into the water. The water meniscus extending toward the thermal sensor when the latter approaches the surface may cause intensive evaporation from the sensor surface and, as a consequence, a higher estimate of $|\Delta T_0|$. In our opinion, this is the reason for the high values of $|\Delta T_0|$ measured, reaching 2 and even 3°C at comparatively small values of the heat flux ($q_0 \simeq -200 \text{ W/m}^2$) reported in [116] and other works of the same authors. Apparently, it is methodologically more correct to carry out probing not from above, but from below, which makes it possible to determine the interface and T_0 more accurately. It is necessary to take into consideration also that a single probing does not give a true idea of the boundary layer cooling due to the fluctuating character of T_0 (the range of fluctuations is approximately equal to $\overline{\Delta T_0}$, see Section 3.3.1); therefore, averaging by the ensemble of measured profiles is necessary to determine $\overline{\Delta T_0}$.

3.5. Practical problems arising in connection with the contribution of the thermal boundary layer to the overall variability of temperature of the near-surface layer of the ocean

The difference in T_0 from T_w and T_h, and the difficulties associated with the methods of measuring the two latter values lead to the question of which one of all these temperatures should be adopted as the ocean surface temperature when solving problems of the ocean upper layer dynamics, interaction of the ocean and the atmosphere, and monitoring and simulation of the climate. It is quite difficult to give an answer to this question because of the temperature variability in the near-surface layer and its complicated structure, which was discussed in general terms in Chapter 1. The picture is simple only in some simple cases.

It is obvious that the small thicknesses of the cold thermal boundary layer and the most typical values of $\overline{\Delta T_0}$ (tenths of a degree) make negligible the direct contribution of this layer to the enthalpy of the whole UQL. On the other hand, determination of the UQL (or near-surface layer) enthalpy on the basis of T_0 instead of T_w (or T_h) may cause considerable errors. The difference between T_0 and T_w (or T_h) may also affect the determination of the components of the heat flux q_T and q_E by known empirical dependences, wherein T_w (or T_h) is usually used, while it is necessary to use T_0. From the estimates of Paulson and Simpson [200], made on the basis of these dependences without consideration of absorption of solar irradiance by the thermal boundary layer and with the application of relationship (3.23) at $\lambda = 6.5$, neglect of ΔT_0 produces systematic overstating of the calculated values $q_E + q_T$ by approximately 5% in the water temperature range from 0 to 30°C. Almost the same result was obtained by Malevsky–Malevich [69]—not more than 2% for the winter months and not more than 7–10% for the summer months. It is obvious that the stated systematic errors are smaller than the overall random error in determining the heat flux by the existing integral methods ($\pm 20\%$). It should be noted that both the cited estimates do not consider the possibility of the existence of a warm boundary layer when $T_0 > T_w$ in some regions and under certain conditions. It should also be taken into consideration that the value $\overline{\Delta T_0}$ (and T_0) itself depends on the local value of the heat flux and T_w [see (3.32) and (3.33)]. Therefore, the errors in determining $q_E + q_T$, obtained by neglecting $\overline{\Delta T_0}$, should be even slightly smaller than the stated ones, all the more so as the constants in the empirical dependences for determining $q_E + q_T$ are chosen just for T_w, but not for T_0. Methods of estimating the flux q_0 based on measuring the vertical temperature gradient dT/dz in water on a linear part of the profile $T(z)$ near the surface, i.e. $q_0 = -\chi_T\, dT/dz$, make it possible to eliminate fully the necessity of considering $\overline{\Delta T_0}$. These methods of contact (vertical probing) and radiometric (with a two-channel radiometer) measurements of q_0 are described, for example, in ref. [116].

It is evident from the aforesaid that it is necessary to adop T_w for the ocean surface temperature (under conditions of correct measuring), while the influence of $\overline{\Delta T_0}$ can be neglected when determining q_0, or when determining climatic trends. Recall in this connection the idea stated by Malevsky–Malevich in ref. [69], namely, consideration of the $\overline{\Delta T_0}$ drop in climatic models without simultaneous consideration of the increase in ocean heat losses under gale conditions due to the formation of water splashes will cause an increase in the total by area and time systematic error in the calculated heat exchange between the ocean and the atmosphere instead of its decrease. However, this does not mean that it is necessary to neglect the $\overline{\Delta T_0}$ drop in general. For example, when solving local problems under sufficiently homogeneous conditions of exchange between the ocean and the atmosphere, it makes sense to determine all the elements of the thermal balance without the known systematic errors, even small ones. It is most important to consider the $\overline{\Delta T_0}$ drop when determining q_T, because it may lead to a change in the direction of this flux component at small water–air temperature differences. Later, when satellite monitoring of SST becomes regular practice, reliable methods will be required to determine correctly and, depending on the type of problem, to use T_0, T_w, or T_h. It is necessary to solve in this connection which particular characteristics, i.e. T_w, T_h, or $\overline{T}_0$, it is advisable to store in databanks relating to the ocean surface in order to satisfy enquiries associated with the solution of all types of problems. No matter what the final choice, it should be made with consideration of the variety of physical regimes of the near-surface layer of the ocean, determining, as shown in Chapter 1, the typical background thermal structures of this layer to which the stated characteristics are attached.

3.6. On the satellite measurement of the ocean surface temperature

The information given in the previous sections of this chapter permits us to express some appropriate ideas on measuring the ocean surface temperature from satellites.

Scanning IR radiometers installed on meteorological satellites are gradually becoming the principal means of obtaining information on the ocean surface temperature (SST) on a global scale. The good space resolution ($\simeq 1$ km) and sensitivity ($\simeq 0.3$ K) [125, 212] of the best modern multi-channel radiometers of type AVHRR, installed on satellites with polar orbits of series NOAA, and the possibility of observing the same surface area twice a day make them an indispensable means for studying the dynamics of the upper layer of the ocean. The single-channel IR radiometers installed on the geostationary METEOSAT satellites are characterized by a slightly poorer resolution and sensitivity (8 km and more than 0.5 K, respectively). However, they give information on the ocean surface temperature with an interval of 0.5 h which allows us to filter effectively the cloud cover and to plot regular (e.g. weekly) maps of the SST on large water areas [196], and to study (in the absence of cloud cover) the comparatively fast processes of the ocean surface thermal field restructuring. As shown by the experience of satellite data analysis, modern satellite IR radiometers record fairly well the diurnal cycle of SST under conditions of strong solar heating [136, 212].

The processing algorithms adopted in recent years, which are based on comparing the radiative temperature in two (3.7 and 11, or 11 and 12 μm) or three (3.7, 11, and 12 μm) IR channels in AVHRR [125, 225], make it possible today to ensure satisfactory accuracy in recovering SST ($\sigma = 0.6$–$0.8°C$ [125, 225]) in certain oceanological problems in regions free of clouds, fog, and volcanic aerosol. For example, comparison of the mean monthly maps of SST, based on AVHRR and ship data in the north-east Pacific, produced a root mean square for the difference $T_{AVHRR} - T_{ship}$ equal to 0.78°C with a mean deviation of 0.25°C [125]. It is quite probable that the individual contact measurements could be a substantial source of random errors because the satellite data are characterized by a 2-fold smaller 'noise' ($\sigma_{AVHRR} = 0.61$°C) as compared to ship data ($\sigma_{ship} = 1.24$°C) [125].

The differences between satellite and point contact and other measurements of the SST in the ocean are inevitable and are conditioned (in the case of reliable calibration of the radiometer and ideal atmospheric correction) by a variable scale of averaging (more than 1 km^2 on a satellite), errors in geographic referencing of satellite data (up to several tens of kilometres in the open ocean), the difference in time of carrying out measurements, the presence of films that change the radiating capacity of the water surface, and mainly the natural space–time variability of the temperature drop ΔT_0 in the thermal boundary layer of the ocean and the character of $T(z)$ in the diurnal thermocline (see Sections 3.3.8 and 4.3). The latter is conditioned by the fact that the radiometer measures the temperature of a layer with a thickness of less than 50 μm, while any other method of contact measurement (towed thermal sensor, upper bathometer, expendable and aerial expendable bathythermographs, CTD probes, sensor in ship water intake, drifting buoys) registers the temperature at a depth from several tens of centimetres to several metres. Therefore, the best results of comparing satellite and contact measurements are obtained in regions of intensive wind or convective mixing with a time difference in the measurements of maximum 1 day, and a difference in the geographic position of the measuring points within 1–3 tens of kilometres [154, 225]. And quite the contrary, the greatest difference can be anticipated either in regions of high dynamic variability (e.g. in frontal regions, upwelling zones, etc.), or in calm or light wind weather under conditions of intensive solar heating when the diurnal temperature cycle (see Fig. 3.9c and ref. [212]) can reach 3–3.5°C. For example, a case is discussed in ref. [212] of recording a patch by AVHRR (TIROS-N satellite) with a temperature 3.5°C higher as compared to the surrounding waters at about 1500 LST which was not detected on the surface temperature map plotted using the ship data. The reality of the satellite image was confirmed in this case by the diurnal SST cycle according to METEOSAT data. The wind velocity map in the region of measurements and interpretation of the TIROS-N visible channel image confirmed light winds over the region with a higher SST.

Similar examples illustrate the fact that satellite data can undoubtedly produce a more realistic synoptic image of the SST and its variability than ship measurements carried out at different times along individual tacks, or along a network of tacks, and more so episodic or random ship measurements. All the points cited above as well as the continuous improvement of the equipment and algorithms of processing radiometric measurements make it possible to believe that SST satellite maps will soon become the basic source of information on SST in various fields of oceanology and the base for creating databanks with the purpose of monitoring climatic variability. The publication of MCSST (multi-channel sea surface temperature) maps in the USA based on AVHRR data and replacing the GOSSTCOMP (global operational sea surface temperature computation) map which was published for about 10 years beginning with 1972 and which was plotted by using data of single-channel radiometers [225] seems very important in this respect as well as the publication of such bulletins in France as *SATMER* [136] and *Veille Climatique Satellitaire* [196].

Obviously, the requirements of comparability of the data of various radiometers on numerous satellites with their wide-scale utilization in problems of studying and monitoring the temperature regime of the ocean surface dictates the necessity of regular control of radiometer calibration on the basis of reliable ground-truth measurements. Measurements of the SST by IR radiometers from a ship or low-flying planes are irregular and therefore cannot be the base of a required ground-truth control network. Evidently, the best means today for obtaining such information are the small and relatively cheap drifting buoys [125, 225], whose data are transmitted to a united centre via special channels of communication. The constants in the algorithms for recovering SST (different for daytime and night) by measurements of the radiative temperature in two or three channels of AVHRR are currently based on these data that are used for control.

Nevertheless, it is necessary to take into consideration that when referencing the satellite measurements to the values of the water temperature, measured on some near-surface horizon, the recovered values of SST actually correspond not to T_0, but to a temperature differing from T_0 due to the space–time variability of the vertical gradient dT/dz near the ocean surface (including the thermal boundary layer). This introduces uncertainty into the physical concept of SST of satellite origin. On the one hand, as a measure of T_0 it contains an error which is determined with certain difficulty, and on the other, it is not an unerring measure of T_w. In the case of full isothermy in the UQL (Fig. 1.1b), the satellite SST is equivalent to T_w and it is possible to use the dependences (3.32) and (3.33) for the transition to $\bar{T}_0 = T_w + T_0$. Obviously, it is necessary to have information on the wind velocity to make a choice between them. The latter and the characteristics of the waves can be measured today from satellites by means of SHF equipment. Images of the surface in one of the visible channels of the spectrum can be an additional source of information on the condition of the ocean surface, making it possible to indicate regions of calm and light wind. There are still no empirical dependences to predict an accurate temperature profile $T(z)$ close to the surface in the case of various regimes of ocean–atmosphere interaction for cases of a warm boundary layer or intensive solar heating (Fig. 1.1c and a). This question demands subsequent methodological elaboration.

Chapter 4

The cyclical nature and space structure of the near-surface layer in connection with solar heating and convection

In the previous chapter we came across the fact that the behaviour of the thermal boundary layer in connection with the origin of primary convective motions in it could be represented as a cyclical one-dimensional process. It turned out later that the one-dimensional idea of cyclical growth (up to critical instability) and destruction of the thermal boundary layer agreed very well with the fact of three-dimensionality of the space structure of the primary convective motions near the water surface detected by more accurate observations. It became clear that the thermal boundary layer in reality had a three-dimensional structure whose space–time variability during observation on an isolated vertical only created the impression of a quasi-periodic (cyclical) process.

It seems that the thermal boundary layer in miniature presents numerous specific features of the behaviour and three-dimensional structure of the entire near-surface layer of the ocean despite the fact that the former is almost always hydrostatically unstable, while the latter fluctuates in the vicinity of an indifferent equilibrium with predominating stability in the daytime. Diurnal solar heating, replaced by nocturnal cooling, is a powerful cyclical factor predetermining the rhythm of its variability. The seasonal cycle is of similar significance. The hierarchy of scales of convective motions [42, 145] based on the growth of effective exchange coefficients, the interaction of the progressive and wavy motions of the water, and also other internal and external factors predetermine new, larger space scales of water circulation and structures of the hydrophysical fields in the near-surface layer of the ocean. It should be noted that this aspect of the near-surface layer of the ocean is the least studied. Nevertheless, it is necessary to imagine and understand well the physical nature and scales of this space–time variability. As in Chapter 3, we will approach this understanding starting from one-dimensional ideas.

4.1. Volume absorption of solar energy in the near-surface layer of the ocean. Irradiance models

The heat flux at the ocean–atmosphere interface is determined by the sum of the components:

$$q_\Sigma(t) = (1 - A_0)q_\odot(t) + q_0 = q_s(0, t) + q_0, \tag{4.1}$$

where $q_0 = q_E + q_T + q_R$ characterizes heat transfer from the surface (see Chapter 3); $q_\odot(t)$ is the total flux of solar irradiance reaching the water surface and depending on the latitude, elevation of the sun (time of day and season), and condition of the atmosphere (cloud cover, air transparency, steam content in the air); A_0 is the sea surface albedo ($A_0 \simeq 0.06$–0.07 can be adopted for practical problems in low and temperate latitudes); and $q_s(0, t)$ is the solar irradiance flux in the water near the surface ($z = 0$), respectively. Its diurnal and seasonal cycles and vertical distribution in the water $q_s(z, t)$ determine the temporal changes in the thermal structure $T(z)$ and heat content (enthalpy)$Q(z)$ of layer 0–z, which can be written down as follows with disregard of vertical turbulent heat exchange and advection:

$$\frac{\partial T(z)}{\partial t} = \frac{1}{\rho c_p} \frac{\partial q_s(z, t)}{\partial z}, \tag{4.2}$$

$$\frac{\partial Q(z)}{\partial t} = \int_z^0 \frac{\partial q_s(z, t)}{\partial z} \, dz = q_s(0, t) - q_s(z, t). \tag{4.3}$$

In turn, $q_s(z)$ and $\partial q_s(z)/\partial z$ depend greatly on the optical transparency of the waters. The absorption of solar energy in waters with greater turbidity is greater per unit of volume as compared to pure water in the 0–10 m layer and is less at $z > 10$ m [72, 84, 242]. The difference in the volume absorption in turbid and pure water is not very great in the upper few metres, while in the first 5–6 cm from the surface it is practically absent owing to the high absorption of IR irradiance directly by the water. The maximum difference (twice as much) is at a depth of 2–5 m; the difference is considerably greater at a depth below 10 m, reaching two orders at a depth of 50 m [84]. The differences in the vertical distribution of the proportion of transmitted solar irradiance $q_s(z)/q_s(0)$ and integral absorption $(1 - q_s)(z)/q_s(0)$ in the upper tens of metres in waters of various optical transparencies can be judged by the data in Tables 4.1 and 4.2, taken from Ivanov (Chapter 5 in ref [72]), and Pelevin and Rutkovskaya [84], respectively. The calculations in ref. [84] were performed with consideration of the strong spectral inhomogeneity of the downward light and absorption by the water, for which purpose the authors divided the spectral range of 320–2700 nm into 60 bands. The parameter $m = 100 \, a_{500}$ in Table 4.2 is the water optical index (a_{500} is the index of vertical attenuation of the downward irradiance at a decimal base in inverse metres for a 500 nm wave associated with a similar a_{500E} index at a base e, used in the work of Yerlov and some other authors, by the relationship $a_{500} = a_{500E} \log e$).

The value $m = 1.2$ corresponds to very pure waters (Sargasso Sea), $m = 1.5$ to pure oceanic waters, $m = 3.0$ to waters of greater turbidity (e.g. in zones of equatorial upwellings), and $m = 7$ to turbid coastal as well as sub-Arctic and sub-Antarctic waters [84]. The Roman and Arabic numbers given in parentheses in Tables 4.1 and 4.2 designate standard types of waters by Yerlov* which roughly correspond to the values of transmission and integral absorption in the column.

---

*Types I–III: for transparent oceanic waters; types 1–9: for more turbid coastal waters.

Table 4.1
Vertical distribution of the proportion of transmitted solar irradiance (1) and integral absorption in a 0–z m layer (2) (per cent) according to the data in ref. [72]

z(m)	Very pure waters (1)		Pure waters (II)		Turbid waters (III-1)	
	1	2	1	2	1	2
0	—	—	100	0	—	—
0.01	—	—	85	15	—	—
0.1	—	—	67	33	—	—
0.2	—	—	59	41	—	—
0.5	—	—	51	49	—	—
1	45	55	45	55	40	60
2	38	62	35	65	28	72
3	35	65	30	70	22	78
5	32	68	24	76	11	89
10	27	73	16	84	2	98
20	20	80	8	92	0	100

Table 4.2
Vertical distribution of the share of transmitted solar irradiance (1) and integral absorption in O–z m layer (2) (per cent) according to the data in ref. [84]

z(m)	$m = 1.2$ (I)		$m = 1.5$ (IA)		$m = 3.0$ (II)		$m = 7(1-3)$	
	1	2	1	2	1	2	1	2
0	100	0	100	0	100	0	100	0
0.001	94.2	5.8	94.2	5.8	94.2	5.8	94.2	5.8
0.005	89.4	10.6	89.4	10.6	89.4	10.6	89.4	10.6
0.01	86.2	13.8	86.2	13.8	86.2	13.8	86.2	13.8
0.05	77.9	22.1	77.9	22.1	77.9	22.1	77.7	22.3
0.1	73.9	26.1	73.9	26.1	73.8	26.2	73.5	26.5
0.5	62.5	37.5	62.0	38.0	59.0	41.0	55.0	45.0
1	57.0	43.0	56.0	44.0	52.0	48.0	47.0	53.0
2	49.0	51.0	48.0	52.0	43.0	57.0	37.0	63.0
5	39.0	61.0	38.0	62.0	31.0	69.0	20.0	80.0
10	30.0	70.0	28.0	72.0	19.0	81.0	7.4	92.6
25	18.0	82.0	14.5	85.5	4.0	96.0	0.5	99.5
50	8.6	91.4	6.2	93.8	0.6	99.4	0.008	99.992
75	4.5	95.5	2.9	97.1	0.09	99.91	0	100
100	2.5	97.5	1.6	98.4	0.014	99.986	0	100

Data of the type in Tables 4.1 and 4.2 or obtained during direct measurements in the ocean can serve as the basis for presenting $q_s(z)$ in analytical form, which is necessary when solving a wide range of model problems associated with climatology of the ocean surface thermal balance, determination of the diurnal cycle of the mixed-layer depth, for forecasting the thermal structure and dynamics of the near-surface layer. Various authors use different $q_s(z)$ irradiance models [82, 92, 93, 134, 200, 217, 239, 242] that more or less take into account the spectral composition of the solar irradiance and the variability of the coefficients of vertical attenuance a_λ for various wavelenths λ from λ, z, and type of waters. The most widespread and physically substantiated is the presentation of $q_s(z)$ in the form of a sum of n exponents:

$$q_s(z) = q_s(0) \sum_{i=1}^{n} R_i e^{-a_i z}, \qquad (4.4)$$

where $\sum_{i=1}^{n} R_i = 1$, and the coefficients R_i and a_i are selected on the basis of greatest conformity with the measured or tabulated values of $q_s(z)$ or $q_s(z)/q_s(0)$. Correspondence of the a_i values to some fixed a_λ values is not absolutely obligatory. Different variants of (4.4) with $n = 1, 2, 3, 9$, and 27, used in models of seasonal and diurnal UQL variability [92, 93, 134, 200, 217, 239, 241], are presented in Table 4.3, where the Roman numerals designate the types of waters by Yerlov and $\gamma_i = 1/a_i$.

As given in ref. [134], the application of approximations 2 and 6 (Table 4.3) in a one-dimensional closure model of UQL evolution gives almost the same pattern of development of the thermal structure and currents in the near-surface layer, which is close to the real pattern, in a wide range of wind velocities, while model 1 with one exponent greatly distorts the pattern at $u_{10} < 20$ m/s. At wind velocities $u_{10} \geqslant 20$ m/s, all three variants are approximately of equal value. Therefore variant 2, as the simplest one, is recommended by Dickey and Simpson [134] for all problems except simulation of exchange processes near the ocean–atmosphere boundary, where it is advisable to apply more complete spectral decomposition. The comparison we made of the values $q_s(z)/q_s(0)$ calculated by models 2, 3, and 6 with the data in Tables 4.1 and 4.2 demonstrated that spectral decomposition 6 gives very great discordance with the tabulated values (up to 17%) just in the very upper 1 cm layer despite the summing up of nine exponents. The best result here is given by variant 3 (discordance within 2%). In the depth range $0.5 < z < 20$ m, the best (of models 2, 3, and 6) conformity with the data in Table 4.1 is given by variant 2 for type I water (discordance within 4%), while calculation by models 3 and 6 gives discordance of 18 and 9% from the tabulated values, respectively. The coefficients in ref. [134] (model 2) for type III water are definitely incorrect, because the calculated values of $q_s(z)/q_s(0)$ at $0 < z < 1$ m for turbid waters turn out to be greater than those for pure waters (type I). The comparatively small discordance (within 5%) at $1 < z < 50$ m gives, according to Woods *et al.* [242] (the coefficients are not published), presentation of $q_s(z)$ in the form of three exponents (model 5), which is widely used by Woods *et al.* [241, 242] when describing the global climatic variability of solar irradiance and in integral models of the upper layer for $z > 1$ m. Similar discordance in the same range of depths is produced, according to our estimates, by the three-exponential approximation of Solovyev (model 4 [93]): within 5.6% for type I water and 4% for type IB. Spectral decomposition with $n = 27$ (model 7) is recommended in refs [239, 242] in the upper 1 m.

The strong dependence of $q_s(z)$ on the optical transparency of the waters necessitates the application of various combinations of R_i and a_i for different types of water. The uncertainty in determining the type of water (I or II) may cause uncertainty in the model calculations of the ocean surface temperature by more than 1°C, which exceeds the inter-annual climatic variability of T_0 in the extratropical latitudes (see ref. [242]). On the other hand, incorrect choice

Table 4.3

Parameters used by different authors in the approximation $q_s(z) = q_s(0) \sum_{i=1}^{n} R_i \exp(-z/\gamma_i)$

No.	Authors	Types of water	Number of terms, n	i	$\lambda(\mu m)$	R_i	$\gamma_i(m)$	Where used
1	Kraus and Turner, 1967 (see ref [242]) Denman, 1973; Alexander and Kim, 1976 (see ref. [217])	—	1	1	—	1	10,20	UQL integral models
		I	1	1	—	1	10	UQL integral models
		II	1	1	—	1	5	UQL integral models
2	Paulson and Simpson, 1977; Simpson and Dickey, 1981, 1983 [134,217]	I	2	1	—	0.58	0.35	UQL closure models, UQL daily variability
				2	—	0.42	23	
		III	2	1	—	0.78	1.4	
				2	—	0.22	7.9	
3	Boguslavsky, 1956; Solovyev, 1979 [92]	—	2	1	—	0.53	0.03	Determination of the change in depth of the convective mixed layer in diurnal cycle
				2	—	0.47	6.66	
4	Solovyev, 1982 [93]	I	3	1	—	0.45	14.084	Determination of the change in depth of the convective mixed layer in diurnal cycle
				2	—	0.27	0.357	
				3	—	0.28	0.014	
		IA	3	1	—	0.45	12.820	
				2	—	0.27	0.357	
				3	—	0.28	0.014	
		IB	3	1	—	0.45	10.0	
				2	—	0.27	0.357	
				3	—	0.28	0.014	
5	Woods *et al.*, 1984 [242]; Woods and Barkmann 1986 [241]	I–III	3		Parameters not stated[a]			Global variability of solar heating; UQL integral models with resolution of the diurnal cycle
6	Paulson and Simpson, 1981 [200]; Simpson and Dickey, 1981 (see ref. [134])	0[b]	9	1	0.2–0.6	0.237	3.48×10	Estimation of the influence of solar irradiance absorption on drop of T_0 in thermal boundary layer; UQL closure models
				2	0.6–0.9	0.360	2.27	
				3	0.9–1.2	0.179	3.15×10^{-2}	
				4	1.2–1.5	0.087	5.48×10^{-3}	
				5	1.5–1.8	0.080	8.32×10^{-4}	
				6	1.8–2.1	0.0246	1.26×10^{-4}	
				7	2.1–2.4	0.025	3.13×10^{-4}	
				8	2.4–2.7	0.007	7.82×10^{-5}	
				9	2.7–3.0	0.0004	1.44×10^{-5}	
7	Woods, 1980; Woods *et al.*, 1984 [239, 242]	0[b]	27		Parameters not stated[a]			Determination of the diurnal cycle of thermal compensation depth; description of $q_s(z)$ profile in upper metre layer of the ocean

[a] In models 5 and 7, all the coefficients γ_i are multiplied by $\cos\theta$, where θ is the zenith angle with account of refraction.

[b] 0 is pure water.

of coefficients R_i and a_i for other types of water may be the reason for incorrect conclusions. For example, as a result of applying the inappropriate, as mentioned above, variant 2 for type III water in the UQL closure model [134], the amplitude of the diurnal cycle of T_0 in turbid waters (type III) turned out to be less than that in pure waters. The latter is at variance with the calculation of solar heat absorption in different types of water (Tables 4.1 and 4.2) and with field observations. According to the data in ref. [182], the difference (diurnal cycle) of SST values, obtained by means of satellite measurements with an interval of 12 h (at 0730 and 1930 LST), is always greater in turbid coastal water than in more transparent oceanic waters.

It is obvious that it is necessary to carry out accurate mapping of the types of water in the World Ocean to ensure the correct choice of coefficients R_i and a_i for the region under study. Apparently, this mapping is possible only on the basis of satellite measurements (using devices of the Coastal Zone Colour Scanner type), because the optical transparency of the waters is determined by processes of significant space and time variability: seasonal blooming of plankton, local upwellings, transport of suspended matter and plankton by currents, eddies, transversal upwelling jets, etc. Widescale utilization of satellite data on the optical colour index will apparently lead to new, more versatile methods of considering the spectral transparency of water in irradiance models based on more detailed optical classification of water [84] as compared to the Yerlov classification.

One of the major problems in connection with the absorption of solar energy in the near-surface layer of the ocean is the problem of estimating the quantity of heat ΔQ_D which is either obtained or irretrievably lost by the near-surface layer toward the end of each diurnal cycle. It is quite evident from most general considerations that ΔQ_D is positive during the hydrological summer when the ocean accumulates heat, and negative during the hydrological winter. In the short periods between the two seasons near the equinocial point, $\Delta Q_D \simeq 0$. It can be concluded from the same considerations that the values of ΔQ_D are maximum in the temperate latitudes in regions with a large seasonal cycle of T_w and diminish in the direction toward the poles and the equator, where the seasonal changes of T_w are relatively small. Observations demonstrate that the actual diurnal values of $\pm \Delta Q_D$ (see Fig. 6.6 in ref. [58]) can reach 8.4–9.2 MJ/m^2 (200–220 cal/cm^2) at the extrema of the annual cycle, but deviate noticeably from the mean value for the given latitude and the given season owing to advection and other fluctuations of the thermal balance of the near-surface layer of the ocean. For example, investigations in a hydrophysical field in the Tropical Atlantic (trade wind zone, 16°30′ N) in 1970 [110, 142] produced the following characteristic values of ΔQ_D:

	MJ/m^2	cal/cm^2
April–June	+0.92	+22
July–August	+1.38	+33
August–September	+2.34	+56

Other researchers (see the discussion in ref. [142]) obtained values of ΔQ_D close to $+3.77$ MJ/m^2 ($+90$ cal/cm^2) in temperate latitudes in summer. The seasonal and latitudinal trend in these examples is in full accordance with the preceding discussions.

The maximum value of enthalpy Q_D in the near-surface layer during the diurnal cycle is another major index relating to diurnal solar heating which is conditioned by the absorption of solar irradiance on cloudless days. The enthalpy maximum is always observed at about 1600 LST along with the maximum of T_h. The value Q_D is a function of the daily total of solar irradiance $Q_s(\varphi, M)$* reaching the ocean surface under serene skies in the given latitude φ in the given month M. Our observations given below demonstrate that it is possible to state roughly for summer conditions in temperate latitudes and for all the seasons in low latitudes that

$$Q_D \simeq 0.6 Q_s. \tag{4.5}$$

The data on which this conclusion is based were obtained at long-term (multi-day, diurnal, and multi-hour) stations in various regions of the World Ocean by multiple (sometimes hourly) probing of the thermal structure of the near-surface layer. The results of these measurements are summarized in Table 4.4.

Actual data on measuring enthalpy $Q(t)$ in the 20 m thick nearsurface layer

Table 4.4

Region, station	Latitude	Date	Q_s MJ/m^2 (cal/cm^2)	Q_D	Q_D/Q_s	Remarks
Tropical Atlantic, 'Polygon-70' station No. 602, research vessel *Akademik Kurchatov*	16°30′ N	30 July–2 August 1970	22.2 (530)	13.4 (320)	0.60	Q_s value obtained from actinometric measurements
Indian Ocean, station No. 450, research vessel *Dmitry Mendeleyev*	0°02′ S	13 February 1972	28.7 (688)	16.7 (400)	0.58	Q_s value obtained from Table 5.2 ref. [72]
Sargasso Sea, POLYMODE, station No. 2749, research vessel *Akademic Kurchatov*	30°50′ N	21 September 1978	26.3 (628)	15.7 (375)	0.60	Q_s value obtained from Table 5.2 in ref. [72]

*Data for $Q_s(\varphi, M)$ are given in Table 5.2 in ref. [72] and in Table 135 in the *Smithsonian Meteorological Tables*, 6th edn, Publ. 4014, Washington, 1958, p. 421.

Table 4.5
Enthalpy of near-surface layer 0–20 m. Sunrise at 0600 LST, sunset at 1800 LST

Time (h) (LST)	Enthalpy $Q(t)$		$Q(t)/Q_D (\%)$
	(MJ/m^2)	(cal/cm^2)	
7	0.63	15	4.0
8	2.09	50	13.3
9	3.77	90	24.0
10	5.44	130	34.7
11	7.12	170	45.3
12	9.00	215	57.3
13	11.30	270	72.0
14	13.39	320	85.3
15	15.28	365	97.3
16	15.70	375	100
17	14.86	355	94.7
18	13.81	330	88.0
19	12.77	305	81.3
20	11.93	285	76.0
21	10.88	260	69.3
22	9.63	230	61.3

Note: According to Table 5.2 in ref. [72], the heat total Q_s reaching the ocean surface was 26.3 MJ/m² (628 cal/cm²) by 1800 LST. The heat losses after 2200 LST were highly distorted by advection.

during the daytime part of diurnal cycle by the data of measurements at the daily station No. 2749 of research vessel *Akademik Kurchatov* on 21 September 1978 in the Sargasso Sea are given as an example in Table 4.5. The measurements (every 3 h) were made by means of a CTD probe 'AIST' under serene skies in a zone free of disturbances from the drifting vessel body (see Section 1.4). Table 4.5 presents data interpolated for each hour and taken from a curve that was smoothed to eliminate small short-term distortions due to advection.

The hourly values of $Q(t)/Q_D\%$, given in Table 4.5, can be used for a rough estimation of the values of $Q(t)$ by Q_D and in other cases of daily heating of the near-surface layer under serene skies, considering, certainly, the actual day length.

4.2. Propagation of cyclical heating through the near-surface layer thickness

The variation of the ocean surface temperature in the diurnal cycle is of quasi-periodic* character with the maximum at about 1500–1600 LST, i.e. 3–4 h after the maximum inflow of solar irradiance, and the minimum at about 0700

---

*Its difference from purely harmonic oscillation is conditioned by the greater rate of water heating at absorption of solar irradiance as compared to the cooling rate [118, 134], and slower propagation of heating into the depth against the buoyancy force as compared to the deepening front of convection during nocturnal cooling.

LST, i.e. soon after sunrise. Propagation of this heat wave from the surface into the depth due to mixing occurs with gradual diminishing of the amplitude $A_T(z) = (T_{max}(z) - T_{min}(z))/2$ and increasing with depth phase lagging $\varphi(z)$. Depending on the intensity of solar heating and the conditions of local turbulent mixing, the depth z_{max}, at which the diurnal temperature cycle practically disappears, changes from several metres to 30–40 and even 50 m [134, 142] (see also Section 4.3). For example, measurements made in the TROPEX experiment in the Tropical Atlantic in August–September 1974 [205] under conditions of appreciable cloudiness and a wind velocity of more than 2 m/s gave $A_T(0) = 0.12°C$, $z_{max} = 15$ m; the time lag of maximum temperature on horizons 5–10 m from the maximum at the surface was 1–2 h and the minium was 1 h. The following lags were recorded in other observations [158] at a wind velocity of 6–8 m/s: 20 min at $z = 3$ m, 40 min at $z = 6.65$ m, and 70 min at $z = 12$ m.

Such regularities of vertical heat wave propagation are reproduced well in modern UQL closure models [134], making it possible to obtain a detailed image of the variation of its thermal structure in the diurnal cycle, including vertical distribution of the coefficient of turbulent heat exchange by the vertical $K_T(z)$. Long before the introduction of these models into oceanological practice, the Fourier–Schmidt method, based on measuring $A_T(z)$ and $\varphi(z)$ on two horizons z_1 and z_2 [118], was widespread:

$$K_T = \frac{\pi}{\tau} \left(\frac{z_2 - z_1}{\ln\left[A_T(z_1)/A_T(z_2)\right]} \right)^2 \tag{4.6}$$

or

$$K_T = \frac{\pi}{\tau} \left(\frac{z_2 - z_1}{\Delta\varphi} \right)^2, \tag{4.7}$$

where τ is the heat wave period, equal to 8.64×10^4 s for the diurnal cycle; $\Delta\varphi = \varphi(z_1) - \varphi(z_2) = 2\pi/\tau\left[t(z_2) - t(z_1)\right]$; and $t(z)$ is the moment of maximum (minium) at horizon z.

Formulae (4.6) and (4.7) were obtained under the assumption of constancy of K_T in the layer considered. Therefore, their application in the upper metres close to the surface, where volume absorption of solar irradiance is maximum and the influence of stratification is strong, can produce considerable errors [118]. To control correct estimation of K_T as some mean value within the layer considered, in the opinion of Shtokman it is necessary to use both (4.6) and (4.7) always simultaneously. Coordination in estimating K_T may testify to the reliability of the results. Formula (4.6) is used more often in practice (see, for example, refs [142, 158]), more rarely a mean value of K_T from both methods [198].

The estimates of K_T obtained by various authors on the basis of field measurements of $T(z)$ under various conditions naturally differ by more than two orders: $K_T = 10^0 - 10^1$ cm^2/s in ref. [118], 100 cm^2/s in refs [142, 198], and 420 cm^2/s in ref. [158]. The small values of K_T correspond to the conditions of light winds and stable stratification near by the surface, and the large values to intensive turbulent mixing of the layer. According to ref. [158], a change in the wind velocity by $\simeq 1 \, \text{m/s}$ results in an increase of the K_T estimate obtained from (4.6) by more than an order. It is necessary to take into consideration also that the effective coefficient of temperature conductivity is not constant with depth even in relatively thin layers (see ref. [217]).

4.3. The diurnal (daily) thermocline and variability of the thermal structure of the near-surface layer of the ocean in a diurnal cycle

The diurnal (or daily) thermocline is one of the main specific features of the thermal structure of the near-surface layer of the ocean (Fig. 4.1a). In the case of well-expressed diurnal cyclicity of solar heating and nocturnal cooling, it develops fully in the afternoon and is finally destroyed in the early morning hours, migrating during the day by depth and changing by contrast and the vertical temperature gradient in it. The diurnal thermocline is best expressed in summer and in low latitudes, while in winter it can be fully absent in the temperate and high latitudes, especially in stormy weather. The characteristic feature of the diurnal thermocline, which distinguishes it from the seasonal and main thermoclines, is the fact that quasi-geostrophic currents have no time to develop in it, and geostrophically balanced slopes of the isothermal and isopycnic surfaces are absent, respectively. This might be the reason why the

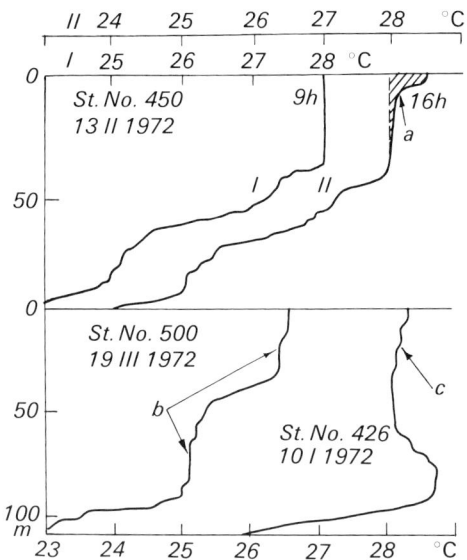

Figure 4.1. Diurnal thermocline and its various manifestations. Observations of K. N. Fedorov during the seventh cruise of the research vessel *Dmitry Mendeleyev*, February–March 1972, Indian Ocean [74]. (a) Common diurnal thermocline at lower boundary of heated layer (dashed) about 6–7 m thick which developed against the background of nocturnal UQL (profile I) by 1600 LST; (b) stepped structure of UQL due to advection; (c) stepped structure of diurnal thermocline due to meteorological conditions changing during the day.

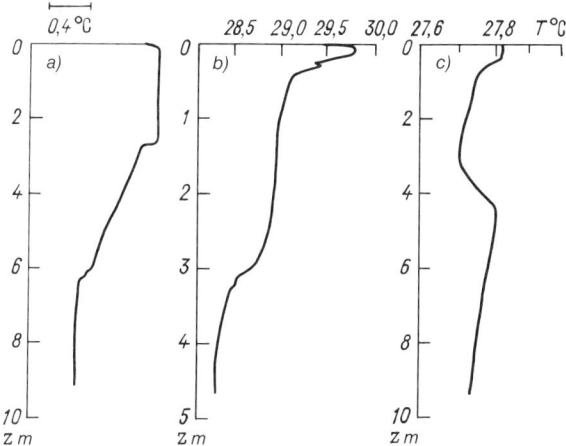

Figure 4.2. Vertical temperature profiles obtained by an emergent profiler in calm and light wind weather. (a) At sharp deepening of the mixed layer at 1936 LST [9]; (b) under intensive solar heating at 1504 LST after a large cloud passed over the vessel [92]; (c) at the beginning of diurnal heating (at 1033 LST) against a background of a freshened and cooled (due to nocturnal convection) 4 m layer [16].

diurnal thermocline has attracted no attention in oceanology until recently. At the same time, the diurnal thermocline can perform a very significant role in many processes associated with the transport of energy from the surface into the oceanic water thickness due to the quite high vertical density gradients developing in it during very intensive solar heating in light wind weather and especially when the near-surface layer is freshened by precipitation.

Sometimes normal development of the diurnal thermocline is interrupted by clouds and intensification of the wind, which is frequently accompanied by abundant precipitation.* Then after the resumption of heating, a new heated layer develops with a new intermediate thermocline on the lower boundary against the background of a shallow mixed layer (Figs 4.2b and 4.2c). The diurnal thermocline may consist of two to three intermediate thermoclines in the case of alternating cloudy and clear weather against the background of a light or moderate wind [74] (see Fig. 4.1c). The steps on the profiles of $T(z)$, resembling alternation of intermediate thermoclines, may also arise as a result of advection, i.e. horizontal displacement of adjacent spatially inhomogeneous layers in relation to one another (Fig. 4.1b). Hence, it follows that it is not always possible to explain correctly the specific features of the diurnal thermocline structure on the basis of single measurements of $T(z)$ at a certain point without knowledge of the prehistory and space variability of the temperature fields, and without comparing the results with the salinity data. Difficulties may also arise in determining its boundaries. At the same time, from the practical point of view the problem of forecasting or estimating the depth

---

*The effects of precipitation are discussed in detail in Section 4.5.3.

of occurrence and the main characteristics (e.g. Väisälä–Brunt frequency N) of the diurnal thermocline is today, apparently, one of the most important problems, especially when considering the requirements of remote sensing (see Chapter 6). It is impossible to solve this problem without knowledge of the rest of the ocean near-surface layer structure and its variability in the diurnal cycle. Comprehensive assistance in interpreting field data has been rendered in recent years by numerical models as a result of which it became possible at last to understand many subtle questions associated with the diversity of combinations of solar irradiance intensities, wind mixing, and convection in the diurnal cycle.

Until recently, all the judgements on the regularities of the thermal structure evolution near the ocean surface in the diurnal cycle were made on the basis of one-dimensional models elaborated in conformity with seasonal variability [72]. Indeed, these models made it possible to obtain quite realistic estimates of the minium thickness of the mixed layer in the daytime and also the amplitudes of the diurnal temperature trend therein [142]. However, it would be difficult to expect to obtain on the basis of such models all the details of the ocean near-surface layer diurnal structure discussed earlier, not to mention the possibility of forecasting the regularities of their variability in the diurnal cycle. Nevertheless, new model developments which have appeared in recent years [92, 93, 134, 239, 241, 242] in the framework of a one-dimensional approach have made it possible to demonstrate a series of the most significant specific features of the diurnal cycle. It is necessary to state the following ones:

(1) the significance for the diurnal cycle of the thermal compensation depth concept* h_c [239] as a horizon at which the solar irradiance absorption rate is equal to the rate of heat losses through the free surface and where $\partial T/\partial z \equiv 0$, respectively;

(2) the regular variation of the thermal compensation depth during the bright part of the day;

(3) the role of convection, which is the basic and, in the absence of wind, the sole factor in the near-surface layer contributing to heat transfer from the ocean to the atmosphere, and performing redistribution of solar heating into the UQL thickness;

(4) the influence of spectral differentiation of solar irradiance volume absorption in waters of various optical transparencies on the structure and evolution of the diurnal heated layer.

The fundamental result of Woods' calculations [239, 241] is that the lower boundary of the layer of convective mixing h'_0 during the greater part of the day, when $h'_0 < H_0$ (Fig. 1.1a), coincides with the depth of thermal compensation h_c (it is practically always slightly smaller). This result reflects the simple physical

---

*Although the concept of thermal compensation depth was introduced into oceanology more than 30 years ago, it started to be really considered only quite recently when simulating the diurnal cycle (see the discussion in ref. [239]).

fact that convection during the day is unable to overcome stable thermal stratification which starts directly beneath h_c. Hence, it is possible to estimate approximately at any time of the day the depth of convection penetration from the diurnal cycle of the thermal compensation depth which is determined by the relationship

$$q_0 = q_s(h_c, t) - q_s(0, t). \tag{4.8}$$

To do this, it is possible to use one or the other approximation $q_s(z)$ from Table 4.3 and the measured (or estimatd by any method) values of q_0 and $q_s(0, t)$ for the given latitude of the area and the given season. An example of h_c variation with time calculated by Woods [239] for the trade wind region ($20°$ N, $45°$ W) with the application of an irradiance model (7) from Table 4.3 is presented in Fig. 4.3a. Solovyev [92, 93] obtained independently a similar dependence for $h_c(t)$ (Fig. 4.3b).

Calculations demonstrate that h_c and h'_0 are not more than 1 m during the greater part of the bright time of the day at low and temperature latitudes when heat transfer from the surface is low (Fig. 4.3). Even with a not very strong wind (force 2–4) it is not convection, but wind-wave mixing that determines at this time the thickness of the diurnal mixed layer h_0. In this case, the convective layer is only a part of the mixed one ($h'_0 < h_0$), and convection does not participate in deepening the latter. This follows most vividly from the numerical model calculations given in refs [239, 241]. They also demonstrated that the contribution of convection at light and moderate wind velocities is determinative during the evening and night hours. The mean climatic variation of h_0 over 1 year with diurnal modulation due to solar irradiance, obtained by Woods and Barkmann with the integral UQL model for the region $41°$ N, $27°$ W, is shown

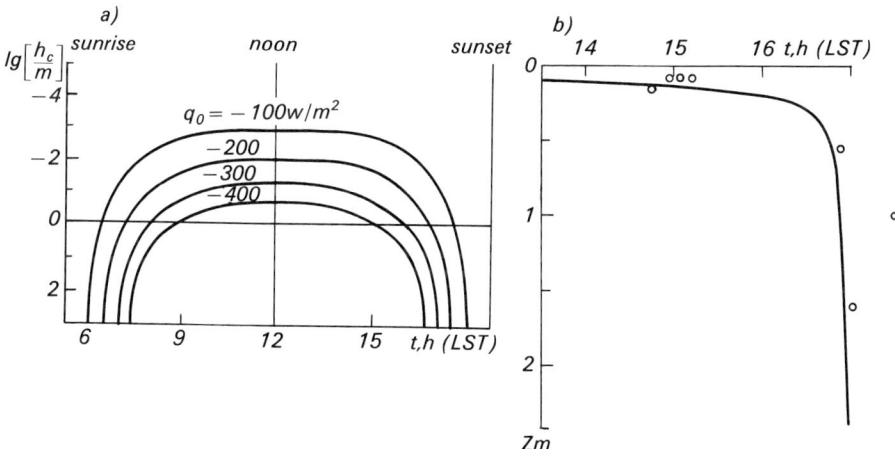

Figure 4.3. Variation of thermal compensation depth during the bright time of the day as calculated by Woods [239] (a) and Solovyev [92] (b). Experimental points correspond to data of probing with an emergent profiler on 1 October 1977, at POLYMODE-77 polygon under conditions close to calm [92].

in Fig. 4.4 [241], from which it follows that the diurnal variation of h_0 is maximum in spring and minimum in autumn. As far as we know, these model calculations are the first ones and represent quite a successful attempt to incorporate the diurnal cycle into a one-dimensional model of UQL seasonal variability.

At relatively small values of the heat flux from the surface ($|q_0| \simeq 100 \, \text{W/m}^2$) in hours close to midday, when h_c is not more than several millimetres thick and limits from below the thickness of the thermal boundary layer δ by the condition $\partial T/\partial z_{h_c} \equiv 0$, the unstable vertical distribution of the density in this layer does not cause convection because the respective boundary Rayleigh number Ra^* (see Section 3.3.3) is not more than $Ra^*_{cr} = Ra'_{cr} = 128$. Indeed, as at $|q_0| = 100 \, \text{W/m}^2$ we have $|\Delta T_0| \simeq 1.2 |\Delta T_0| \simeq 0.5$ C (see Table 3.4), then $Ra' < 51$ even at $\delta \simeq 2 \, \text{mm}$. However, in the time interval of 2–3 h, both ways from midday, the value h_c reaches the centimetre scale and does not limit the thickness δ. It is sufficient for the latter to reach 3 mm so that Ra' should exceed 128, and a diurnal mixed layer should begin to form under the thermal boundary layer due to the penetration of primary convective elements ('thermals') from above.

The thermal structure of the near-surface layer between the free surface and the developing diurnal thermocline at each time moment is determined by the heat losses q_0 from the surface, the volume absorption of solar energy in the several upper metres of the water, convection, and wind mixing, and is quite complicated in the general case (Fig. 1.1a). The diurnal near-surface mixed layer with temperature T_h and thickness $h_0 - \delta$ is disposed directly under the thermal boundary layer with thickness δ, containing an inverse rise of temperature with depth ($T_h > T_0$). The diurnal near-surface mixed layer is formed by the combined effect of solar heating and convective and wind-wave mixing, and is one of the most variable elements of the near-surface layer in the diurnal cycle. This layer is practically non-existent at dead calm, intensive solar irradiance, and small flux q_0, when there is no wind mixing and convection is light, and the diurnal thermocline commences directly below the thermal boundary layer. When the wind is moderate in the hours close to midday and convection is relatively light, it is possible, in accordance with refs [239, 241] on the basis of (4.8), to single out a thinner layer of convective mixing within the diurnal mixed layer with the lower boundary on horizon $h'_0 < h_0$. In other cases, when the wind is light or there is no wind at all, and the heat losses from the surface are great, the whole diurnal mixed layer is determined by convection, and $h'_0 = h_0$. The thickness of the layer of convective mixing $h'_0 - \delta$ is minimum in all cases at local midday; it increases rapidly in the late afternoon (after 1600 LST) and in the early evening hours (Fig. 4.3). In this case, the proportion of convection in the total mixing increases gradually and from some time, when h'_0 approaches h_0, convection becomes the determinative factor of mixing. Subsequent deepening of the boundary $h'_0 = h_0$ continues up to convective mixing of the entire UQL in the early morning hours when $h'_0 = h_0 = H_0$. The temperature contrast $\Delta T_D = T_h - T_w$ in the diurnal thermocline reaches the maximum approximately at 1600 LST and then diminishes gradually on deepening of the mixed layer,

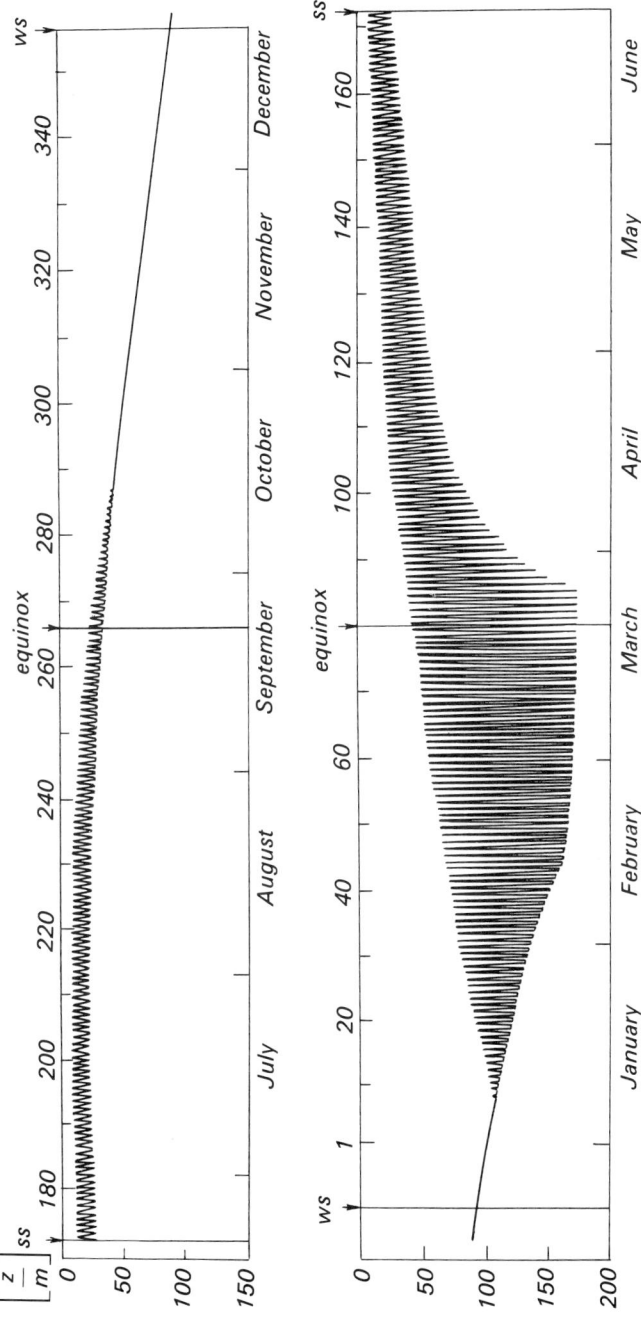

Figure 4.4. Seasonal modulation of the diurnal mixed layer h_0 (depth of turbocline occurrence [241] over 1 year calculated by mean climatic data for 41° N and 27° W in conformity with the model of Woods and Barkmann [241]. SS (summer solstice); WS (winter solstice)

and completely disappears by the time the UQL temperature levels by the vertical. The sharp growth of h'_0 after $\simeq 1600$ LST is frequently accompanied by a sharp temperature jump near the upper boundary of the diurnal thermocline (Fig. 4.2a) with a temperature contrast reaching 16% of the total temperature drop in it [9]. According to the model of Barenblatt [7], this discontinuity is a consequence of unsteady heat exchange in the thermocline at its rapid deepening. The rates of deepening at which the temperature jump was observed reached, according to estimates based on measurements with an emergent profiler [9], 0.05–0.10 cm/s, which is in good agreement with the model.

Temperature inversion with maximum T_{max} at the depth of thermal compensation develops sometimes in calm weather beneath the diurnal mixed layer whose boundary is determined in this case by convection (Fig. 1.1a). These inversions are most likely unsteady and short-living, with the exception of cases when horizon h_0 coincides with the boundary of the near-surface freshened layer and the temperature inversion is hydrostatistically steady. In the absence of a convectively mixed layer, the temperature maximum and the depth of thermal compensation are located directly beneath the temperature inversion of the thermal boundary layer. Examples of near-surface temperature inversions, recorded by direct measurements, are given in Figs 3.9c and 4.2b. They can also be found in Fig. 1 in ref. [117] and Fig. 1 in Ref. [102].

The variations during the day in the values h_0 and H_D, which are the upper and and lower boundaries of the diurnal thermocline, respectively, and in its temperature contrast ΔT_D are of special interest from the point of view of potential estimation and forecasting of the depth of occurrence and of the vertical temperature gradient of the diurnal thermocline. The temperature contrast is determined mainly by the diurnal trend of T_h, as in the absence of advection the value T_w in the diurnal cycle should not be changed in principle by more than $\Delta Q_D/(\rho c_p H_0)$; that at H_0 of about 40–50 m should not exceed $\simeq 0.05°C$. The diurnal trend of T_h equal to the diurnal trend of T_0 within the limits of accuracy determined by diurnal variation of $\overline{\Delta T_0}$ and identified commonly with the diurnal trend of SST, depends on combining the intensities of daytime heating and mixing and therefore is always coordinated with the diurnal trend of h_0. Depending on the factors stated above, the diurnal trend of ΔT_D can vary in different situations from several hundredths of a degree to 3.5°C (see Section 1.1, Fig. 3.9c, and also refs [19, 102, 134, 212]). The sharpness of the diurnal thermocline $(\partial T/\partial z)_D$ and the depth of occurrence of its lower boundary H_D also depend on the mixing intensity.

The diurnal cycle of T_0 and the daily variations of T_h, as well as the sharpness of stratification in the diurnal thermocline, are also influenced by the time shift between the approaching maxima of wind velocity and insolation. The greatest growth of T_0 and T_h in the daytime and the greatest contrast of ΔT_D can be anticipated at a time shift of 12 h (the strongest wind during the night and calm or light wind at noon). And vice versa, coincidence of the wind velocity maximum with the period of maximum heating results in a maximum amplitude of the diurnal heat wave and a light temperature contrast in the diurnal thermocline [134]. Alternation of intensification and abatement of the wind during the day

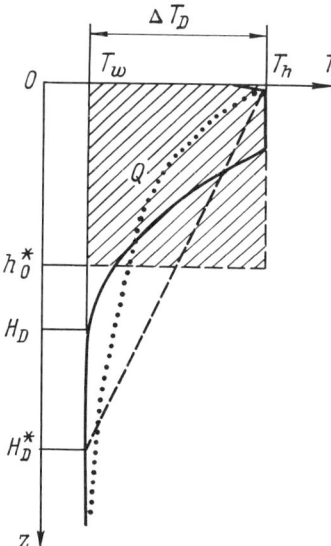

Figure 4.5. Diagrammatic representation of the limit (dashed line) and actual (solid and dotted line) profiles of $T(z)$. The dashed area corresponds to the enthalpy of the diurnal heated layer Q. See text for other explanations.

in the presence of continuous insolation under clear skies may give rise to the appearance of several intermediate steps on the profile $T(z)$ between δ and H_D of the afore-cited type (see Fig. 4.1c). In any case, with the same enthalpy $Q(t)$, characteristic of the given hour (see Table 4.5), all the vertical pofiles $T(z)$ corresponding to the regimes of moderate and intensive mixing should be within the limits given in Fig. 4.5 in the absence of advection and at a sufficiently correct and stable diurnal cycle of T_h. The dashed lines in this figure show the limit forms of the profiles that are never practically attained, and the heavy solid line indicates schematically the actual profile of $T(z)$. Accordingly, horizons h_0^* and H_D^*, which can be considered as estimates of the actual value of H_D from above and below, are determined by the following expression at the given enthalpy $Q(t)$ accumulated during the day or part of the day:

$$\frac{Q(t)}{\rho c_p \Delta T_D(t)} = h_0^* < H_D < H_D^* = \frac{2Q(t)}{\rho c_p \Delta T_D(t)} \tag{4.9}$$

and differ only by a factor of 2. To the point, the left-hand limit in (4.9) corresponds to the case when h_0 is the lower boundary of the totally mixed diurnal heated layer, while the right-hand limit represents the case when there is no mixed layer proper and the diurnal thermocline practically starts from the level of thermal compensation. In this case, horizon H_0^* should be adopted for the lower boundary of the heated (but unmixed!) layer as well as the diurnal thermocline. As a rule, at least 80% of ΔT_D is concentrated within the limits from the surface down to H_D^* in the case of intensive heating without mixing, and the layer with the sharpest vertical temperature gradient (dotted curve in

Fig. 4.5) is within the same limits. Expression (4.9) can be used to estimate the depth of occurrence of the diurnal thermocline lower boundary using the data of the SST diurnal cycle obtained on the basis of satellite or shipboard data, as well as to forecast this depth on the basis of climatic information. It is possible to use to this end routine daily values of $\Delta T_D(t)$ as well as the spread of the diurnal cycle ΔT_{Dmax} equal to $\Delta T_{Dmax} = 2A_T(0)$. Accordingly, it is necessary to have estimates of $Q(t)$ or Q_D which can be obtained on the basis of (4.5), finding $Q_s(\varphi, M)$ in Table 5.2 in ref. [72] and applying in this case to approximate estimation of $Q(t)$ as a certain share of Q_D by analogy with Table 4.5. To obtain $Q_s(\varphi, M)$, it is also possible to use empirical formulae, considering the sun elevation and the atmosphere transmission as has been done in the model in ref. [134].

It is interesting to make several estimates of h_0^* and H_D^* by using available satellite and shipboard data on the variation of SST with the application of expression (4.9). The necessary initial data and results of the estimates are given in Table 4.6.

The estimates in Table 4.6, judging by the orders of the values and as compared with the results of measurements (example IV, 1400 and 1900 LST), are quite verisimilar and produce values of h_0^* from 0.8 to 2.9 m and of H_D^* from 1.6 to 5.8 m under conditions of practically calm (wind less than 1.5 m/s) to light wind of 3 m/s at different times of the day. In some cases (III and IV), the diurnal cycle of h_0^* is observed with an increase of the heated layer thickness by the end of the day. In case I, where anomalously intensive heating is clearly bound with freshening of the near-surface layer by the discharge of the Elbe, some seeming disturbance of the diurnal cycle of h_0^* is most likely associated with the errors (± 0.5 K) in satellite measurement of the radiative temperature at the beginning of the day when ΔT_D is small, and with respective errors in SST recovering. The decrease of h_0^* by 1900 LST in case IV is caused by displacement of the vessel during the day, when the measurements were made, to another region (we made the measurements along a S–N tack 30 miles long, including two stations, Nos 2698 and 2699). It can be recognized on the basis of the results mentioned above that the suggested method of approximate estimation of the depth of occurrence of the diurnal thermocline lower boundary is quite useful for widescale application when analysing satellite and other data on the diurnal cycle of SST.

The afore-cited estimates and multiple probing of the near-surface layer under conditions of solar irradiance, made by us in the Sargasso Sea in 1978, indicate that the Väisälä–Brunt frequency N in the diurnal thermocline in the lower part of the heated layer (between horizons 2 and 5 m) can reach $(2–4) \times 10^{-2} \, \text{s}^{-1}$ in common cases in light wind weather. According to ref. [20], $N \simeq 8 \times 10^{-3} \, \text{s}^{-1}$ in a 7–11 m thick heated layer under a moderate wind ($\simeq 5$ m/s). In the case of extremum intensive heating (I in Table 4.6) $\bar{N} \simeq 5.9 \times 10^{-2} \, \text{s}^{-1}$ even on the average at $H_D^* \simeq 2$ m, and if it is considered that the diurnal thermocline thickness is not more than 1 m, then actually $N \simeq 8 \times 10^{-2} \, \text{s}^{-1}$. Thus, the diurnal thermocline is often substantially sharper than the seasonal and main ones [81, 102]. It can be expressed more sharply when the non-linear effect of advective

Estimation of h_0^* and H_D^* using data of satellite and shipboard measurements of diurnal trend of the ocean surface temperature

Time (h) (LST)	I SST (°C)	ΔT_D (°C)	Q, MJ/m² (cal/cm²)	h_0^*(m) / H_D^*(m)	II SST (°C)	ΔT_D (°C)	Q, MJ/m² (cal/cm²)	h_0^*(m) / H_D^*(m)	III SST (°C)	ΔT_D (°C)	Q, MJ/m² (cal/cm²)	h_0^*(m) / H_D^*(m)	IV SST (°C)	ΔT_D (°C)	Q, MJ/m² (cal/cm²)	h_0^*(m) / H_D^*(m)
6	13.0	0	0	0	11.1	0	0	0	20.4	0	0.5 (12)		28.0	0	0	0
7																
9	14.0	1.0	4.4 (104)	1.0 / 2.0												
10									21.0	0.6	4.6 (110)	1.8 / 3.6				
11													29.1	1.1	8.2 (195)	1.8 / 3.6
12	16.0	3.0	10.4 (248)	0.8 / 1.6												
13									21.5	1.1	9.5 (228)	2.1 / 4.2				
14					12.5	1.5	7.3 (174)	1.2 / 2.3					29.7	1.7	15.3 (366)	2.2 / 4.3 (3.0)
15									21.9	1.5	12.9 (308)	2.1 / 4.2				
16	16.5	3.5	18.1 (433)	1.2 / 2.4			8.5 (204)		21.8	1.4	13.2 (316)	2.3 / 4.6	29.5	1.5	18.0 (430)	2.9 / 5.8
: 19													29.3	1.3	14.6 (350)	2.7 / 5.4 (4.5)

I—North Sea, 55° N, 12 July 1979, SST by METEOSAT data, wind <1.5 m/s [212]; data on SST confirmed by shipboard measurements; II—SE of Cook Strait, 40° S, 1–2 May 1983, SST by data of NOAA-7 and NOAA-8 satellites, wind conditions unknown [124]; III—Bay of Biscay, 44° N, 17 September 1985, SST by METEOSAT data, wind conditions unknown [136]; IV—Sargasso Sea, 28° N, 21 July 1978, T_h measured by a towed sensor from the research vessel *Akademik Kurchatov* in motion and by a CTD probe at stations. Measured values of H_D are given in the right-hand column in parentheses. Values of Q_D (underlined) were determined with consideration of (4.5) by Q_s in Table 5.2 in ref. [72]. The similarity with the data in Table 4.5 was used to determine the hourly values of $Q(t)$.

redistribution of diurnal totals of absorbed solar irradiance on account of convergent currents in the frontal zones, or convergent drift currents bound with the specific meteorological conditions, is added to the natural diurnal temperature trend. In this case, the values of ΔT_{Dmax} can be especially high and the calculated values of the diurnal increment of the near-surface layer enthalpy may exceed the normal diurnal values of Q_D [see expression (4.5) and Table 4.4 on p. 111].

Cases are possible under very intensive wind–wave mixing and/or light heating when the diurnal thermocline is practically indistinguishable and difficulties arise in determining the diurnal mixed layer boundary h_0 against the background of UQL. Woods and Barkmann [241] suggest the horizon where the commonly high values of the dissipation rates of the turbulence kinetic energy ε, according to their estimates, diminish by 2–3 orders*, and which they called 'turbocline' as the lower boundary of this layer (h_0). However, special measurements of the turbulent pulsations (preferably, rates) are necessary to determine the depth of occurrence of the turbocline. These measurements are quite realistic currently and can be carried out operatively [81, 152, 214], but difficulties connected with the choice of the threshold value ε under conditions of natural alternation of turbulence are also evident [241]. It should also be taken into consideration that under certain special conditions (e.g. when additional sources of turbulence kinetic energy, such as the vertical shear of velocity, are available at some depth), a sharp change of ε with the depth may not be observed at the lower boundary of the diurnal mixed layer h_0. These conditions are characteristic, for example, of the near-surface layer above the upper boundary of equatorial counter-currents [152].

Single measurements of $T(z)$ may produce dissimilar estimates of h_0 and H_D for other reasons as well. As in any other pycnocline, internal waves can be generated in the diurnal thermocline. Their generation can be bound, in particular, with the penetration of turbulent eddies from the diurnal mixed layer [20]. Accordingly, periodic displacement by depth can be observed in the diurnal thermocline of isotherms as well as the thermocline boundaries h_0 and H_D. The h_0 and H_D values can be modulated by Langmuir circulations (see also Section 4.5.4).

4.4. The blocking role of the diurnal thermocline with regard to turbulence in the near-surface layer of the ocean

Hydrostatically stable stratification in the near-surface layer, suppressing turbulent pulsations of velocity, creates a certain 'blocking effect' [81, 102], preventing the propagation of turbulence from the surface, where it is generated

---

*These estimates are based, most likely, on the traditional concept of stationary equilibrium ('dissipation' is equal to 'generation') turbulence, which is inapplicable in the given case because dissipation should prevail greatly over generation near by the pycnocline which limits the UQL from below.

by wind–wave mixing and convection, into the UQL thickness. This is evidenced by indirect indications [e.g. the character of the $T(z)$ variation with the change of heating conditions], as well as by the results of direct measurements of turbulent pulsations of velocity in the ocean. One of the major indirect indications is the rapid destruction from below of the abrupt boundary of UQL (H_0), or the diurnal mixed layer (h_0), which is manifested in slackening of the temperature vertical gradient and even in the formation of small steps on this boundary [74, 81, 110]. This effect is best noticed in calm weather immediately after the beginning of diurnal solar irradiance against the background of the layer of the density jump that intensified during the night below the UQL. It can be observed during the daytime in windy weather immediately after appreciable abatement of the wind which supported before that the mixed layer boundary (h_0 or H_0) in a sharpened state despite solar heating. Observations demonstrate that the near-surface 'blocking effect' tells extremely rapidly (within 1 h) on the structure of the jump layer below the UQL even if the boundary of the latter is quite deep, e.g. 40 m [74], or even 80–90 m deep [110]. It should be noted that the lower boundary of the UQL recovers its abruptness just as rapidly after subsequent intensification of the wind.

The association of the depth of turbulence propagation with thermal stratification and the diurnal cycle of UQL turbulization are demonstrated in the work of Paka and Fedorov [81] based on direct measurements of velocity pulsations by the microturbulent probe 'Baklan' in the equatorial area of the Pacific ($2°$ N–$2°$ S, 163 and $167°$ W), i.e. in a region where it is possible to anticipate additional turbulization of the UQL from below due to intensive vertical shear of velocity at the upper boundary of the equatorial subsurface counter-current (Cromwell current). It turned out that the thickness of the near-surface turbulized layer is determined not by this shear, but by the intensity of thermal stratification and the time of day, respectively. Intensive turbulence penetrated during the night to a depth of 60–80 m, and in the daytime (1300–1600 LST) it was observed only in a layer of $\simeq 15$–20 m even at a wind velocity of 7 m/s. The obvious correlation of the degree of the near-surface layer turbulization with its thermal stratification even at intensive shear of velocity in the lower part of the UQL led Paka and Fedorov [81] to the conclusion that turbulization of the UQL under conditions of moderate winds and in the absence of precipitation* should experience a diurnal cycle everywhere with the exception of areas in high latitudes.

New direct measurements of turbulence [152] also in the equatorial part of the Pacific with a subsurface counter-current ($0°$ N, $139°50'$ W) confirmed the conclusion on the diurnal cycle of turbulence in the upper 30 m mixed layer. The recorded variation of ε in the diurnal cycle, coinciding in phase with the coming solar irradiance, reached two orders. Besides, a diurnal cycle of

---

*Intensive precipitation also creates a 'blocking effect' (see Section 4.5.3) comparable to the effect of thermal stratification, but the precipitation lacks a vividly manifested diurnal cycle in the majority of climatic zones.

turbulence energy was discovered with a 10- to 100-fold variation in the density stratified layer (30–90 m) with a velocity shear wherein the diurnal cycle of Väisälä–Brunt frequencies was not traced. The minimum of the turbulence intensity in this layer progressively lagged behind in time with depth in regard to the maximum of insolation (up to 3–6 h at the horizon 65 m). The cause of the diurnal cycle of ε in the stratified layer is not yet clear. It should also be noted that the diurnal cycle of ε and N lead to respective changes in the Ozmidov scale $L_0 = \varepsilon^{1/2} N^{-3/2}$. According to estimates in ref. [152], L_0 is equal to several tens of centimetres at noon and to several metres during the night in the layer below 30 m.

The attenuation of turbulence in the diurnal thermocline results in the formation in the UQL thickness of layers with a thickness of several tens of metres wherein turbulence is not practically observed. It is probable that these layers in daytime are in an intermediate condition between turbulent and laminar regimes. The introduction of the slightest disturbance into these layers (e.g. local rise of shear or penetration of convective motion from above) apparently results in their rapid returbulization. This explains evidently the rapid recovery of the abrupt lower boundary of the UQL after the disappearance of the heated layer near the surface, or at intensification of the wind, which was noted in refs [74, 81, 110].

The suppression of turbulence by thermal stratification affects rapidly the vertical distribution of suspended matter (decrease of its concentration) in the near-surface layer of the ocean. Most likely, this is the reason for the effect of 'clarification' of the $\simeq 10$ m thick heated layer described in ref. [15] and consisting in a decrease of the attenuation index ($\Delta a/a \simeq 0.04$–0.2) soon after the beginning of diurnal heating at light wind waves.

An important parameter determining the diurnal mixed layer rate of deepening into the stratified thickness is the coefficient of turbulent entrainment k on its lower boundary h_0. It reflects the proportion of total energy available for the generation and maintenance of turbulence (convective and/or wave–wind), which is used to increase the potential energy of the mixed layer solely due to entrainment of colder volumes of water from the underlying stratified thickness of the diurnal thermocline. In the case of free convection (no wind) [14],

$$k \equiv \frac{\langle T'w' \rangle_{h_0}}{|q_0/(\rho c_{\mathrm{p}})|} < 1, \tag{4.10}$$

where T' and w' are the turbulent pulsations of temperature and the vertical velocities, respectively.

In the presence of convection and wind, it is necessary to consider also the inflow of energy transmitted to the water by the wind. In the case of developed waves, when the main portion of the wind energy is transmitted to the water through tangential stress, this inflow of energy G can be expressed by the wind velocity u_{10} at a height of 10 m in the near-water layer of the atmosphere, the orbital velocity of water particles u_{w} in the developed waves, and the mean coefficient of friction $c_D \simeq 10^{-3}$ [20]:

$$G = \rho_a c_D u_{10}^2 u_w, \tag{4.11}$$

where ρ_a is the air density.

Woods and Barkmann [241] adopted $k = 0.15$ for the conditions of free convection in their integral UQL model. The laboratory experiments of Varfolomeyev and Sutyrin [14] carried out under the guidance of one of us with the special purpose of studying the behaviour of k depending on q_0 and $\partial T/\partial z = \Gamma_T$ under conditions of free penetrating convection have demonstrated that the asymptotic regime of this convection is determined by two dimensionless parameters, i.e. $\tilde{Q} = q_0/(\rho c_p k_T \Gamma_T)$ and the Prandtl number Pr. Values from 0.17 to 0.27 were obtained in the experiments at $\tilde{Q}$ from 30 to 50, respectively, for the coefficient of entrainment k. The overall trend of the dependence of k on $\tilde{Q}$ is in good agreement with another known result, i.e. $k = 0.07$ at $\tilde{Q} \simeq 10$ [14]. The problem of the limit value of k for free penetrating convection in these experiments remains obscure. On the other hand, the value $k = 0.5$ is in satisfactory agreement with field observations [20] under conditions of a mixed regime in the presence of a moderate wind with a velocity $u_{10} \simeq 5$ m/s. Hence, it can be concluded that the limit value of k for free convection is somewhere between 0.3 and 0.5.

The work mentioned above [20], wherein the case of wind mixing under conditions of intensive solar heating in the Equatorial Atlantic is discussed in detail on the basis of field data, is interesting also because the depth h_* of wind–wave turbulence penetration into the diurnal thermocline is estimated on the basis of ref. [6]. In the afternoon, when $q_0 = -210$ W/m^2 and $N \simeq 8 \times 10^{-3}$ s^{-1} ($\Gamma_T \simeq -0.03°$C/m), $h_* \simeq 1.7$ m, demonstrating that the turbulence generated by the wind of velocity 5 m/s is unable to overcome even such a relatively light diurnal thermocline whose thickness reached 10 m in the given case.

4.5. Typical scales and characteristics of space structure inhomogeneities of the near-surface layer of the ocean

The near-surface layer of the ocean is under a continuous and extremely variable, spatially inhomogeneous influence of the atmosphere. It is logical to expect under these conditions that the space scales of the near-surface layer structural inhomogeneities will be conditioned mainly by the space scales of the atmospheric influences. However, observations in the ocean demonstrate that the near-surface layer also contains internal scales within which certain specific structural inhomogeneities, e.g. Langmuir circulations, convective cells, etc., form as a result of external forces. The instability of the oceanic fronts, which, as a rule, are manifested most vividly just in the near-surface layer, generates vortical disturbances in its structure in a wide spectrum of horizontal scales from several metres to tens and hundreds of kilometres. Other scales are conditioned 'from below', e.g. the scale of the internal waves of the seasonal thermocline, which is perceptible up to the surface proper, or the scale of synoptic eddies associated with the baroclinic Rossby radius of deformation Ro depending on the characteristics of the main thermocline. Vertical motions and vortical disturbances of various scales can also be generated at the interaction of the currents with the bottom topography. Finally, there are also such cases (see Section 6.3) when the scales of inhomogeneities of the atmosphere near-water layer are conditioned by specific, in the given case frontal, space scales of the SST field. In principle, it is enough to look through a dozen satellite images of the ocean surface in the IR scale to see how complicated, different in scale, and variable the space structure of the SST field in the frontal zones is, and during the daytime in summer even in relatively 'calm' regions in the open ocean.

The reaction of the near-surface layer, as a result of its small thickness, to external forces is practically 'instantaneous' if it is compared to typical time scales of the main physical processes in the ocean. The most rapid reactions of the near-surface layer are the following: wind–wave and convective mixing; freshening, salinization, and solar irradiance; development and attenuation of drift currents, wind waves, Langmuir circulations, and thermal inhomogeneities of calm weather. The typical time of reconstruction of the near-surface layer regimes never exceeds the limits of a day, although, as given below, the subsequent thermal and dynamic reactions of the deep layers of the ocean to certain atmospheric influences (e.g. typhoons and tropical cyclones) can be manifested for many days and even months. Besides, the diverse secondary effects, generated in the ocean as a result of atmospheric and other effects on the near-surface layer, require different times for development and attenuation. Typical time scales of some of the latter exceed greatly the characteristic time of the effects per se, e.g. the time scales of geostrophic adaptation of motions roused in the ocean and the mass fields, and also the typical time of relaxation of the aroused geostrophic currents.

Unfortunately, non-stationary problems, which are necessary for analytical and model approaches, are still set and solved vary rarely in geophysical hydrodynamics. Therefore, subsequent consideration of a series of regularities determined in this section will be carried out by analysing field data. Some of the model quantitative estimates, bound with local wind effects on the ocean, will be discussed in Chapter 5.

4.5.1. *Thermal effect of strong atmospheric formations and processes*

Tropical cyclones, hurricanes, and typhoons exert the strongest effect on the near-surface layer of the ocean. Their form is close to circular with typical diameters ranging from 100 to 1000 km. The maximum wind velocities are often more than 50 and even 100 m/s. Moving over the ocean with a rate of several tens of kilometres per hour, these formations are accompanied by force 8–9 storm waves, abundant precipitation, upwelling of water from deep horizons to the surface, intensive wind–wave mixing, and extreme intensification of heat and mass exchange with the atmosphere. Under certain conditions, all these processes may leave an extremely noticeable and long-living trace in the upper layer of the ocean, penetrating several hundred metres into the depth. The

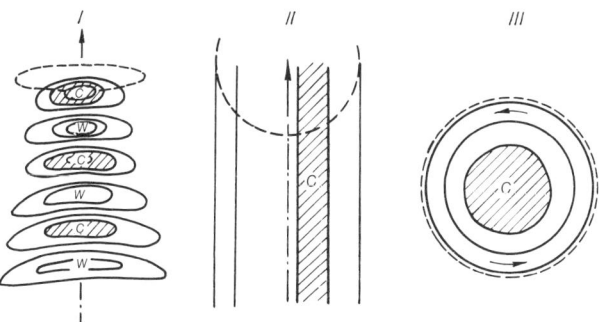

Figure 4.6. Three types of trace left behind by typhoons and tropical cyclones in the upper layer of the ocean. (I) Wavy trace (by Geisler) of rapidly moving tropical cyclone ($V > c_1$); (II) cold band left behind by slowly moving tropical cyclone ($V < c_1$); (III) vortical cyclonic disturbance arising under the effect of a quasi-stationary typhoon ($V = 0$). Areas of maximum cooling are designated by the letter C and are dashed. Warm areas are designated by the letter W. The direction of cyclone displacement is shown by arrows, and the trajectory of their centres by a dash-dotted line. The dotted line shows schematically the areas of effect of the maximum winds. The isotherms are indicated by solid lines.

character of this trace depends on the hurricane displacement rate V. As the thermohaline and dynamic disturbance, arising in the ocean beneath the hurricane, radiates in the process of geostrophic adjustment inertial-gravity waves, whose parameters are determined by UQL thickness H_0 and by stratification of the jump layer beneath it, it is convenient to compare the hurricane displacement rate V with the phase rate of the first mode of the inertical-gravity waves $c_1 = \sqrt{g'H_0}$, where $g' = g\Delta\rho/\rho$ is the reduced acceleration due to gravity and, in its turn, $\Delta\rho$ is the density drop of ρ in the jump beneath the UQL. If $V > c_1$, then the disturbances in the UQL thickness and the pattern of the water temperature anomalies in the near-surface layer, respectively, are of periodic (wavy) character. If $V < c_1$, then the trace of the hurricane is a continuous band of cooled water extending along the trajectory. If a typhoon or a tropical cyclone remains in the same place for 2–3 days, it swirls beneath itself a real cyclonic eddy of synoptic scale with a penetration depth of about 1000 m or more and with respective thermal disturbance. These three types of disturbance are represented schematically in Fig. 4.6. Disturbances of types II and III are the longest-living ones. The space scales of the temperature anomalies in the near-surface layer of all three types of disturbance are determined by combining the scales of the atmospheric effect with the characteristic scales of the dynamic reaction of the ocean (see Section 5.1). The results of the studies of Geisler, Fedorov, Ivanov, Pudov, and others, laid down in the basis of grouping such disturbances, have been systematized and further developed by Sutyrin in Part II of monograph [113]. Disturbances relating to type II ($V < c_1$) are the best studied ones. They are accompanied by a 1.5–6°C decrease of SST in the trace as compared with the background depending on the force of the hurricane, while the arising temperature anomalies remain for several weeks and even months. Even after such a small hurricane as 'Ella', whose trace was examined in detail by us in the Sargasso Sea in 1978, the temperature in the trace dropped against the background by 1.8–2.0°C, and the pattern of thermal anomalies remained for more than 3 weeks. Detailed data on the thermal structure of such disturbances are found in refs [101, 105, 113].

Common atmospheric cyclones leave a noticeable, but essentially shorter-living trace (not more than several days) in the temperature field of the ocean. A sharp increase in the wind velocity and a decrease in the air temperature ΔT_a at the passage of the cold front, which is accompanied by an increase in the heat flux from the surface $|q_0|$, result in a drop in the ocean surface temperature and an increase in the mixed layer thickness H_0. An interesting case of such local interaction of the atmosphere and the ocean in the warm ring of the Gulf Stream was discussed recently in ref. [214]. The atmospheric cyclone, crossing the region of investigations, was accompanied by two storms with wind velocities over 20 m/s separated by an interval of 3 days. The temperature inversion of about 4°C from the surface to horizon 170 m was an interesting specific feature of the initial vertical structure of the waters near the surface. Despite compensation of the inversion by vertical distribution of salinity, the hydrostatic stability of this layer was weak, especially between 30 and 150 m. The water temperature close to the surface should be increased due to mixing of the inversion layer by the vertical.

However, the latter temperature rise was compensated, at least in the initial period, by cooling as a result of heat losses from the surface. Thus, 3 days after the first storm, despite an inflow of cold air with $\Delta T_a \simeq 5°C$, the ocean surface temperature remained practically invariable (about 12°C). The UQL thickness increased at the same time from 15 to 50 m, but decreased to 25 m by the time of the second storm. After the second storm, the mixing started to penetrate into the least stable layer and, as a result of rapid deepening of the UQL lower boundary, the water temperature at the surface began to rise despite the compensating effect of cooling. The water–air temperature difference reached 15°C approximately 3–4 days after the beginning of the second storm; the heat flux from the surface increased to 900 W/m²; and the UQL thickness increased to 170 m. The water temperature near the surface reached 13.5°C, which is only 0.5°C lower than the mean temperature of the inversion layer 0–170 m before the beginning of mixing. Hence, according to our estimate, cooling of the 25–170 m layer by 0.5°C in the 4 days after the beginning of the second storm corresponds to a total loss of heat of about 303 MJ/m² (7250 cal/cm²) or to a mean heat flux from the surface of about 880 W/m², which is very close to the value stated in ref. [214] and obtained on the basis of known integral formulae. The formed powerful convective-mixed layer manifested itself well on the vertical profiles $T(z)$ and $S(z)$, and on the vertical profile of the turbulent energy dissipation rate ε estimated on the basis of direct turbulent pulsation measurements of the rate. The fact that the boundary of the mixed layer on profile ε was 15 m deeper than on profiles $T(z)$ and $S(z)$ demonstrated active turbulent entrainment from the underlying stratified thickness. The cause of this intensive growth of UQL thickness was convection and not wind mixing, as shown by the estimates of Shay and Gregg [214] of the Monin–Obukhov scale L, and their comparison with the UQL thickness ($5 < |H_0/L| < 10$).

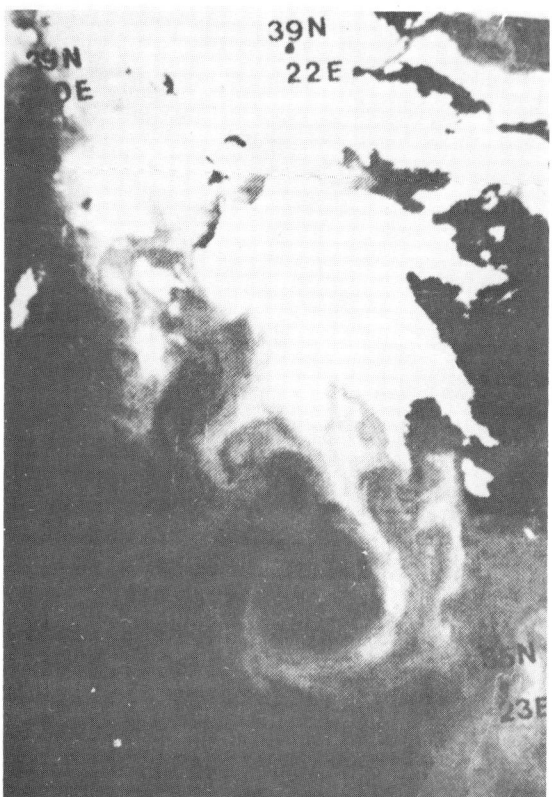

Figure 4.7. Cold near-surface jet drawn into orbital movement by an anticyclonic eddy southward of the Peloponnese Peninsula. IR image transmitted from NOAA-9 satellite, orbit 2717, 23 June 1985 [136]. See interpretation diagram corresponding to this image in Fig. 6.8.

An instructive example of the combined effect of coastal cooling of waters (as a result of upwelling or cold air outflow from the shore behind the cold atmospheric front) and entrainment of the cooled waters in the orbital movement by the anticyclonic eddy, which was not manifested initially in the SST field, is demonstrated in a succession of six IR images of a region in the Mediterranean Sea southward of the Peloponnese Peninsula obtained from the NOAA-9 satellite (orbits 2449–2816) from 4 to 30 June 1985 (see pp. 8–9 in issue 22 of ref. [136]). The waters, cooled in the Messiniakos Kolpos down to 20°C, were drawn into the orbital movement of the eddy of $\simeq 100$ km diameter southward of the Akritas Cape and, against the background of the surrounding warm waters with a temperature of 23°C, spread out, slowly heating, at its periphery at a velocity of $\simeq 35$ cm/s*, as estimated by us. An almost closed circular band of relatively cold waters formed gradually in the near-surface layer on the periphery of the eddy with a width up to 20–30 km and length (by circumference)

---

*See the discussion on this estimate in Section 6.2 and Fig. 6.8.

of about 300 km, limited on both sides by temperature contrasts of 1.5–3.0°C, clearly distinguishable in the IR images (Fig. 4.7). The SST contrast was conditioned in this case by the atmospheric processes, and the space scales of inhomogeneity by the scales preset 'from below' from the water thickness by the eddy of synoptic scale developed therein.

4.5.2. Thermal effect of atmospheric convective cells

The effect of atmospheric convective cells, having a typical diameter of several tens of kilometres, on the thermal structure of the ocean near-surface layer is conditioned by the space inhomogeneities of the near-water wind velocity u_a and heat losses q_0 from the ocean surface, respectively. In closed convective cells, ascending air flows occur in the central areas, while air descends on the periphery. The pattern is just the opposite in open cells. Correspondingly, convergence of the horizontal air flows under the clouds and divergence in the space between the clouds are observed in the near-water layer in all cases. As convective cells commonly displace over the ocean together with average transfer of air in the lower troposphere, the field of the near-water wind is, in this case, of periodic space structure, wherein a certain background wind velocity u_b in the near-water layer is modulated by the periodically varying convective component u_c, so that $u_a = u_b + u_c$. In this case, measurements at a fixed point produce a wind velocity modulated by a harmonic addition with a period depending on the diameter of the cell and the velocity of its transfer. The opinion was expressed in some works [55, 229] that periodic intensification of cooling due to intensification of the wind velocity should be manifested as a periodic temperature signal penetrating into the water to a depth of several dozen metres. According to data in ref. [229], modulation of the velocity of average transfer $u_b = 8$ m/s by the convective component u_c reached ± 0.5 m/s for intensive atmospheric convection at the transfer of cold and dry continental air in winter over the warm waters of the East China Sea. Correspondingly the oscillations of q_0 had a period of approximately 1 h and an amplitude of about ± 35 W/m^2 (or $\simeq 6\%$ from the average level of heat losses), which gave an additional total heat loss of -0.125 MJ/m^2 (-3 cal/cm^2) during the passage of each cell ($\simeq 1$ h). According to our calculations, the periodic temperature signal induced by this addition could not be recorded by equipment with a sensitivity worse than 0.01°C on horizons deeper than 3 m. The hourly fluctuations of the water temperature that, nevertheless, were perceived at all levels down to the bottom (75 m) were most likely induced by quasi-periodic convective breakthroughs caused by the overall continuous cooling of the water from the surface. It was periodic intensification of cooling due to atmospheric convective cells that could be, in this case, the 'trigger mechanism' for these breakthroughs. Numerical simulation of convection [12] demonstrated the potential effect of the periodic processes on its space and time scales. About 1 h (45 min by the estimate in ref. [229]) was necessary in the given case to form the convective breakthrough. Therefore, when $\bar{q}_0 = -560$ W/m^2, the cooled water in each such breakthrough could not have a deficit of heat of more than 2.1 MJ/m^2 (50 cal/cm^2), and the temperature fluctuations at the bottom associated with the breakthroughs could not exceed 0.007°C. Only highly localized breakthroughs collecting water from an area greatly exceeding the cross-section of the sinking region could induce local fluctuations of several hundredths of a degree at the bottom. In any case, the limit cooling of the entire 75 m water thickness per day with the given thermal balance could not exceed 0.13–0.14°C, or, again, 0.006°C per atmospheric cell passing through the observation point. The difference of the diurnal phase shear of correlation between the convective breakthroughs and the atmospheric index (75 min) from the nocturnal phase shear (about 25 min), given in ref. [229], is of certain interest. It is the result of the effect of diurnal solar heating 8–12 MJ/m² (200–285 cal/cm²) that retarded the onset of convective instability in the near-surface layer. The time (about 5 min) stated in ref. [229] within which highly correlated temperature fluctuations were perceived at all the horizons from the surface to the bottom corresponds to the vertical sinking rate of large convective elements equal to 25 cm/s.

It is necessary to note in conclusion that the descending flows of cellular convective circulation in the near-water layer of the atmosphere during rain give, according to the measurements and estimates of Gautier [148], at least no less contribution to the ocean surface temperature decrease than the precipitation per se (see Section 4.5.3). When the convective cells are in the same place for a long time, the descending flows may result in considerable cooling of the near-surface layer of the ocean which was freshened by precipitation.

4.5.3. *Effects of precipitation in the near-surface layer of the ocean*

Rains precipitating on the ocean surface freshen the near-surface layer and change the water temperature near the surface. The thickness h_w of the freshened layer, the values ΔS and ΔT of freshening and temperature change, respectively, and the deepening rate of the freshened layer lower boundary in the process of precipitation depend on the intensity and duration of the rain and on the mixing intensity. If the values of salinity S_1 and S_2 near the surface are accurately measured immediately before and after the rain, it is easy to calculate the thickness of the freshened layer h_w in metres by the quantity of precipitation which is commonly measured in millimetres:

$$h_w = \frac{\delta_w \times 10^{-3} S_2}{S_1 - S_2}. \tag{4.12}$$

On the other hand, if h_w is known, then it is easy to determine $\Delta S = S_1 - S_2$ and S_2, respectively, by S_1 and δ_w:

$$\Delta S \simeq S_1 \delta_w \times 10^{-3}/h_w, \tag{4.13}$$

$$S_2 \simeq S_1(1 - \delta_w \times 10^{-3}/h_w). \tag{4.14}$$

As δ_w is one of the most difficult values to measure in the ocean, it is possible to obtain sufficiently accurate estimates of δ_w on the basis of correctly measured

S_1, and S_2 and h_w:

$$\delta_w \simeq \frac{\Delta S h_w}{S_1 \times 10^{-3}} \text{(mm)}. \tag{4.15}$$

Estimates made on the basis of (4.12)–(4.15) and confirmed by numerous observations indicate that heavy showers with a duration of 1–2 h and intensity 20 mm/h and more can freshen a layer up to 1 m thick by 0.5–1.0‰ in calm weather and a light wind, which exceeds greatly the typical amplitude of local variability of salinity in the near-surface layer in the absence of rain. Short-term light rains reduce the salinity near by the surface by 0.2–0.3‰.

The freshened areas (patches) resemble saucers in shape and have a characteristic scale corresponding to the scales of individual cumulo-nimbuses or their typical clusters, i.e. from 1 to 10 km. The freshened patches can remain for several days in calm or light wind weather, while the freshening ΔS can increase due to numerous heavy showers.

In the case of strong winds, the rain freshening penetrates to deeper horizons due to more intensive mixing. In this case, however, according to (4.13), the values of ΔS are essentially smaller and are practically not different from amplitudes of the background salinity variability. An interesting effect is observed in the case of very heavy showers accompanied by a strong wind, namely, the addition of positive buoyancy together with rain water into the near-surface layer practically neutralizes the effect of turbulence generated by the wind. As a result, while it rains, the initial mixed layer does not deepen at all or deepens very slowly. According to observations in ref. [204], carried out at a wind velocity of 10 m/s during rain that lasted for about 2 h,* the mixed layer, wherein freshening took place, remained invariable in thickness $h_w = 7$ m at least for 1 h. During this time, the freshening reached $\Delta S = 0.25$‰ which, according to (4.15), is equivalent to precipitation of 48 mm. After that the rain most likely continued for some time ($\simeq 20$–30 min), because the quantity of precipitate, estimated at the end of the observations by the value ΔS at a maximum h_w, was 60 mm. Seven hours after the rain, h_w reached a thickness of 20 m at $\Delta S = 0.11$‰. The average rate of deepening of the mixed layer lower boundary was 5.1×10^{-2} m/s and apparently would have been greater if not for the preliminary freshening due to the rain.

The freshened areas (patches) were preserved for a much shorter period of time under a strong wind than in calm and light wind weather. According to ref. [204], maximum freshening $\Delta S = 0.25$‰ remained for only slightly longer than 1 h, and the salinity in the observed region almost recovered its initial values 12 h after the end of the rain.

A drop in the water temperature is observed after rain more often near the surface [19, 24]. It rarely exceeds 0.2–0.3°C (Fig. 4.8a) and is determined by the joint effect of the precipitation temperature, which, as a rule, is lower than the water temperature, and the changes in the conditions of heat exchange with

---

*The times of the beginning and end of the rain were not recorded.

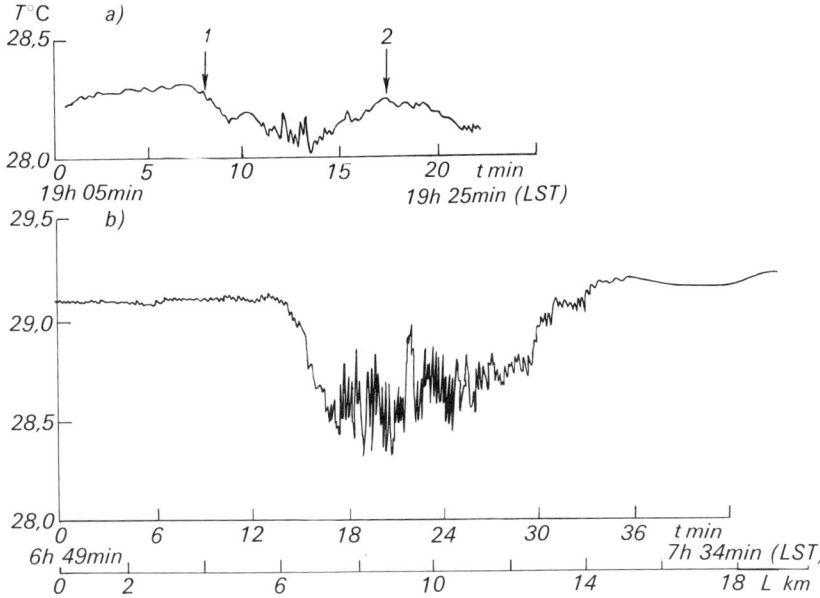

Figure 4.8. Characteristic variation of ocean surface temperature ($z \simeq 0.15\,\mathrm{m}$) recorded by a towed thermal sensor when the vessel crossed the rain 'trace' (vessel speed 15 knots). Our observations in the Sargasso Sea, 27th cruise of the research vessel *Akademik Kurchatov*, August–September 1978. (a) During rain (1: vessel enters rain band; 2: vessel leaves rain band); (b) in morning hours (before solar heating) after precipitation during the night.

the atmosphere due to intensification of the wind and air temperature drop. Tha water temperature decreases with its continuous recording at a horizon close to the surface, when the vessel crosses the rain zone, and is of such characteristic shape (Fig. 4.8a) that it is possible for the purpose of description to speak of a thermal 'trace' of the rain. Sharp small-scale temperature fluctuations with a characteristic horizontal scale of 50–100 m, clearly visible in Fig. 4.8a, are observed in the rain traces recorded in conditions close to calm in the vicinity of the patch's centre. According to available data, similar fluctuations are observed in the salinity field. The internal waves on the lower boundary of the freshened layer, where N can reach $(1-3) \times 10^{-2}\,\mathrm{s}^{-1}$, are the most probable cause of these fluctuations [102]. Areas freshened by rain remain near the surface for a long time and, as a result, the rain trace can often be found in absolutely dry, clear weather when no trace of the rainy weather can be detected even in the atmosphere. By similarity with the term 'fossil (or relic) turbulence', we called these rain traces 'fossil rain traces' [24]. In Fig. 4.8b it is shown that similar small-scale temperature fluctuations occur there.

The precipitation of rain is equivalent to intensive solar heating by its effect on turbulent exchange in the near-surface layer of the ocean. Positive buoyancy is added in both cases near the surface and a substantial amount of turbulence energy, generated by the wind, is spent to overcome it. Thus, in the case discussed in ref. [24], the 2 h rain, during which the rainfall on the ocean surface amounted to 60 mm, was equivalent to the addition of about 21 MJ/m^2 (500 cal/cm^2) of solar heat to the near-surface layer which corresponded to 9–10 h of solar heating in summer under clear skies. This cannot but cause suppression of turbulence generated by the wind (see Section 4.4), which, in its turn, causes easily predicted consequences.

In the daytime, during the period of intensive solar heating, lenses of freshened water on the ocean surface turn into peculiar solar heat traps. We repeatedly observed in the Sargasso Sea during the POLYMODE experiment in 1977 and 1978 cases of a sharp rise in the temperature in the near-surface layer of the water immediately after abundant rains. Figure 4.9, showing a series of vertical temperature profiles in the near-surface layer, obtained by an emergent profiler, before, during, and after a rain, demonstrates that 1.5 h after the rain the temperature in the upper 2 m layer directly adjoining the surface rose by 0.4°C. And it happened after quite a short heavy shower that lasted not more than 15 min and brought only 5–7 mm of precipitation, whereas heating occurred with intermittent cloudiness, gradually reducing from force 7–8 to 1–2. Freshening near the surface did not exceed 0.05–0.10‰. Superheating of the near-surface layer in the case of more intensive freshening and under clear skies reached 1.5–3.0°C in the afternoon as compared to the temperature on the horizon 3–4 m, i.e. below the freshening boundary (see Fig. 4.10 and ref. [102]).

As cited above, freshening of the near-surface layer can 'accumulate', and freshened lenses can merge if the intermittent heavy showers are observed for several days in calm weather. The diurnal temperature of the near-surface layer in such regions of the ocean as the Sargasso Sea, where calm weather is frequent in summer, can reach 30–32°C in some places, while in the neighbourhood, where there was no rain, the temperature of the water near the surface is within its common range of 27.5–28.5°C. Our observations in the Sargasso Sea have shown that regions with a near-surface layer freshened by 0.2–0.3‰ and with a high diurnal temperature sometimes have a diameter of 150–200 km. The formation of such superheated regions, in turn, can contribute to the formation of large clusters of cumulo-nimbus and to the origin of tropical cyclones and typhoons [24].

At night, the density jump on the lower boundary of the freshened lens prevents propagation of convection into the deeper layers. As a consequence, the near-surface layer in the rain trace is subjected during the night to more intensive radiation cooling than in the neighbouring areas, where there was no rain and where convective mixing penetrated to a greater depth. As a result, the 'fossil rain trace' can be colder in the morning hours than immediately after the fall of precipitation (Fig. 4.8b). Taking into account that the ocean loses during the night on the average about 4.2 MJ/m^2 (100 cal/cm^2) due to radiation and contact heat exchange [239], it is quite sufficient in the absence of wind of the deficit of salinity $\Delta S = 0.1$‰ for the freshened 3–4 m thick layer to preserve during the night its lower boundary, cooling down by 0.10–0.30°C in relation to the surrounding waters wherein the temperature only equalizes by depth within the limits of the UQL as a result of convection. A case of such cooling which was preserved by the beginning of diurnal heating is shown in

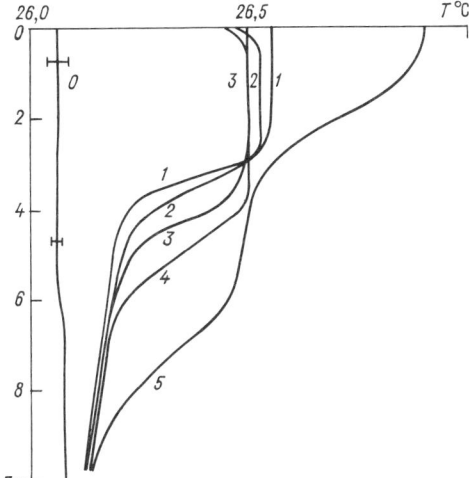

Figure 4.9. Development of the temperature vertical profile in the near-surface layer during solar heating in light wind weather 5 h after sunrise, including a brief heavy shower from 0940 to 0950 LST on 22 September 1978. Sargasso Sea, station No. 2749, research vessel *Akademik Kurchatov*. Profiles $T(z)$: (0) 0652; (1) 0938; (2) 0942; (3) 0945; (4) 1009; (5) 1127 (LST everywhere). Initial profile (0) obtained by the CTD probe 'AIST', others by an emergent profiler.

Fig. 4.2c. Heating of the freshened lens in the previous day will only contribute to an additional rise of the hydrostatic stability of the lens's lower boundary and its capacity to withstand nocturnal convective mixing with the underlying waters. This effect is the reason for the long 'life' of the rain traces in the ocean.

One of the consequences of rain precipitating on the sea surface should be the sign-variable T, S correlation of space thermohaline inhomogeneities in the near-surface layer. Immediately after heavy rains, which are intermittent in

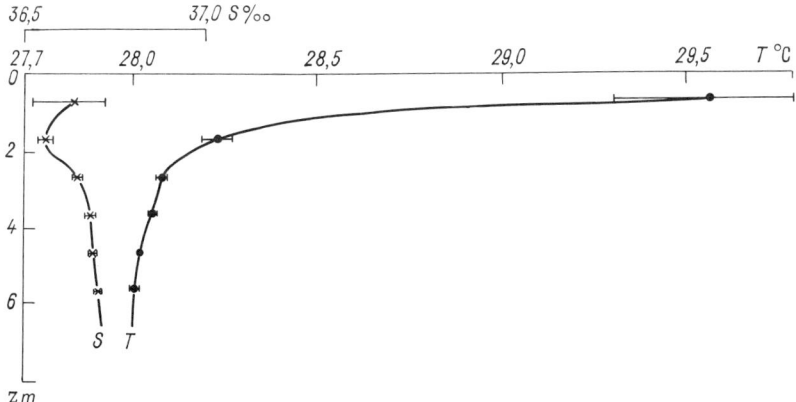

Figure 4.10. Vertical temperature and salinity profiles recorded by the 'AIST' probe on a calm day during intensive solar heating in the Sargasso Sea (2000 LST, 25 August 1978).

space, the developing inhomogeneities should have a positive T, S correlation which should change into a negative one in the course of solar heating of the near-surface layer. The positive T, S correlation should manifest itself again in the early morning hours as a result of cooling of the freshened patches during the night.

A number of interesting methodological problems arise when applying (4.15) for estimating the amount of precipitation based on measuring the temperature and salinity in the near-surface layer of the ocean. For example, Fig. 4.9 shows that it is possible to use the vertical temperature profile, obtained by an emergent profiler or by some other method, 1–1.5 h after the end of the rain, for approximate estimation of the freshened layer thickness in the absence of profiles $S(z)$ if there is sufficiently intensive solar irradiance at the given time. The lower boundary of the newly formed heated layer will be the approximate boundary of freshening h_w, and the values of S_1 and S_2 can be obtained by sampling from the surface with a bucket. However, it is necessary to take into account that the space structure of the near-surface layer could be inhomogeneous by temperature and salinity before the rain. Estimates made without consideration of this effect may contain serious errors. This is clearly demonstrated by measurements made by us during the 27th cruise of the research vessel *Akademik Kurchatov* in the Sargasso Sea (see Figs 4.11a–4.11c). The first probing [$S_1(z)$ and $T_1(z)$, Fig. 4.11a], made by means of the CTD probe 'AIST' according to the method expounded in Section 1.3, started 47 min (at 1633) before the beginning of the rain, and the second [$S_2(z)$ and $T_2(z)$, Fig. 4.11c] started 5 min after the end of the rain (at 1758). From the beginning of the rain and until its end, T and S on horizon 0.7–1.0 m (Fig. 4.11b) were recorded continuously by the same probe. The ship's rain gauge recorded 13 mm of precipitation in 33 min, and 20 mm in a special tray fixed in another place on the vessel. If δ_w was determined only by the profiles $S_1(z)$ and $S_2(z)$ on the basis of averaging the salinity in the 5 m freshened layer, then the freshening would be $\Delta S = \bar{S}_1 - \bar{S}_2 = 0.85‰$, and $\delta_w = 119$ mm from (4.15), which clearly lacks correspondence with reality. Meanwhile, estimation by continuous recording of the salinity variation on horizon 0.7–1.0 m (Fig. 4.11b) produces quite a different result ($\Delta S = 0.37‰$). It is obvious that freshening from 36.37 to 36.00‰ did not start immediately and was stepwise at 1738, i.e. from the time the freshening front passed through the sensors of the probe. Then the salinity decreased gradually as a result of turbulent mixing of the freshened layer, which was evidenced by considerable salinity fluctuations recorded by the probe. The rate of deepening of the freshening front before 1738 was about 3.9–5.6 cm/min and could be extrapolated to the time of the rain end, which gave an estimate of h_w of about 1.5 m. It should be noted that this rate, conditioned by turbulization of the near-surface layer owing to a heavy shower with brief intensification of the wind up to 3–5 m/s, is equal to the rate of deepening of the mixed layer in the case [204] discussed above, where it was conditioned by a 10 m/s wind already after the end of the rain. Estimation of δ_w by the values of S_1, S_2, and h_w, obtained on the basis of continuous recording of the salinity, gives $\delta_w = 10.4$–20.0 mm, which is in good agreement with the direct measurement of the amount of precipitation, but

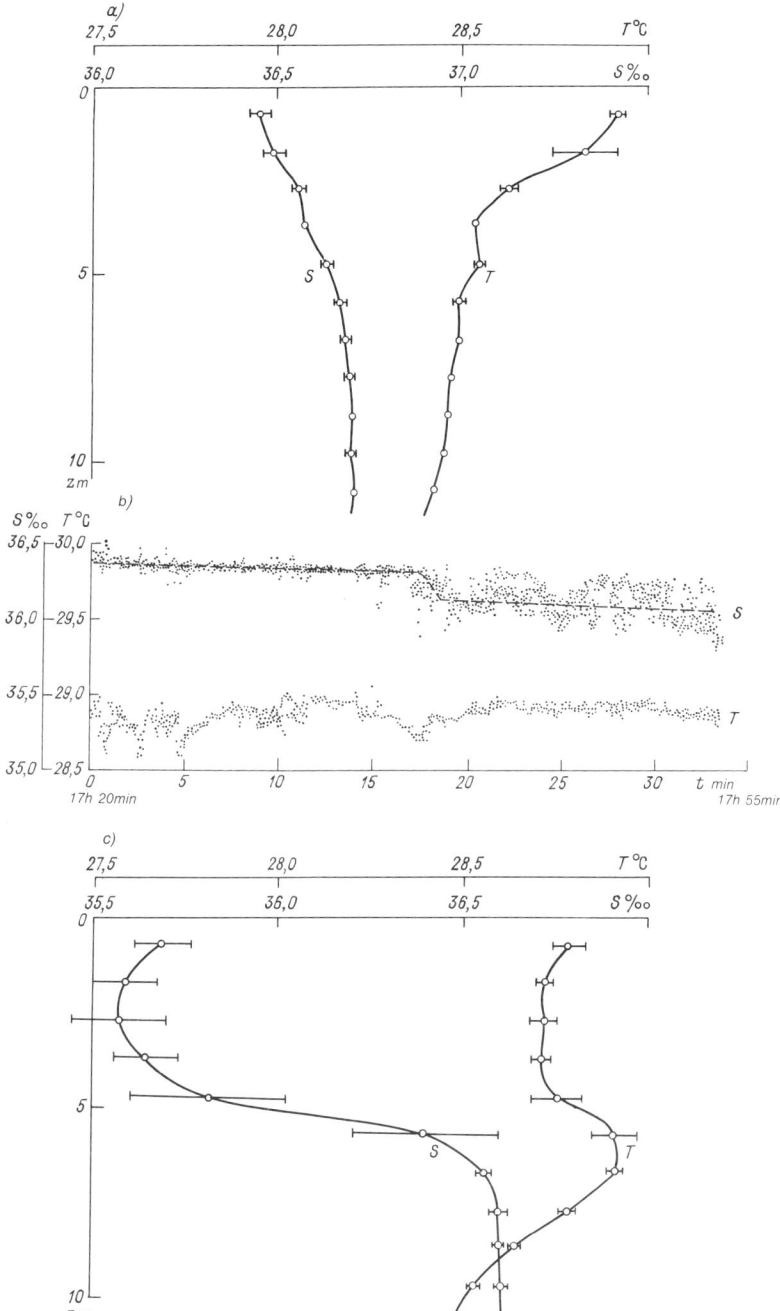

Figure 4.11. Results of recording temperature and salinity by the 'AIST' probe, 23 August 1978, in the Sargasso Sea. (a) Vertical profiles before rain (1633 LST); (b) changes on horizon 0.7–1.0 m during rain from 1720 to 1753 LST; (c) vertical profiles after rain (1758 LST).

differs greatly from the estimate of δ_w by the profiles (119 mm). The differences between the two estimates are caused by the fact that the freshening of the 5 m layer, observed in Fig. 4.11c, is a background structure associated with freshening by the previous heavy rains in the neighbouring region, where to the vessel drifted in the time between measurements of the two profiles $S(z)$. This is also confirmed by the temperature inversion on profile $T_2(z)$ below the freshened layer which is not found on profile $T_1(z)$. These effects explain well why some authors do not find freshening of the near-surface layer even after heavy rains, or obtain a higher estimate of precipitation by formula (4.15) on the basis of single profiles obtained from the drifting vessel (see the discussion in ref. [24]).

4.5.4. Modulation of the thermal structure of the near-surface layer of the ocean by internal waves and Langmuir circulations

Internal waves penetrate into the near-surface layer of the ocean from the underlying stratified thickness and create a system of quasi-horizontal convergent–divergent motions, redistributing the heat accumulated during the day [49, 102, 103, 109]. The near-surface layer of the water grows thicker in the areas of convergence and grows thinner in the areas of divergence, which causes horizontal differences in the diurnal profiles $T(z)$ and temperature fluctuations when it is recorded by a thermal sensor towed on a fixed horizon close to the surface (commonly not more than 0.5 m). These fluctuations with amplitudes of 0.5–1.5°C and with a horizontal scale of about 1–5 km are

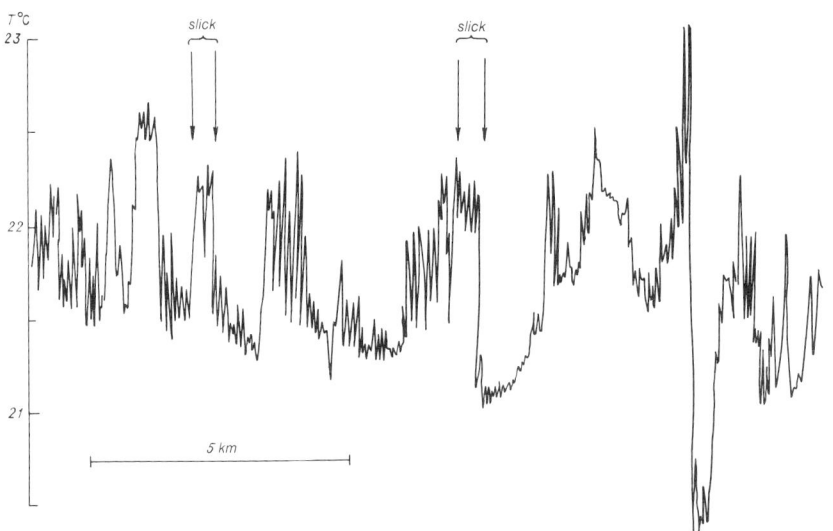

Figure 4.12. Fragment of recording surface temperature ($z \simeq 0.15$ m) carried out by V. E. Sklyarov with a towed thermal sensor when the vessel passed near the Walters bank (33°00′ S, 43°50′ E) at 1300–1600 LST in fair weather with some clouds and a wind less than 3 m/s. Arrows indicate boundaries of slick bands. Fifth cruise of the research vessel *Akademik Mstislav Keldysh*, January 1983.

manifested most clearly together with the overlapping small-scale fluctuations (50–100 m) in calm weather with intensive solar irradiance (Fig. 4.12); therefore, they are called 'thermal inhomogeneities of calm weather' [109]. The regularities of their formation and evolution in the diurnal cycle in connection with a random field of internal waves are discussed in detail in refs [103, 109] and have been confirmed recently by new simultaneous measurements on two horizons [114].

The first special measurements, making possible the establishment of a direct link of the temperature field near the surface with the observed quasi-periodic pattern of the internal waves of the seasonal pycnocline and their visible manifestations on the surface, were carried out during the 34th cruise of the research vessel *Akademik Kurchatov* [49] in the region of the Peruvian coastal upwelling where the sloping thermocline was close to the surface. The measurements demonstrated that the temperature close to the surface was 0.3–0.6°C higher in the slick bands, corresponding to the areas of convergence and located over the rear slopes of the wave crests, than in the bands of ripples (Fig. 4.13). Flying fish, dolphins, and even whales were encountered in abundance in the slick bands where foam and surfactants accumulated. The higher temperature in the slick bands is also demonstrated in Fig. 5 in ref. [102] and in Fig. 4 in ref. [49]. The record of the thermal inhomogeneities of calm weather with a spread up to 2.7°C (Fig. 4.12) obtained by Sklyarov when the research

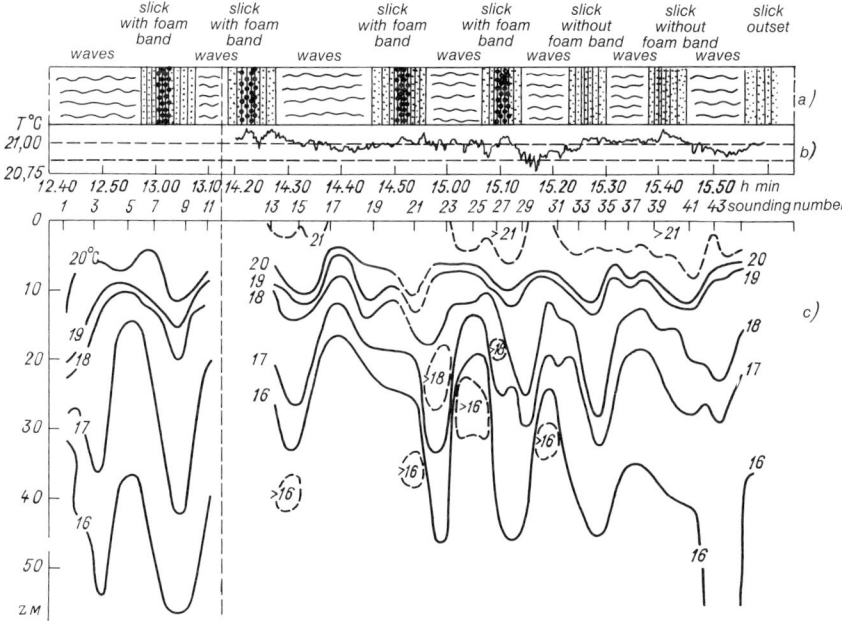

Figure 4.13. Surface phenomena (a), record of temperature by thermal sensor on horizon $z \simeq 0.15$ m (b) and isotherm in layer 0–60 m (c) at a station in the region of the Peruvian upwelling during the period of observations from 1240 to 1600 LST, 15 February 1982 [49].

vessel *Akademik Mstislav Keldysh* (fifth cruise) passed near the Walters bank (Madagascar range) with an abrupt fall of the depth from 18 to 4000 m is extremely interesting. Attention should be given to the unusually sharp frontal temperature gradients at alternation of warm and cold bands, reaching 2–3°C per 100 m. The slopes of the underwater mountain most likely contribute to the generation of very steep and high internal waves of kilometre length.

The small-scale temperature fluctuations with amplitudes of 0.1–0.4°C on the 'inhomogeneities of calm weather' are not an artefact, although the hunting of the sensor in a layer with a sharp temperature gradient can add a high-frequency contribution. Such fluctuations are most likely conditioned by comparatively short internal waves on the lower boundary of the diurnal thermocline as it is in the rain 'trace' (the scale of fluctuation is the same and the values of the Väisälä–Brunt frequency, as mentioned before, are sufficiently high in both cases). The other potential sources of such fluctuations are the inhomogeneities caused by convection [109] and orbital motions in water associated with waves or swells [114]. According to the observations discussed in ref. [49], a smoother character of the small-scale fluctuations would really correspond to the slick areas. However, this correlation was not observed in other cases (Fig. 4.12).

As shown in ref. [109], the general spread of fluctuations in the 'inhomogeneities of calm weather' reflect the really existing horizontal patchiness of the temperature field near the surface and the differences in the vertical thermal structure in the neighbouring areas of the near-surface layer. The depth of this patchiness penetration depends on the thickness of the heated layer and the diurnal thermocline [see Section 4.3, relationship (4.9) and estimates by field data in Table 4.6]. According to ref. [114], characteristic inhomogeneities of kilometre scale were recorded simultaneously on horizons 0.1–0.15 and 5 m, beginning at 1300 LST at wind velocities $u_{10} = 2$–3 m/s, small diurnal SST cycle ($\Delta T_{Dmax} = 0.7°C$), and temperature fluctuations in the patches of not more than 0.1–0.2°C. These observations were made in the tropical zone of the Indian Ocean in October and, therefore, the value Q_D could not be less than 16–17 MJ/m^2 (400–420 cal/cm^2), whence on the basis of (4.9) it can be concluded that the thickness of the diurnal mixed and heated layer H_D could not be less than 6 m. In another case, as shown by a comparison of the continuous records of temperature obtained by us in the Sargasso Sea on 24 August 1978, on horizons 0.15 and 3 m (T_{15} and T_{300} in Fig. 4.14), the inhomogeneities of calm weather under lighter winds, the greater diurnal cycle of T_{15} ($\Delta T_{Dmax} \simeq 1.6°C$) and the greater spread of fluctuations T_{15} in the patches (up to 0.5–0.6°C) were restricted to a layer of 2–2.5 m up to 1400–1500 LST and started to manifest themselves on the 3 m horizon only after 1530 LST. By this time, $\Delta T_D = T_{15} - T_w$ decreased to 0.9–1.0°C, and the boundary H_D deepened to 3.5–4.5 m.

The convergent and divergent motions in Langmuir circulations [72, 102], arising due to interaction of the drift currents and waves at winds from 3 to 10 m/s, represent another mechanism redistributing the diurnal heating and conditioning the horizontal differences in the thermal structure. The distance between the band of convergence is more often equal to the double thickness

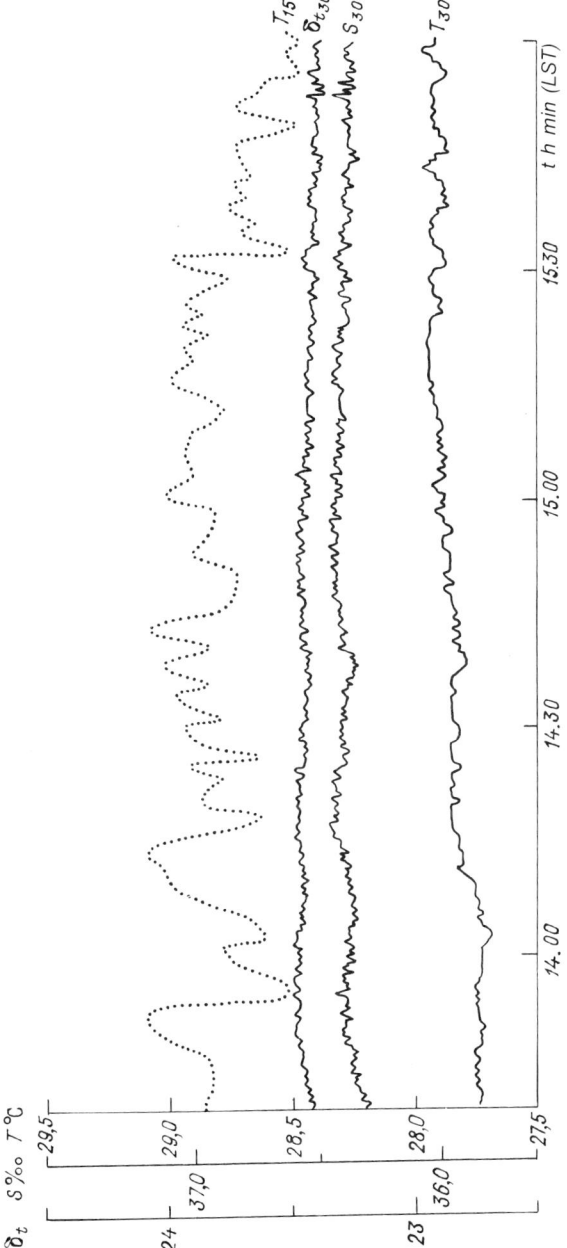

Figure 4.14. Records of temperature T_{15} near the surface ($z \simeq 0.15$ m) and temperature T_{300}, salinity S_{300}, and density σ_{t300} on horizon 3 m obtained by us during the 27th cruise of the research vessel *Akademik Kurchatov*, 24 August 1978, in the Sargasso Sea by a towed thermal sensor and 'AIST' probe in the flow system from 1340 to 1600 LST.

of the mixed layer which is the closest to the surface and can fluctuate from several metres to several tens of metres, and their visibility on the surface is usually due to the presence of some tracer (debris, algae, foam).

The heat entrained by the downwelling flows creates in the layer embraced by circulatory motions a band-like thermal structure, but as a result of wind–wave mixing at the wind velocities mentioned above, the horizontal differences of the temperature in the latter band-like structures are much smaller than in calm weather inhomogeneities, and commonly are not more than 0.5–0.6°C. The enthalpy of a layer several metres thick may be different in the afternoon between the areas of convergence and divergence by 0.3–0.4 of the daily sum of absorbed solar irradiance [102]. At a stronger wind, the Langmuir circulations contribute to rapid spreading of the diurnal heating into the entire UQL thickness. The diurnal thermocline may be very faint in the process of its development under these conditions, and rapidly merges with the seasonal one.

A perfect case of downwelling warm water along the line of convergence, characterized by a slick band on the surface of the ocean, was recorded recently under the guidance of Paka by means of a near-surface thermal trawl during the tenth cruise of the research vessel *Akademik Mstislav Keldysh* in the Tropical Atlantic (Fig. 2.2). The small-scale fluctuation close to the surface ($z = 0.5$ m) did not spread below the horizon $z = 1.5$ m. However, downwelling warm waters under the slick band were observed to horizon 3.5 m by the temperature signal with a contrast of 1.0–1.5°C in relation to the surrounding waters and was still perceived as a slight 0.05–0.1°C temperature rise on the horizon 5.5 m. The width of the band of downwelling warm waters was about 30 m. The general pattern of the temperature variation on horizons 1.5 and 3.5 m (see Fig. 2.2) is qualitatively similar to that observed during convection in the laboratory basin (Fig. 3.2b) and in the ocean (see Section 4.5.5 and Fig. 4.16 below), the difference being that the sign of the temperature contrast at convection is different.

4.5.5. Manifestation of convection in the thermal structure of the near-surface layer of the ocean. Hierarchy of convection scales

The system of convergent and divergent motions arising in the cold thermal boundary layer of water in a laboratory basin and in the ocean due to the formation of primary convective elements of centimetre scale has already been discussed in Section 3.3.1. However, the depth of penetration of the primary convective elements, conditioned by the physical parameters of the water ($\alpha, \nu, \rho, c_p, k_T$) and the thermal flux q_0 from the surface ($z_c \sim q_0^{-1/4}$ [145]), is restricted to tens of centimetres owing to the diffusion of heat, while actual convective mixing in the ocean involves layers with a thickness of tens and hundreds, and in areas of deep convection, especially at high latitudes, even thousands of metres. Obviously, there exist larger than the primary centimetre, space scales of convection, although the process of scale enlargement is not yet completely clear.

One of the potential variants of this process is considered by Foster [145], i.e. the thin mixed layer formed by the primary elements of convection acts as a source of larger convective elements at the subsequent stage of convection whose scales are determined not by molecular (coefficient k_T), but by turbulent (coefficient K_T) heat exchange; in its turn, the new layer is a source of still larger convective elements for the next stage of mixing, and so on, and the value of K_T is determined for each new stage by the limit thickness of the previously mixed layer. So a real 'hierarchy of convection scales' arises in the liquid [145]. The verisimilitude of this concept has been qualitatively confirmed in the laboratory experiments of Dikarev and Zatsepin [42] with turning over of the initially hydrostatically stable two-layer liquid system in a flat vertical basin of small width whose shape was a good approximation to the conditions of two-dimensionality (x, z). In the course of convection development, enlargement of its scales from the minimum (1 cm and less) to several tens of centimetres (2–3 convective elements over the 1.2 m length of the basin) takes place, as given in the succession of images (Fig. 4.15). In this case, the size of the basin does not only restrict, but also determines the characteristic space scales of the last stages of convection.

It is not excluded that in the real ocean, where there are no lateral walls, the internal waves and the Langmuir circulations, causing horizontal convergent–divergent motions with scales from several metres to several kilometres, may assign a space scale of free convection, arising at surface cooling of the ocean. The results of two-dimensional digital simulation of convection in ref. [12] are evidence in favour of this assumption. It is interesting to note in this respect that the manifestations of Langmuir circulations in the SST field, the 'inhomogeneities of calm weather' induced in the ocean by internal waves, and the convective fluctuations of the water temperature near the surface in the laboratory basin are very similar in character (see Fig. 5 in ref. [102], Fig. 1 in ref. [109], and Figs 3.2a and 3.2b), although their space and time scales are quite different. The similarity of the manifestations' character of all these circulation effects of different scales in the velocity and temperature fields is a good reason for assuming that larger scales of convection, developing after the primary one, also form convergence and divergence areas in the near-surface layer of the ocean, as well as a respective space inhomogeneous structure of the temperature field.

However, observations in the ocean give extremely little irrefutable evidence as to the existence of thermal or some other inhomogeneities near the surface bound with secondary scales of convection. A mosaic of densely packed convective cells, having a diameter of 6–30 m with warmer waters in the centre of each cell and colder waters at the edges, can be mentioned as the most reliable case detected from an aeroplane by means of an IR radiometer during a calm night (Gulf of Mexico) [184]. Arkhipova and Rzheplinsky [5] observed a cellular pattern with a cell diameter of about 15 cm on the surface of a pond in calm weather. Ordinary chaff was used as an improvised tracer. The convective nature of the 'patterns' observed in calm weather on the ocean surface in some other known cases is less evident. For example, the reason is unclear for the formation and lengthy existence of zooplankton bands under intensive solar heating in

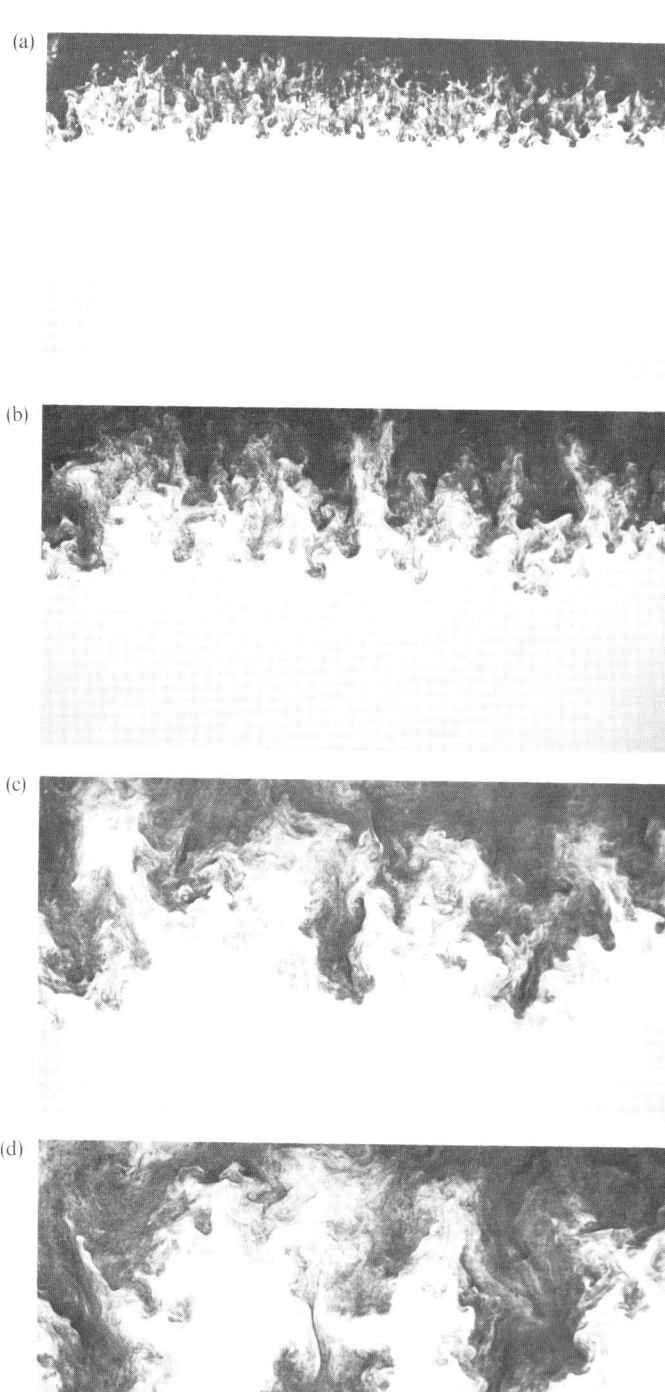

calm weather and light winds, having a length of more than 70 km, a width of 2–8 cm, and $\simeq 1.5$ m spacing between the bands as reported by Owen [199]. It is hard to explain from the positions of convection the preservation of a constant spacing between the bands from 1300 to 2000 LST when the thermal balance of the near-surface layer changed greatly owing to a decrease in insolation and the appearance of a 2–3 m/s wind, all the more so as the depth of convective mixing in this time interval should have gradually increased to several metres (if there was no sharp salinity stratification near the surface).

The absence of such a convenient tracer as cumuli in the atmosphere, which themselves are a consequence of convection, makes it difficult to determine the characteristic space scales of the convective cells in the ocean. Besides, convection in the ocean is maximum during the night, when it is difficult to carry out shipboard visual observations or to apply the use of aeroplanes and helicopters. Satellite IR radiometry possesses insufficiently high space resolution and lacks the necessary sensitivity to detect thermal manifestations of convection. Apparently, the thermal convection does not spread beyond kilometre scales by the horizontal and is accompanied by temperature inhomogeneities whose amplitudes are not in excess of $\pm 0.5°C$. It may be that the only means of detecting such manifestations of convection are accurately made measurements of the temperature in the near-surface layer of the ocean by means of sufficiently sensitive sensors in the regime of towing simultaneously on several horizons under conditions close to calm in the evening, night, and early morning hours. We made such measurements in the Sargasso Sea in 1978 from the research vessel *Akademik Kurchatov*. They consisted in continuous (lasting for several hours) recording of the temperatures T_{15} and T_{300} on two horizons, namely, (1) 0.15 m, by means of a towed sensor with a constant time $\simeq 15$ s and a space resolution of about 100 m;* and (2) 3 m, by means of the CTD probe 'AIST' in a flow system aboard the vessels with pumping water from the given horizon and with worse space resolution, respectively (see ref. [108]). It seems to us that thorough analysis of the results produced important and, in certain cases, unique information on the development of convection in the near-surface layer of the ocean and on the character of its thermal manifestations during night hours.

The most complete and interesting measurements cover the period from 23 to 28 August 1978. Fair weather with some clouds and winds within 0.5–2.0 m/s, rarely intensifying to more than 3 m/s, prevailed from 23 to 26 August, while periods of increased cloudiness up to force 7–8 and intensification of the wind to 4–7 m/s were observed on 27 and 28 August. The diurnal values of T_{15} rose to 29.0–29.5°C as a result of intensive solar irradiance (Fig. 4.14), and the temperature near the surface remained at 28.8–29.0°C until 0100–0200 LST

Figure 4.15. Development of convective motion at an initial deficit of buoyancy $I_0 = 12\,\text{cm}^2/\text{s}^2$ in consecutive time moments t after turning over the basin. Experiments by S. N. Dikarev and A. G Zatsepin [42]. (a) $t = 10$ s; (b) $t = 25$ s; (c) $t = 40$ s; (d) $t = 54$ s.

---

*The method of continuous recording of T_{15} by a towed sensor, applied in these measurements, is discussed in refs [103, 109].

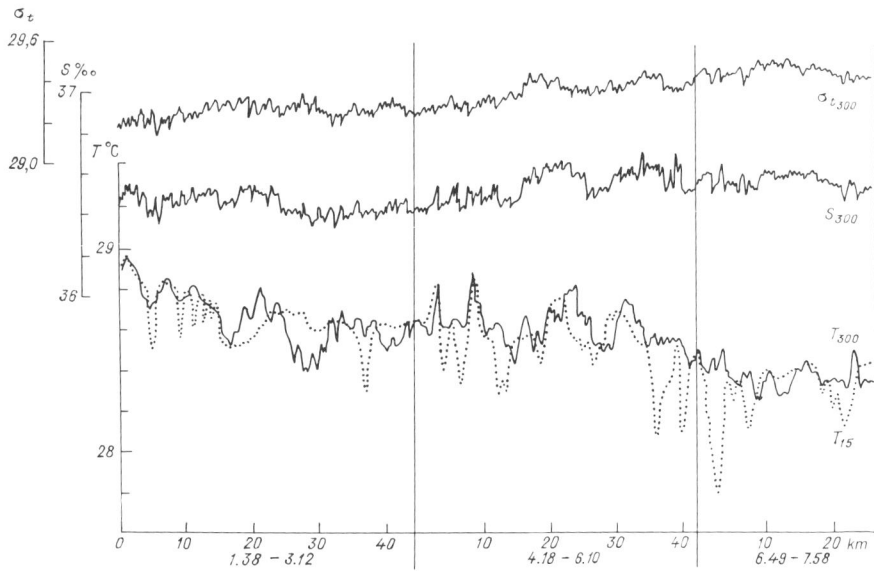

Figure 4.16. The same as in Fig. 4.14; fragments of records obtained by the authors during night and early morning hours, 23 August 1978.

(Fig. 4.16). The temperature difference $\delta T = T_{15} - T_{300}$ even at 2200–2330 LST (Fig. 4.17) still reached 0.5–0.8°C, and the first significant changes of the sign in this difference (cooling from the surface to $\delta T \simeq -0.2°C$, shown by an arrow in Fig. 4.17) were recorded only at about 2330 LST. Such cold 'outbursts'[†] as related to the general level of $T_{15} \simeq T_{300}$ are quite frequent between 0100 and 0700 LST and reach 0.3–0.6°C with a characteristic spacing along the route of the vessel from 1 to 6–7 km (Fig. 4.16), and extremum temperature drops ($\simeq 0.6°C$) at a maximum 3–4 km width of the cooled sections are observed close to 0600–0700 LST. Many of the cold outbursts are well correlated in time with smaller cold 'outbursts' (up to 0.05–0.10°C) when recording T_{300}. Apparently, this is the way the largest scale of convective inhomogeneities is manifested, although it is necessary to make the reservation that we had a definite mixture of space and time scales due to the movement of the vessel which it is difficult to separate.

Still smaller fluctuations having a space scale of about 100 m have also been detected by our measurements. They started to develop in the near-surface layer at about 2230–2300 LST and reached the maximum (up to 0.5°C) during the night and in the early morning hours. The association of these fluctuations with convection was indisputably established by comparing their intensity with hourly meteorological information. Unsmoothed analogue records of T_{15}, patterns of which were obtained during the night of 26–27 August 1978 and

---

[†]Curves of the T_{15} variation, obtained by means of a towed sensor, are smoothed out in Figs 4.14, 4.16, and 4.17 by better comparability with the curves of the T_{300} variation.

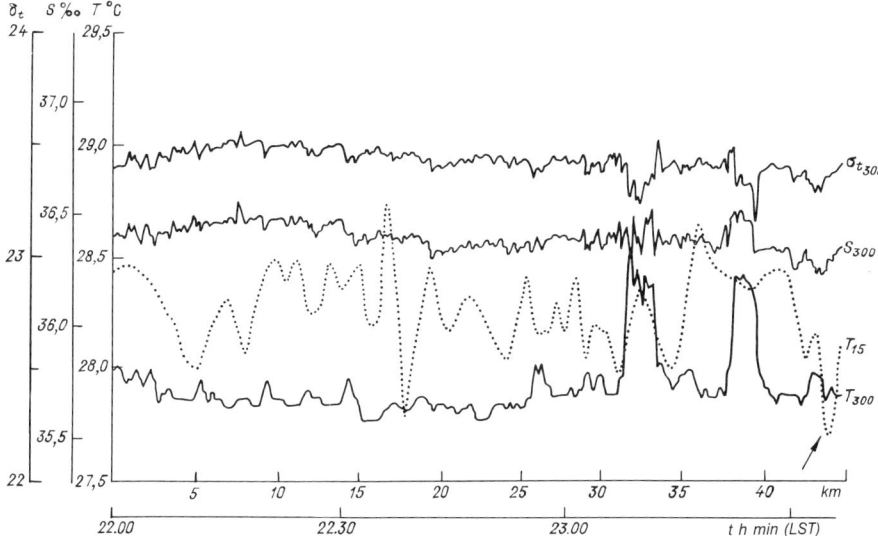

Figure 4.17. The same as in Fig. 4.14; recording made during the night at 2200–2340 LST, 24 August 1978.

are presented in Fig. 4.18, are used for this analysis. The 'calmest' one (without small-scale fluctuations) is fragment 'a' in the record of T_{15} in Fig. 4.18 (1), made at about 2230 LST on 26 August 1978 with a cloud cover of force 7, a wind slightly less than 4 m/s, and 77% relative humidity. Another fragment ('b') in the same hours of 24 August 1978, corresponding to the smoothed curve of T_{15} in Fig. 4.17 and given here for comparison, shows small fluctuations (up to 0.05°C) with a horizontal scale of 100 m against a background of very sharp residual inhomogeneities of diurnal calm weather. This fragment was obtained under practically clear skies, a wind of 1.5 m/s, and 70% relative humidity. In our opinion, the presence of small-scale fluctuations is associated in this case with increased effective radiation in calm weather, and corresponds to the beginning of convection in comparatively large-scale (100 m) cells, which is impeded in the previous case by the slightly fresher wind homogenizing the near-surface layer at comparatively low effective radiation. The convective temperature fluctuations of 100 m scale demonstrated a close to maximum spread of $\simeq 0.5$°C at 0300 LST on 27 August 1978 (Figs 4.18, 2). At the time there was no cloud cover, the wind reached 5 m/s, and the total heat losses from the surface, including effective radiation, were close to the maximum values in the given concrete conditions ($|q_0| > 380$ W/m^2). The cloudiness increased to force 4 by 0620 LST, the wind abated slightly, and flux $|q_0|$ decreased to 340 W/m^2. The spread of T_{15} fluctuations reduced to 0.25–0.30°C (Fig. 4.18, 3). A sharp drop in the wind velocity from 5 to 1 m/s and an increase in the cloud cover from force 4 to 7 were observed simultaneously at about 0715 LST. The total heat losses dropped to 180–190 W/m^2, which led to a rapid reduction in the spread of the convective fluctuations to 0.1°C and a general increase in T_{15}

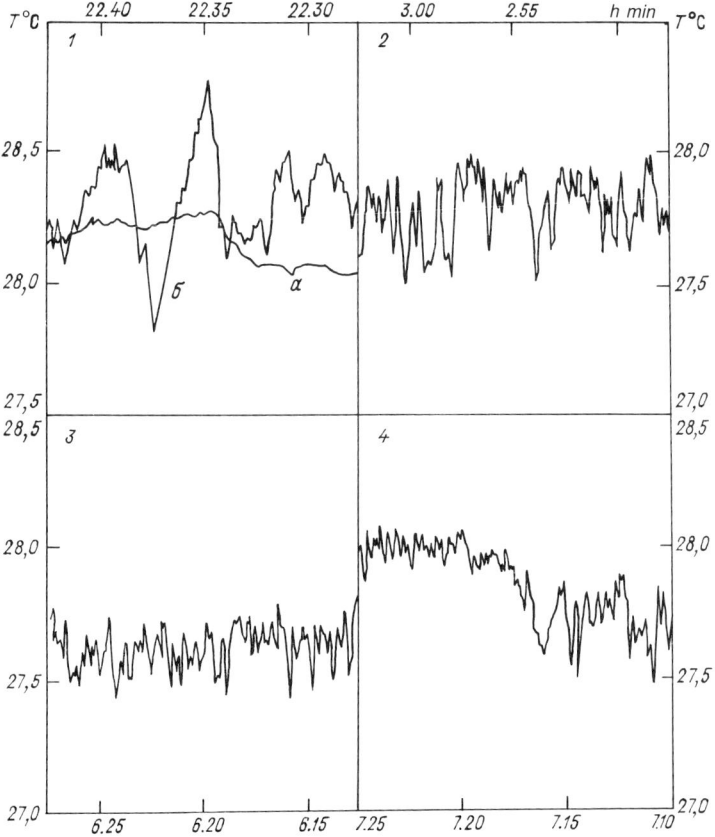

Figure 4.18. Fragments of record with different character of T_{15} variability during the night of 26–27 August 1978 (explanations are given in the text).

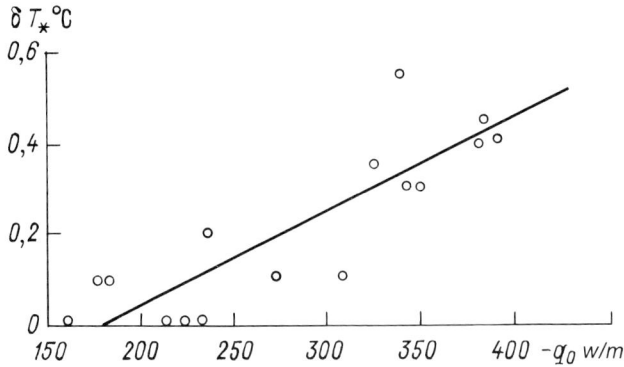

Figure 4.19. Dependence of the spread of δT_* fluctuations, similar to those in Fig. 4.18, on the total heat flux to the atmosphere.

by about 0.2°C (Fig. 4.18, 4). As the connection of the fluctuation spread of T_{15} on the horizontal scale 100–150 m with the total heat losses from the surface is obvious, their convective origin is beyond all question. For further proof, Fig. 4.19 shows the dependence between the spread of δT_* fluctuations on the 100 m scale and q_0 using the data of our night observations in different hours from 23 August to 1 September 1978. The regression line with the coefficient of correlation 0.85 indicates that small-scale fluctuations were not displayed on the records of T_{15} in general up to q_0 values of about -175 W/m^2. They reached a maximum spread of $\simeq 0.5$–0.6°C at q_0 of about -400 W/m^2 and then again reduced if subsequent increase of heat losses from the surface occurred due to intensification of the wind velocity beyond the limits of 6–7 m/s. The last result is in agreement with our conclusions on the transition from free to forced convection just at these wind velocities (see Section 3.3.5).

It is possible that the manifestations of kilometre scale convections discovered by us are influenced by the earth's rotation so that the latter manifestations in the morning hours gain a vortical nature. The experiments of Dikarev [41] have shown that the convective motions in a homogeneous liquid gain a stable vortical structure under the effect of rotation (Fig. 4.20), the eddies always being of cyclonic character, and the ordering of their arrangement (typical in-between spacing) at the given angular velocity of basin rotation depends on the heat flux q_0 from the surface of the fluid. In the case of deep convection in layers 1000 m and more thick, the convective eddies have typical diameters of about 10 km [41]. It is likely that these eddies can gain a kilometre scale during nocturnal convection with limited UQL thicknesses of 10–100 m and at in-between spacings of about 5–10 km. But there is, as yet, no proof of this assumption.

4.5.6. *Thermal regime in the vicinity of fronts, shoals, banks, and coasts*

Sloping frontal interfaces result in various stratification on both sides of the front. The lower boundary of the UQL may be very deep on the cold side, while the density stratification on the warm side, connected with the deepening frontal interface, forms a jump on the horizons of only several metres in the zone extending along the front and having a width from several hundred metres to several kilometres, depending on the slope of the frontal surface. It is logical to anticipate that the amplitude of the near-surface layer temperature diurnal cycle on both sides of the front should be different in this case owing to more intensive diurnal heating and greater nocturnal cooling on the warm side of the front as compared to the cold one. Although such studies have not been carried out in the ocean, the analysis of IR information in ref. [182] gives evidence in favour of this assumption, i.e. the amplitude of the diurnal cycle on the warm side of the upwelling front off the Californian coast exceeded 0.5°C and was less than 0.5°C on the cold side. A still more striking example can be found in satellite IR images of the Mediterranean Sea of 17 September 1985 published in ref. [136]. According to METEOSAT data, a relatively cold band 15–20 km wide and about 100 km long was observed in the morning hours

(a)

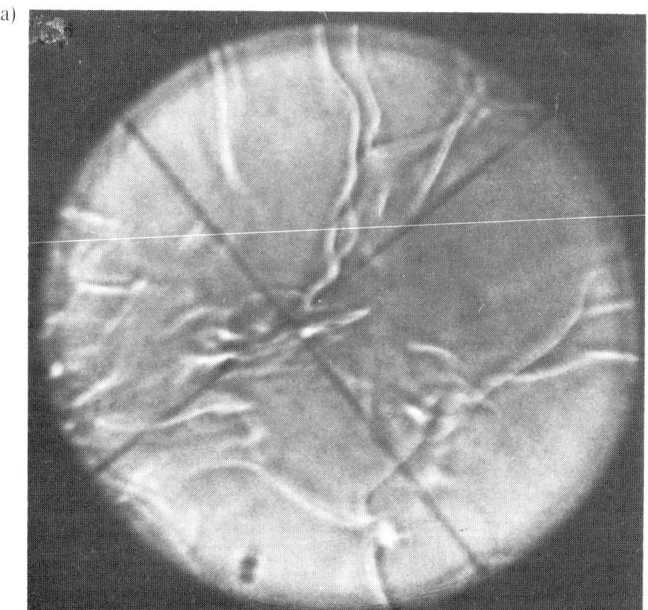

(b)

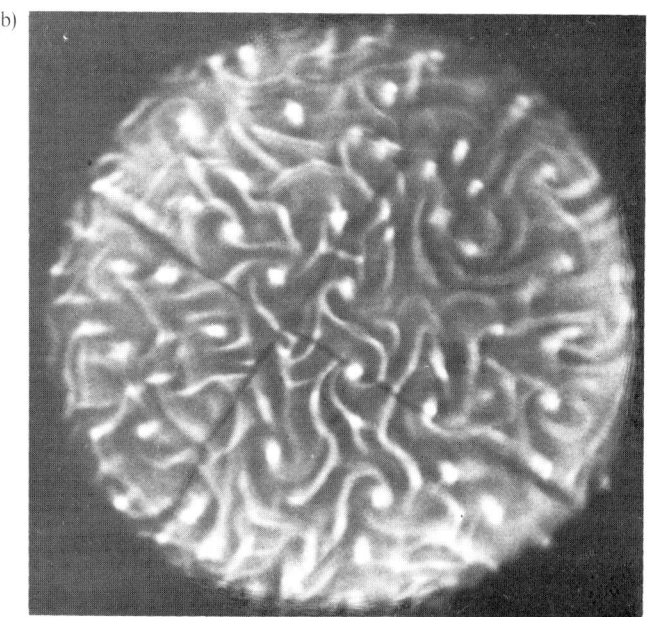

Figure 4.20. Horizontal distribution of convective elements obtained by S. N. Dikarev [41] by means of a shadow visualization device. (a) Without rotation; (b) with rotation ($\Omega = 1.5$ rad/s). Clockwise revolution of fluid, $q_0 = -1050$ W/m^2, water layer thickness 10 cm, observation area $\simeq 150$ cm^2.

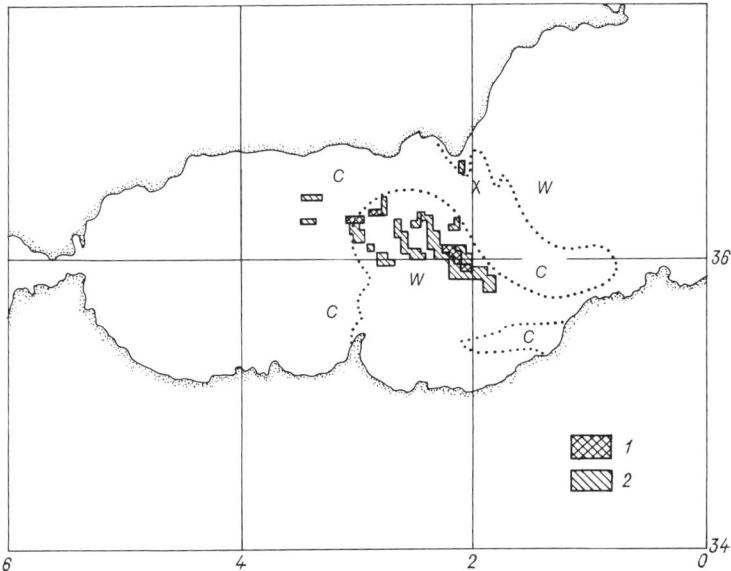

Figure 4.21. Intensive diurnal heating on the warm side of the front in the Mediterranean Sea on 17 September 1985 at 1600 LST (by data of a METEOSAT-2 IR radiometer [136]). Approximate position of front between warm (W) and cold (C) waters by averaged data for 13–19 September is shown by a dotted line, the area of anomalously high heating is dashed: (1) $\Delta T_D > 2.5°C$; (2) ΔT_D from 2.5 to 1.5°C.

along the frontal interface with a temperature contrast of about 2–2.5°C eastward of the Alboran Sea and extending SE–NW. Maximum heating was observed in this band at 1600 LST as compared to the surrounding waters. The difference in the SST values in this band between 1030 and 1600 LST reached 2.5°C, while it was not more than 0.5–1.5°C in the surrounding waters. The given situation is represented schematically in Fig. 4.21 for clearness.

Similar effects should be observed over banks and coastal shoals where the temperature diurnal cycle should have a greater amplitude owing to the propagation of diurnal heating and nocturnal cooling to a smaller depth limited by the bottom as compared to the adjoining deep waters. Short-living thermal fronts can form on the outer boundaries of the shoals and banks. A similar effect is manifested in the seasonal temperature cycle. This effect in large lakes at moderate latitudes in combination with the specific features of the equation of state of freshwater results in the formation of the so-called thermobar, i.e. a very sharp thermal front, delimiting the coastal warmed waters and the waters in the central part of the lake, having a temperature of maximum density (about +4°C) in spring.

Generally speaking, the water regimes over banks and coastal shoals are diverse and depend on the combination of the total depth over the bank or shoal and the mixing conditions due to wind and local currents. The above-cited effect of anomalous heating on shoals is connected with the conditions of calm or light wind weather and total depths of 3–5 m. In the case of depths in excess of 10–20 m and intensive turbulization due to wind or strong (e.g. tidal) currents, cold areas mixed to the bottom over banks and shoals may form, contrasting sharply with the surrounding or neighbouring stratified waters in the deeper parts of the ocean. The formation of fronts under these conditions in the seasonal cycle is described in ref. [103] on the basis of the well-known studies of Simpson, Hunter, Pingree, and others. The thermal regime of the near-surface layer in coastal areas can be subjected to the strong influence of upwelling. Certain aspects of this problem are discussed in Chapter 5.

Chapter 5

Characteristic features of water motion
in the near-surface layer of the ocean

The problems of dynamics of sea currents have gained recently new actuality owing to the development of satellite methods of measuring the physical characteristics of the ocean and, in particular, the relief of its level surface and near-surface currents. We have already discussed in Section 3.6 the inevitable and regular differences that are revealed by direct comparison of SST values, obtained by satellite measurements, with traditional data obtained as a result of shipboard contact measurements. Satellite methods will give information on currents, which, if certain preliminary conditions at comparing are not observed, will just as inevitably and regularly differ from the information collected by traditional methods or obtained by means of numerical models. In order that this circumstance should not surprise anyone or should not be attributed to the errors of one or the other new methods, it is useful to know what the actual velocities of the currents in the near-surface layer of the ocean are composed of. The answer to this question will help future users of satellite information to understand what is actually measured by one or the other method, and what procedure of processing is necessary to produce comparable values when measuring by different methods. We intend to discuss below some specific features of water motion in the near-surface layer of the ocean from this untraditional point of view.

5.1. What are the components of the currents that are recorded in the near-surface layer of the ocean?

If we proceed from established and well-known facts, and if we ignore tides (and tidal currents), internal waves and associated currents, and wind waves (and Stokes transport), then the following answer could be given to the question in the title. The current vector in a stationary regime far from the coast and bottom is composed of two basic components, namely, the geostrophic current and the wind drift.* The former is determined at each point of the ocean by the balance between Coriolis acceleration (due to the earth's rotation) and the local gradient of pressure that characterizes the field of masses setting in under the integral effect of the wind field and other external factors in a stratified ocean

---

*Often called the 'Ekman' wind drift owing to the well-known theory of wind currents elaborated by Ekman, a Swedish geophysicist [138].

confined between shores under conditions of hydrostatic equilibrium. The latter is an additive engendered by the direct local effect of stationary tangential stress of the wind on the water surface. Generally speaking, both components are not independent of one another and are bound into an integral equilibrium (or averaged) state by the boundary conditions and the respective stationary (or averaged) fields (wind, pressure, water density, height of level) in the scales of the whole ocean.

It should be noted that the following conditions, practically unrealizable in the ocean, have been mentioned above intentionally: (1) realization of a stationary state, (2) the absence of tidal currents, and (3) the absence of wind and internal waves. These inevitable reservations are necessary in order, for simplicity, to start our discussion on a most elementary basis, gradually adding pertinent complications.

The geostrophic balance and the hydrostatic equilibrium are expressed by the following simple equations:

$$-fu_g = \frac{1}{\rho}\frac{\partial p}{\partial y}; \tag{5.1}$$

$$fv_g = \frac{1}{\rho}\frac{\partial p}{\partial x}; \tag{5.2}$$

$$g\rho = \partial p/\partial z, \tag{5.3}$$

where u_g and v_g are the horizontal components of the geostrophic component of velocity directed east- and northward, respectively (axis z is directed vertically downward); f is the Coriolis parameter equal to $2\Omega \sin \varphi$ ($\Omega = 7.29 \times 10^{-5}\,\mathrm{s}^{-1}$ is the angular velocity of the Earth's rotation; φ is the latitude of the area); p is the pressure; g is the acceleration due to gravity; and ρ is the water density.

Combination of (5.1) and (5.2) with (5.3) and integration by the vertical from some in-depth horizon z_0 to horizon z in which we are interested give equations of 'thermal wind' (well known in dynamic meteorology) that determine the values $u(x, y, z)$ and $v(x, y, z)$, and are the basis of the 'dynamic method' of calculating currents which is used widely in oceanology:

$$-u_g(x, y, z) = \frac{g}{f}\int_{z_0}^{z} \frac{\partial \rho(x, y, z)}{\partial y}\,\mathrm{d}z + u_0(x, y, z_0);$$

$$v_g(x, y, z) = \frac{g}{f}\int_{z_0}^{z} \frac{\partial \rho(x, y, z)}{\partial x}\,\mathrm{d}z + v_0(x, y, z_0). \tag{5.4}$$

The components $u_0(x, y, z_0)$ and $v_0(x, y, z_0)$ determine the velocity vector on the reference horizon z_0. It is quite obvious that it is necessary to know the three-dimensional density field in the ocean and the field of currents on horizon z_0 to use (5.4). While we know the former in the most rough approximation as some average state, for which seasonal detailed elaboration is possible only in some regions and layers, the latter is totally unknown and, for simplicity, is most often assumed equal to zero at some sufficiently deep horizon z_0

of 1500–2000 m:

$$u_0(x, y, z_0) = v_0(x, y, z_0) \equiv 0. \tag{5.5}$$

In this case, horizon z_0 is called a no-motion horizon, or a 'zero horizon'. Oceanologists made much effort in the first half of this century to discover the zero horizon, the 'philosophers' stone' of oceanology, and to elaborate methods of its 'reliable' determination. But no proof was found of the existence of such a common and continuous zero horizon for all the geostrophic currents in the ocean. Progress has been taking shape in recent years in the sphere of applying a geostrophic approximation which is due to the fact that regular rotation of the geostrophic velocity vector with depth has been found, and inverse methods have been elaborated [224, 246], making it possible to recover the profile of an absolute current velocity in limited regions in the absence of motion transverse the isopycnic surfaces by the measured three-dimensional density field, using the combination of conditions for preserving the potential density and the potential vorticity. The limits in applying the method are due to the necessity of meeting in the chosen region all the conditions in the basis of the method. Therefore, it is too early yet to speak of eliminating on a global scale the uncertainty held in (5.4).

If, as before, it is considered that condition (5.5) is carried out somewhere on horizon z_0, then integration of (5.4) from z_0 up to the free surface elevation $\zeta(x, y)$ gives the following expressions for the components of the geostrophic current velocity on the surface:

$$u_g(0) \equiv u(x, y, \zeta) = \frac{g}{f} \frac{\partial \zeta(x, y)}{\partial y}$$

and $\tag{5.6}$

$$v_g(0) \equiv v(x, y, \zeta) = -\frac{g}{f} \frac{\partial \zeta(x, y)}{\partial x}.$$

Expressions (5.6) reflect a fact which is highly advantageous for remote sensing, namely, that the free surface slopes $d\zeta/dx$ and $d\zeta/dy$ contain an integral effect of the density distribution in the ocean thickness from ζ to z_0, but at the same time these expressions do not solve the whole problem because (it is time to recall it!) the geostrophic component is not the full velocity on the ocean surface. The drift 'additive' was mentioned above. It is an additive only in the sense that within the framework of a linear stationary problem one may speak of its independent determination by the local characteristics of the wind and of simple superposition of the geostrophic and drift components as the final result. It is advantageous for the subsequent discussion to preserve the possibility of speaking in this sense of the drift component as an 'independent' additive.

In the classical theory of Ekman [138], the ocean was considered to be continuous in all directions and homogeneous by density ρ and by the value of the coefficient of turbulent viscosity A_z. In this case, neither slopes of the surface $d\zeta/dx$ and $d\zeta/dy$, nor a geostrophic component of the current could develop in general in the case of a homogeneous and infinite wind field. The question was of the purely drift current induced by the field of the wind tangential stress τ with constant components τ_x and τ_y over the whole ocean. Indeed, when setting the problem of a purely drift current in a real ocean, it is necessary to consider the following factors besides the aforesaid complications:

(1) the presence of density stratification, whose characteristics within the layer of wind force are associated with the flow of buoyancy through the surface and, therefore, possess a seasonal and diurnal cycle with superimposed variations of random or episodical character;
(2) the presence of waves on the sea surface;
(3) the inconstancy by depth and in time of the coefficient of turbulent viscosity and other characteristics of turbulence that depend, in particular, on the variable characteristics of density stratification, on the intensity of the wind force (through waves), and on direct turbulization due to convection;
(4) the presence of natural (and not coinciding) boundaries in oceanic water basins and wind systems; and
(5) the limited nature of wind forces in time up to the possibility of very short-term (impulse) wind forces on a limited area of the ocean surface.

Many efforts have been made in the last two to three decades to consider the above factors in the problem of purely drift currents in the ocean (see, for example, refs [77, 183, 209, 243, 244]). Some of the most important results obtained in the cited works will be discussed below. But it is important to note that none of these results is universal yet, i.e. can be applied in all cases as a standard method for calculating the drift additive. Therefore, we are still forced to illustrate our subsequent reasoning by the quite simple results of the classical theory of Ekman which despite extreme idealization provide some basis for general conclusions.

Ekman's formulation of the problem makes it possible to take into account the bottom friction at depth $H(x, y)$, making use there, for example, of the no-slip condition

$$u(x, y, H) = v(x, y, H) = 0, \tag{5.7}$$

thereby gaining the possibility of considering the influence of the shores where the current velocity should become zero. The corresponding expressions for the velocity of the total (geostrophic + drift) current on the surface [96] are

$$u(0) = \frac{D}{2\pi\rho A_z}(\tau_x m_0 + \tau_y n_0) + \frac{g}{f}\left[\frac{\partial \zeta}{\partial x}\alpha_0 + \frac{\partial \zeta}{\partial y}(1 - \beta_0)\right],$$

$$v(0) = \frac{D}{2\pi\rho A_z}(\tau_y m_0 - \tau_x n_0) + \frac{g}{f}\left[\frac{\partial \zeta}{\partial y}\alpha_0 - \frac{\partial \zeta}{\partial x}(1 - \beta_0)\right], \tag{5.8}$$

where $D = \pi/\sqrt{f/(2A_z)}$, and m_0, n_0, α_0, and β_0 are coefficients representing combinations of trigonometric and hyperbolic functions of argument $(\pi/D)H$ (see details in ref. [96]). In an infinitely deep sea $[(\pi/D)H \to \infty]$, i.e. actually far away from the coast, $m_0 = n_0 = 1$, $\alpha_0 = \beta_0 = 0$, and expressions (5.8) transform

Table 5.1
Comparison of Stokes transport and Ekman drift

Main elements of the problem	Stokes transport	Ekman drift
Source of energy	Kinetic energy of wind	Kinetic energy of wind
Method of energy transfer to water	Normal component of wind pressure	Wind tangential stress
Basic physical effect which is the cause	Non-linear effect of unbounded orbits of particle motion in potential waves	Viscous transfer of motion with shear through adjoining air and water boundary layers
Consideration of wind field characteristics	Indirect. Through local characteristics of waves	Direct. Through measured or assigned characteristics of stationary wind field
Other factors considered:		
• Earth's rotation	Not considered	Considered
• Turbulent viscosity	Not considered	Considered either as A_z = constant or as $A_z = A_z(z)$
• Fetch and time of wind action	Indirectly considered through characteristics of waves	Not considered; wind field should be stationary
• Sea depth	Considered; renders effect on shape of orbits	Considered; renders effect through near-bottom Ekman boundary layer in summary current
• Surfactant films	Empirical consideration is possible; effect is great	Empirical consideration is possible through roughness parameter and formula for calculating τ; effect is insufficiently studied

by adjusting A_z in (5.8) as well as the coefficient c_D included in the dependence of τ on the wind velocity, or by selecting the wind coefficient.

There is another widespread point of view (which we and many scientists [73, 173] consider erroneous) that linear superposition of the Stokes transport and Ekman drift is a natural and most simple solution of the problem. This solution is suggested in some studies [52] without any analysis of all the physical aspects of the problem. It follows from other studies [171], made on the basis of statistical analysis of field data on the drift of floats, that this superposition is possible only with mutual adjustment of respective coefficients, including the so-called 'Stokes coefficient' which is used to estimate Stokes transport by the wind velocity with the use of the Pierson–Moscovits spectrum for developed waves (see Section 6.3). The work [171] is generally very instructive, as it shows that the purely wind (drift) current* in the near-surface layer of the deep ocean far away from the coast can be presented with a different but sufficient degree of statistical quality† either as a linear function of the wind velocity (i.e. prevalent Stokes transport; maximum quality), or as a function of the wind velocity square (i.e. prevalent Ekman drift; minimum quality), or as superposition of the square-law Ekman drift and linear Stokes transport (quality close to maximum). In all the cases considered, the coefficients (k, A_z, c_D, 'Stokes coefficient') obtained

---

*Without the geostrophic component.
†Relation of model dispersion of the studied row to field dispersion.

empirically by minimizing the discrepancies between the model and field rows by the method of least squares proved to be different, but fitted well within the limits known from the literature. The drift current angle of deviation from the wind direction either was assigned 45° to the right or was determined itself by the method of least squares. The angle was on the average 15° (to the right) in the latter case for the linear and square-law models, which coincides well with the prediction of the most realistic generalization of the Ekman theory for the variable by depth coefficient of turbulent viscosity $A_z(z)$ [183].

Nevertheless, the problem of the relationship of the drift current and Stokes transport can hardly be solved correctly in the linear formulation even on the basis of accurate determination of the quantities of kinetic energy transported by the wind to the water separately through the truly tangential stress of the wind and through the normal to the surface component of wind waves, developing due to waves. It is useful to turn to the discussion in refs [132, 244] in this connection. In our opinion, the interpretation of the problem in ref. [244] is incorrect for the following reasons. The semi-empiric parametrization suggested by Wu [244] is based on separate determination of the coefficients of wind resistance c_D and wave resistance c_v with consideration of fetch and under the assumption that the total flux of the kinetic energy from the wind to the water, characterized by full* wind stress $\tau = c_D \rho_a u_a^2$, consists of the energy absorbed directly by the drift current ($\tau_c = u_{*w}^2 \rho$) and the energy transmitted directly to the waves ($\tau_v = c_v \rho_a u_a^2$):

$$\tau = \tau_c + \tau_v. \tag{5.12}$$

Here, ρ_a and ρ are the air and water densities, respectively; and u_a is the wind velocity at the height of the anemometer a. It follows from (5.12) that the friction velocity in water u_{*w} connected with the purely drift current is determined as

$$u_{*w} = (c_D - c_v)^{1/2} u_a (\rho_a/\rho)^{1/2}. \tag{5.13}$$

As Wu uses the empirical dependence of the purely drift current velocity V_n on u_{*w} obtained in laboratory experiments

$$V_n = 22 u_{*w}, \tag{5.14}$$

and as the dependence of the Stokes transport velocity V_v on u_a, used in ref. [244], is practically linear even with consideration of fetch, this allegedly permits Wu to express V as a simple superposition of V_n and V_v:

$$V = V_n + V_v, \tag{5.15}$$

linearly depending on u_a. An interesting dependence of V, V_n, and V_v on fetch is found as presented in Fig. 5.2. According to Wu's data, the summary drift and the purely drift component of the current diminish rapidly with an increase of fetch L, unlike Stokes transport which increases slowly at an increase of L as $L^{0.03}$. This behaviour of V and V_n is associated with the actually observed

---

*Strictly speaking, it is impossible to call the stress here 'tangential'.

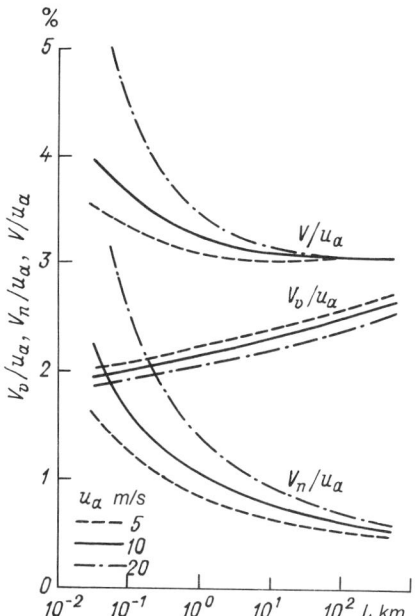

Figure 5.2. Dependence of the drift V_n/u_a and wave (Stokes) V_v/u_a coefficients as well as their sum V/u_a on fetch L by Wu [244].

decrease of the wind resistance coefficient c_D with an increase of L, and with the fact that by the very essence of parametrization, assigned in (5.13), the purely drift component V_n is actually obtained as a remainder after subtracting the Stokes transport V_v from the summary drift V. It is interesting to note that the latter ceases to decrease at $L \simeq 100$ km and remains constant at a level of about 3% of u_a, while V_n continues to decrease slowly as the fetch and V_v increase. This behaviour of the purely drift component is neither in conformity with the classical concept of Ekman (see Table 5.1), nor could be derived from it. It is not without reason that the deviation of the purely drift component from the wind direction was generally not taken into consideration in the calculations of Wu. The shortcoming of this approximation is that τ_c and τ_v in (5.12) are determined not independently of each other, but as summands of an empirically determined whole τ, which in the given text can be called 'pseudotangential' wind stress because it also includes wind pressure components normal to the surface. The decrease of c_D with an increase of fetch just reflects this fact. It would be more correct to determine separately the coefficients of resistance relating individually to τ_c and τ_v directly from the fluxes of energy. In principle, the ideology of this step has been prepared by the discussion of the Stokes transport problem in ref. [132] in the context of considering the turbulent shear layer near the free surface. It is apparently incorrect in this connection to calculate Stokes transport over the entire spectrum of wind waves and swell. Both from the point of view of physics and from the point of view of the dominating contribution to Stokes transport, it is necessary to use the parameters of the main energy-carrying long waves [132, 244], which can be considered potential, in the calculations. The short-period components of the wave spectrum are deliberately not them and should be considered as roughness elements. The effect of the true wind tangential stress and the turbulent flow with shear, induced by it, which creates similarity with the near-wall boundary layer, inherent of the Ekman concept, is connected with the latter roughness elements. Non-linear interaction of the two respective components of the current is manifested in the development of Langmuir circulations* in a plane normal to the wind direction. It is possible to consider from the same positions intensification of Stokes transport when suppressing the high-frequency components of the wave spectrum by surfactant films (see ref. [73]) when a greater proportion of the wind energy is transported directly to the energy-carrying waves due to wind pressure components normal to the surface.

As mentioned before, the wind fields over the ocean are distinguished by extreme space inhomogeneity and time variability. The chaotic state of the ocean surface, induced often in the case of moderate and strong winds as a result of the stochasticity of the wind waves, is characterized by small-scale turbulent eddies in the near-surface layer, developing at destruction of the high-frequency components in the wave spectrum and playing the role of viscosity. The application of similarity with a turbulent near-wall layer is natural under these conditions [132], especially if the purpose is to study the evolution of the vertical profile of the current velocity in the near-surface layer (i.e. a problem with initial conditions). When solving such problems, assigned in the spirit of the Ekman theory, the equations of hydrodynamics represent a logical base for analytical description or numerical simulation of drift effects associated with the non-stationary disturbing effect of the wind on the surface. However, when the near-surface layer is separated from the underlying layers by an abrupt density jump (e.g. due to the diurnal thermocline or near-surface freshening), the turbulent viscosity and its effects can be slackened or, in general, inoperative in the layer of the jump. The problem here is not necessarily in the lower UQL boundary, as it was shown many times before (see Section 4.3, also refs [74, 177]) that a fine structure may exist in the UQL, in which case the uppermost homogeneous layer, adjoining the surface from below, may be many times thinner than the UQL. A simple model [177] based on observations demonstrates that at the development of an abrupt density jump on horizon 8 m in the daytime due to solar heating a homogeneous and invariable tangential wind stress of $0.1 \, H/m^2$ will create in 12 h a drift current with an average velocity of 27 cm/s in the direction of the wind in the mixed 8 m layer. The same wind at night, when convection deepens the homogeneuous layer to 100 m, will create a drift current with an average velocity of only 2 cm/s. The 25 cm/s difference is in good agreement with the intensification of the near-surface current in the daytime, observed by Montgomery and Stroup [189], as compared to the nocturnal current under similar conditions with a difference of 29 cm/s and an

---

*See details on Langmuir circulations (cells) in the monograph of Monin and Krasitsky [73], Section 5.2.

invariable wind. However, the situation is far from being so simple as is implied in the above-cited simplified model. All the deviations from the stationary regime, e.g. the diurnal variations of the stratification parameters and the dependence of $A_z(z)$ on the depth, or the impulse character of the wind force, give rise to another component of the current, namely, the inertial component [149, 209, 243]. Its addition and, probably, non-linear interaction with the drift component can induce such a phenomenon as post-noon intensification of the current below the diurnal thermocline (in the lower part of the UQL). Woods *et al.* [243] called this intermediate maximum of velocity the 'diurnal jet current'* by similarity with the nocturnal jet flow in the atmosphere, induced by the same causes as in the ocean. Although the existence of inertial currents in the ocean has been known since the time of Ekman, their origin due to the diurnal rhythm and the necessity of their consideration in one-dimensional numerical models of the upper layer of the ocean have been realized only recently [243]. It is quite obvious that any instantaneous direct measurements from a vessel, buoy, or satellite (if the latter is possible) of the full vector of the current velocity, including the drift and inertial components, would give values that differ greatly from those that could be recovered either by the dynamic method of hydrological data or by processing the data of only one satellite radio altimeter. This difference should be still greater due to tides and the tidal component of the current, which will not be discussed here. On the other hand, calculation of the drift and inertial 'additives' to the geostrophic component (5.6) recovered by satellite data requires in each concrete case the solution of a problem with initial conditions on the basis of realistic evolution models of the stratification conditions in the near-surface layer and detailed knowledge of the wind field variability. However, the problems associated with currents in the near-surface layer are still not exhausted by this.

It becomes increasingly obvious during the studies that the lateral boundaries, e.g. the coasts, oceanic fronts, natural limits of atmospheric wind systems, oceanic eddies, ice edges, etc., introduce complicating effects into the dynamics and structure of the currents in the near-surface layer of the ocean. As given below in this chapter, satellite images have made possible the discovery of new forms of non-stationary motions of the near-surface layer waters such as mushroom-like currents (see Section 5.3.1) and systems of transversal filaments in the coastal upwelling zones (see Section 5.3.2). Besides, intensive local systems of currents have been found when carrying out detailed complex studies. The latter currents develop:

(1) when atmospheric fronts, tropical cyclones, typhoons, and powerful convective cells of air circulation pass over the ocean under clusters of cumulo-nimbus;

(2) when ice melts in the marginal ice zones;

---

*In other works, the above-described intensification of the near-surface current in the daytime due to layering, induced by solar heating, is called the 'diurnal jet current'.

(3) at relaxation or instability of complex, dynamically equilibrium systems and states;
(4) owing to local dynamics close to sharp oceanic fronts, etc.; and
(5) owing to such oscillating motions as Kelvin and shelf waves.

In some cases, it is still impossible to calculate and even to explain satisfactorily the uncommon characteristics of the observed currents. No above-cited modified variant of the Ekman theory can be used for the solution of such problems because all these variants are based on a one-dimensional approach to the problem of stationary near-surface currents. It is necessary to formulate problems on non-stationary currents with lateral boundary conditions. Intuitively, and not without well-grounded reasons, obtained from observations in the ocean, it can be supposed that many of these problems will be reduced to local geostrophic adaptation of the initial disturbance, induced by the external effect with a space scale L and time scale Δt. The inertial oscillations and associated currents mentioned above stand out in this case as an integral part of the mechanism of local geostrophic adaptation. From the point of view of interpreting field measurements, including remote sensing, the difficulty lies in the fact that we deal now not only with the wind, but also with the limited, in a certain manner in space and time, geostrophic "additive" to the stationary (or mean) geostrophic component in the total circulation of the ocean waters. It would be more correct to speak of at least two such 'additives' of different scales, one of which has been studied very intensively in recent years. These are eddies of synoptic scale ($L \simeq 100$–$300\,\mathrm{km}$, $\Delta t \simeq 2$ months) which are deliberately geostrophic or quasi-geostrophic formations. Although in many cases the latter are induced by the internal dynamics of the ocean (e.g. Gulf Stream rings) and not by external forces, they cannot be considered of general circulation because their contribution to the system of permanent currents at averaging for a sufficiently long period is, most likely, equal to zero, just as the contribution of the second 'additive', developing as a result of localized external forces, is also equal to zero. Let us imagine for clarity the geostrophic and other effects altering the ocean level in the form of deviations of the surface level from the equilibrium geoid corresponding to the ocean resting on the rotating earth. Then, within the limits of a linear approach, it is possible to consider schematically the full deviation of level $\zeta(x, y, t)$ at each moment of time as the sum of several components:

$$\zeta = \bar{\zeta}_g + \zeta'_{g1} + \zeta'_{g2} + \zeta'_t + \zeta'_s + \zeta'_b + \zeta'_w, \tag{5.16}$$

where the prime is attributed to components that should disappear upon averaging in each point (x, y) in a sufficiently long period of time * t_∞:

$$\frac{1}{t_\infty} \int_0^{t_\infty} (\zeta'_{g1} + \zeta'_{g2} + \zeta'_t + \zeta'_s + \zeta'_b + \zeta'_w)\mathrm{d}t = 0 \tag{5.17}$$

---

*For ζ'_{g1} and ζ'_{g2}, it is practically several years or tens of years and much less for the other components.

Here, $\overline{\zeta}_g$ is the deviation of the level from the equilibrium geoid due to stationary (or mean) global circulation; ζ'_{g1} is the geostrophic disturbances of the level of synoptic scale; ζ'_{g2} is the geostrophic deviations of the level due to local short-term external effects lasting from 1 to 10 days and restricted to areas with a scale of 10–100 km by the horizontal; ζ'_t is the tidal excess of the level over the geoid; ζ'_s is the wind-effected level deviations; ζ'_s is the non-stationary ('instantaneous') part of the barometric effect which is not included in ζ'_{g2}; and ζ'_w is the increases of the level connected with long waves.

Only three terms in (5.16), with the subscript 'g', can be used in calculations by the dynamic method on the basis of (5.6). It stands to reason that when measuring $\zeta(x, y, t)$ from a satellite, the problem arises of separating all the components in (5.16), and a no-motion horizon z_{oi} should correspond to each of the three geostrophic components when calculating disturbances of the steric level. Therefore, it is important to learn to estimate correctly the depth of penetration of locally induced currents into the stratified thickness of the ocean waters depending on the duration, intensity, and horizontal scales of the surface disturbance (e.g. a storm), and on the local parameters of stratification. Orlanski and Polinsky [195] concentrated their attention just on this problem with the aim of demonstrating that geostrophic effects of synoptic scale in the ocean may be (other considerations apart) the consequence of localized atmospheric effects.* They determined that if the storm horizontal scale L is essentially smaller than the Rossby barotropic radius of deformation $R_B = \sqrt{gH}/f$, where H is the total depth of the ocean, then the geostrophic reaction of the ocean envelops the upper layer, whose depth h_g depends only on stratification, which is characterized by the Väisälä–Brunt frequency N, and on L, so that

$$h_g = O|fL/N|. \qquad (5.18)$$

A more precise expression for h_g should naturally depend on a criterion assigned at will with the purpose of determining the boundary of the layer reacting to external disturbance (e.g. *e*-times diminishing of velocity as compared to its value on the surface). As for a deep ocean $H = O|10^3|\,\mathrm{m}$ and $R_B = O|10^3|\,\mathrm{km}$, this result should be applicable not only for disturbances, designated in (5.16) as ζ'_{g1}, but also for ζ'_{g2}, i.e. for all storms associated with the passing of common atmospheric cyclones and fronts, tropical cyclones and typhoons, and all smaller surface disturbances of various nature mentioned above.

Let us consider in greater detail the results of ref. [195]. An analysis is given in the work of the non-stationary linearized Navier–Stokes equations for disturbances induced in the ocean by localized atmospheric forcing in relation to the initial state of calm. The principal idea of this analysis is that the rapidly developing and then attenuating atmospheric forcing with a time scale Δt induces in the ocean, besides non-stationary (oscillatory or wave) reaction, also a long-term quasi-stationary disturbance† which attenuates very slowly owing to

---

*A similar problem is also considered in ref. [113] as applied to tropical cyclones.

†Below we will call this the quasi-stationary reaction of the ocean and designate it as ζ'_g, bearing in mind potential possibilities of cases when $\zeta'_g = \zeta'_{g2}$ or $\zeta'_g = \zeta'_{g1} + \zeta'_{g2}$.

the practical absence of viscous turbulent dissipation in the stratified thickness of the ocean. This idea is confirmed in numerous observations in the ocean. The extremely slow relaxation (several months) of stratification disturbances in the trace left behind by tropical hurricanes and typhoons [101, 105] (see Section 5.1), whose effect on the ocean surface lasts for only 1–2 days, is an obvious case. The basic results obtained in ref. [195] can be summarized as follows:

(1) the space scale of stationary disturbance induced in the ocean (diameter of area of maximum level slopes; width of jet current) is equal to the space scale of localized atmospheric forcing (diameter of storm area, diameter of typhoon, etc.) in the case of barotropic and baroclinic reactions of the ocean;

(2) the time scale (Δt) sufficient for localized atmospheric forcing to induce a quasi-stationary geostrophically balanced local reaction of the ocean is of the order of 1 day;

(3) the intensity of the local geostrophic reaction of the ocean (current velocity) v_g, slopes of level $\partial \zeta'_g/\partial x$ or lower boundary of upper quasi-homogeneous layer (UQL) $\partial \eta'_g/\partial x$ is proportional to the product of the atmospheric forcing time scale (Δt) and its intensity (τ, rot$_z \tau$). For example.

$$(\partial \zeta'_g/\partial x)_{max} = - G_0 (R_N/R_B)^2; \tag{5.19}$$

$$(\partial \eta'_g/\partial x)_{max} = G_0 \cdot 2(R_N/L)^2, \tag{5.20}$$

where $R_N = NH/f$ is the Rossby baroclinic radius of deformation and G_0 is the dimensionless amplitude of the space–time structure of localized atmospheric forcing: $G_0 = \tau \Delta t/(2\pi\rho R_N Hf)$;

(4) the quasi-stationary local reaction of the UQL is subordinated to the law of potential eddy conservation;

(5) the relation of the penetration depth h_g of the ocean reaction to its full depth H is equal to the relation of scale L to the Rossby baroclinic radius of deformation:

$$h_g/H = L/R_N = Lf/(NH), \tag{5.21}$$

whence (5.18) follows. Accordingly, if $L \to R_N$, then $h_g \to H$, and the reaction of the ocean is barotropic. If it is assigned that $h_g = (\partial v_g/\partial z)/(\partial^2 v_g/\partial z^2)$ (h_g is the thickness of a layer with an essentially different from zero vertical shear of the horizontal velocity), then at $L \ll R_N$ (or $L/R_N \ll 1$)

$$h_g = \frac{H}{2}\left(\frac{L}{R_N}\right)^2; \tag{5.22}$$

(6) the relation of the geostrophically balanced disturbances $\zeta'_g = \zeta'_{g1} + \zeta'_{g2}$ of the free surface to the disturbances η'_g of the lower UQL boundary for all the scales $L < R_B$ is a small value

$$\left| \frac{\zeta'_g}{\eta'_g} \right| \simeq O|L^2/R_B^2|, \tag{5.23}$$

which makes it possible to apply the known approximation of the solid cover in numerical simulation of such situations (i.e. to consider $\zeta'_g = 0$ in the continuity equation);

(7)   the barometric effect, whose consequence in (5.16) is ζ'_b, contributes to the non-stationary currents of the near-surface layer only in the case of extreme localization $(L < 20\,\text{km})$ on time scales less than 12 h, and does not contribute to the quasi-stationary reaction of the ocean.

From the point of view of ref. [195], the role of the UQL is important only when transmitting the tangential stress of the wind downward to the stratified thickness of the ocean. However, the characteristics of the UQL do not participate in forming the typical scales of the ocean geostrophic reaction L and h_g, at least in cases where h_g exceeds greatly the UQL thickness. Similarly, the relationship L/R_B does not participate in forming the scales L and h_g. This relationship, as it is known, determines the ratio of potential PE_g and kinetic KE_g energy of the geostrophically balanced flow in a homogeneous fluid:

$$PE_g/KE_g = L^2/R_B^2. \tag{5.24}$$

The simple explanation of the geostrophic adaptation process is based in ref. [195] just on this fact. If the initial localized effect consists in a considerable variation of the level height ζ' so that $PE_0 \gg KE_0$, then only insignificant adaptation to the geostrophic regime is necessary and PE_g will differ from PE_0 only very little in the final geostrophic equilibrium, and the relation PE_g/KE_g will remain a large value. But if the initial effect is mainly of the character of localized pumping of kinetic energy proportional to Δt^2 (e.g. by a storm) and $PE_0 \ll KE_0$, then the subsequent geostrophic adaptation will reduce greatly the local levels of energy, making $PE_g \ll PE_0$ and $KE_g \ll KE_0$, leaving, nevertheless, $PE_g/KE_g \ll 1$. The greater share of initial disturbance will go in this case into inertial oscillations and will dissipate.

Let us now consider certain concrete situations to which different elements of the aforementioned analysis are applicable.

Influence of typhoons and tropical cyclones on the ocean. Let us consider, for example, the case of the very compact tropical cyclone 'Ella' in the Atlantic studied by us during the 27th cruise of the research vessel *Akademik Kurchatov* in 1978 [101, 105] (see also Section 4.5.1). The diameter of this cyclone was $L = 100\text{--}150\,\text{km}$, the velocity of motion 20–30 km/h, the time of forcing $\Delta t = 5$ h, the wind velocity at a height of 10 m $u_{10} = 45$ m/s, the ocean depth $H = 5000$ m, and $\bar{N} \simeq 10^{-3}\,\text{s}^{-1}$. Correspondingly, $R_B \simeq 3000\,\text{km}$ and $R_N \simeq 70\,\text{km}$. Then $R_N < L \ll R_B$, and the quasi-stationary geostrophic reaction of the ocean should have been barotropic. This was confirmed by direct measurements on a buoy with current meters on the horizons 90, 700, 3120, and 4170 m, which were practically on the axis of the hurricane trace. When Yegorikhin, Zubin, and Ivanov processed the records of the current meters (rotors), an increase of current velocity due to the passing of the hurricane was discovered against the background of all the horizons stated. The maximum recorded additives to the background velocity were 10–13 cm/s on horizon 90 m, about 10 cm/s on horizon

700 m, and about 7 cm/s on horizons 3120 and 4170 m. The hurricane passed over the buoy from about 1722 LST on 31 August 1978. The maximum velocities on horizon 90 m were observed at about 2000–2400 LST on 31 August, while the maximum velocity on horizon 700 m was displaced approximately by 1 day (by 2400 LST on 1 September), and on horizon 4170 m by another day (i.e. by 2400 LST on 2 September). It was impossible to determine this phase shear more accurately because the matter was of gradual variation of the smoothed values of velocity, while the hourly values demonstrated intensive inertial oscillations with a period of about 1 day and an amplitude of up to 20 cm/s, whose intensification, which was most conspicuous on horizon 700 m, coincided with the time of the passing hurricane (31 August) and was observed until 5–6 September. A powerful single peak of velocity up to 1 knot was observed on horizon 90 m between 2000 and 2400 LST on 31 August.

The velocities of the drift current, calculated by the wind coefficient ($k = 0.03$ [244]), reached 1.35 m/s at the time of the passing hurricane. The maximum geostrophically balanced slope of the level $\partial \zeta'_g / \partial x$, calculated using (5.19), was equal to 6×10^{-7}, and the velocity of the corresponding geostrophic current was about 8 cm/s, which was in good conformity with the results of the measurements discussed above. The maximum slope of the jump layer $\partial \eta'_g / \partial x$, calculated using (5.20), was equal to 10^{-3}, while it reached 1.1×10^{-3} on the cross-section of the trace given in ref. [105]. Everything mentioned above taken together demonstrates the reality of the approach discussed in ref. [195].

Influence of atmospheric convective cells. The thermal effect of localized wind systems connected with atmospheric convective cells and large clusters of cumulo-nimbus on the near-surface layer of the ocean was discussed in Section 4.5.2 using the data of observations [148, 229]. In the context of this section it is useful to estimate the characteristics and depth of the penetration currents induced by these effects.

The disturbances described in ref. [229] were of clearly barotropic character, because the East China Sea in winter in the region of the observations (about $30°$ N, $125°$ E) is well mixed ($N = 0$) to the very bottom (70–80 m). Accordingly, $R_B \simeq 370$ km and $R_N \simeq 0$, which gives $L/R_B \simeq 6.7 \times 10^{-2}$ at a diameter of the cells $L \simeq 24$–30 km. Indeed, disturbances of the current velocity, representing a reaction of the sea to disturbance by atmospheric cells, recurring with a period of Δt for about 50 min (velocity of cell displacement in relation to buoy—about 8 m/s), were sensed by the data of measurements down to horizon 70 m [229]. The solution of the problem for the barotropic case, discussed in ref. [195], gives under the above-described conditions for maximum deviation (amplitude) of the level ζ'_0 a value of 1.3 cm, which the mean slope of the level $\zeta'_0/L = 5.4 \times 10^{-7}$ corresponds to within the limits of the disturbance diameter L. The respective average geostrophic velocity on surface $\bar{v}'_g$ should have reached 7 cm/s, and the drift currents at the peak of disturbance 24 cm/s (at a wind coefficient of 0.03). Unfortunately, it is impossible to compare these estimates with the actually measured disturbances in the current field because the necessary digital data are not published in ref. [229].

The above estimate demonstrates that in the absence of stratification and at small scales of moderate wind disturbance the quasi-stationary geostrophically balanced reaction of a rather shallow sea can be of the same order as the reaction of the deep baroclinic ocean to a strong tropical cyclone.

Local cyclonic vorticity on the near-surface fronts in the ocean. Substantial horizontal velocity gradients at the fronts are a characteristic feature of the near-surface layer of the ocean. It is noteworthy that these gradients are always characterized by cyclonic vorticity. This fact is expressed by the well-known formula of Margules (see p. 79 in ref. [103]) connecting the frontal surface slope tangent with the drops in velocity and water density across the front. Margules's formula is easily derived from the equation of geostrophic motion along the front under the condition of no breaks in the pressure field and pressure gradients at the front. This fact is considered in ref. [103] mainly by analogy with atmospheric fronts where the cyclonic vorticity of the wind at the earth's surface is connected with the wind convergence in the friction layer and with the ascending motions along the frontal surface. The conditions in the near-surface layer of the ocean are not in this respect an exact copy of the conditions in the near-ground layer of the atmosphere. Therefore, cyclonic vorticity of motion occurring everywhere near the oceanic fronts requires a special explanation. The latter can be given from the point of view of the condition for conservation of potential vorticity, including when the fronts develop as a result of localized atmospheric forcing (e.g. due to wind mixing of the waters on the shelf), and, as given in ref. [195], when the local reaction of the UQL is subordinated to this principle. We assume that the front is the boundary between two homogeneous masses of water (or ocean areas) characterized by different constant values of a potential vorticity $\Omega_1 = $ constant and $\Omega_3 = $ constant (Fig. 5.3). $\Omega_1 = (f + \omega_1)/h_1(x)$, where ω_1 is the relative vorticity of motion in area 1, characterizes motion in a relatively thin layer of variable thickness $h_1(x)$, whose lower boundary is sloping and emerges at the surface as a front. The area where $\Omega_1 = $ constant spreads to one side (see the cross-section in Fig. 5.3) to infinity, where $h_1 = h_\infty = $ constant and motion fades. According to Stommel's Gulf Stream theory (see the discussion in ref. [103]), $\Omega_1 = \Omega_2 = f/h_\infty$. At the same time, we assume that the value of the potential vorticity $\Omega_4 = \Omega_4(x)$ in a relatively

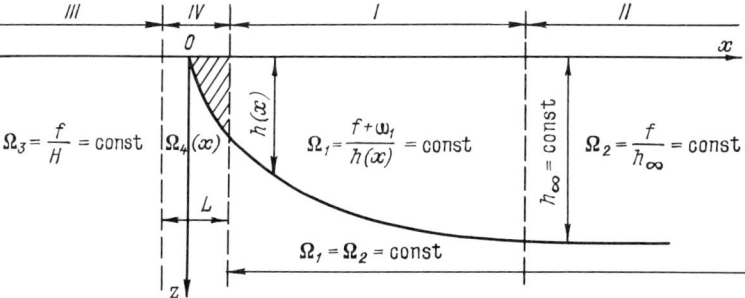

Figure 5.3. Schematic representation of the oceanic front as a boundary between two regions characterized by different constant values of a potential vorticity $\Omega_1 = $ constant and $\Omega_3 = $ constant.

narrow frontal zone of width L, i.e. it is not constant and varies from Ω_1 to Ω_3 without any breaks* so that $\bar{\Omega}_4 = (\Omega_2 + \Omega_3)/2$. If the motion on the other side of the front and at some distance from it is immaterial and $\Omega_3 = f/H$, then simple calculations show that

$$\bar{\Omega}_4 = (f + \bar{\omega}_4)/H, \tag{5.25}$$

where, in turn,

$$\bar{\omega}_4 = \frac{1}{2}\left(\frac{H}{h_\infty} - 1\right)f. \tag{5.26}$$

At all values of $H/h_\infty > 1$ (and this was the initial condition), $\bar{\omega}_4 > 0$, i.e. vorticity in the frontal zone should be cyclonic.

It is convenient to consider the Gulf Stream front as a numerical example, although in Stommel's interpretation the matter is not in the upper layer, but in the layer of 18° water. This makes no difference, and if we adopt the values $H = 5000$ m, $h_\infty = 800$ m, and $f = 9 \times 10^{-5}$ for latitude 38° used in Stommel's example, then we obtain $\bar{\omega}_4 = +2.35 \times 10^{-4}\,\text{s}^{-1}$, which corresponds to strong *cyclonic vorticity*, i.e. a sharp decrease in velocity to the left of the current axis ($dv_g/dx > 0$). To attain this value of $\bar{\omega}_4$ at a velocity drop of $\Delta v_g = 4$ m/s from $v_{gmax} \simeq 3$ m/s on the Gulf Stream axis to $v_g \simeq -1$ m/s on the other side of the front, the width of the frontal zone should be $\simeq 17$ km, which is approximately 3–5 times smaller than the width of the zone of diminishing velocity to the right of the Gulf Stream axis where the vorticity ω_1 has a minus sign (anticyclonic). This asymmetry agrees well with all the known measurements of the transversal profile of the Gulf Stream velocity and with our own observations during the 25th cruise of the research vessel *Akademik Kurchatov* in 1977 along the meridian 72°30' W. The considerations given above represent in a certain sense a supplement to the theory of Stommel, who reproduced in his simple model only the right-hand (anticyclonically vortiginous) part of the transversal profile of the Gulf Stream velocity. It is easy to show that from the equation $\Omega_1 = \Omega_2$, which can be written as

$$(f + \bar{\omega}_1)/h_1 = f/h_\infty,$$

it follows that

$$\bar{\omega}_1 = f(\bar{h}_1 - h_\infty)/h_\infty. \tag{5.27}$$

If it is assumed that $\bar{h}_1 \simeq h_\infty/2$, then from (5.27) we obtain $\bar{\omega}_1 \simeq -0.5\,f$, which confirms the fact well known from observations (see H. Stommel, *The Gulf Stream*, 2nd edition, p. 139).

Although the latter case relates to the Gulf Stream, there are no reasons for doubting that the principle discussed should also be true for fronts developing as a result of local atmospheric forcing on the near-surface layer of the ocean.

---

*The specific point of the front emergence to the surface is not considered here.

5.2. Marginal ice zones

Quite specific conditions contributing to the origin of local jet currents, eddies, fronts, and upwellings characterize the marginal ice zones of the oceans and seas adjoining the ice cover. These zones may have a width up to 50–70 km toward the open sea and several tens of kilometres under the ice. The cause of origin of these boundary zones is evident, namely, the presence of a natural boundary between the open water and the ice cover. The conditions of interaction between the ocean and the atmosphere in the area of the ice edge undergo stepwise changes. It is necessary to add certain factors of purely ice origin, e.g. freshening during the ice-melting period or salinization when it freezes up. The presence of a solid ice mass on one side of the edge, which is accompanied by a temperature contrast between the underlying surfaces, also modifies the wind system in the bottom layer of the atmosphere, adding to the 'purely atmospheric' background also an ice breeze [18], which always blows from the ice to the open water. Although the velocity of this component is low ($\simeq 0.5$ m/s), its constancy ensures quite predictable effects. In particular, the breeze is sufficiently strong to drive ice away from the edge; therefore, a quite wide band of thinned out floating ice is observed under a light wind near the ice edge where the ice cover is incessantly broken up by waves and swell.

Screening of part of the sea surface by ice from the direct effects of wind and a considerable amount of the penetrating solar irradiance is the cause of interesting dynamic and thermal effects in the marginal ice zone and large polynias. Strong boundary currents can be accompanied by the origin of horizontal temperature contrasts and fine vertical layering [126]. Hydrostatically stable temperature inversions can form in calm weather directly near the melting ice in the upper 0–1 m layers of the water [10]. The conditions in the marginal ice zone can be greatly complicated in regions of the ocean where various processes occur beyond the limits of local ones. Thus, the edge of the Arctic ice cover in the Spitsbergen area is close to the region of downwelling Atlantic waters. As demonstrated by observations [130, 165], frontal processes, connected with the presence of these waters and their downwelling or emergence to the surface as a result of upwelling, can render a great influence on the hydrological structure of the near-surface layer in the marginal ice zone.

All the effects mentioned above make the marginal ice zone a dynamically active one, whose frontal character was noted by Vize [18], a well-known Soviet polar explorer, almost 50 years ago. New programmes of observations in the marginal ice zones, realized recently by Soviet [78] and foreign scientists (MIZEX* experiment [191]), have made a comprehensive contribution to our knowledge of the marginal ice zone. But the exploration of this zone is not yet complete. Today, with the launching of research vessels of the icebreaker class of *Otto Schmidt* type, and especially the opportunities of wide-scale utilization of satellite information (floating ice is a very good passive tracer!) the prospects

---

*MIZEX—Marginal Ice Zone Experiment, 1982–1987; participants: USA, Great Britain, Norway, France, Canada, FRG, Denmark, Finland, Ireland, and Sweden.

of exploring the dynamics of the waters in the marginal ice zone are more promising and interesting than ever before.

Let us examine in greater detail the conditions that contribute to the origin of specific systems of currents in the near-surface layer of the marginal ice zone. First of all, it is necessary to mark out the sharp near-surface *layering* of the waters. Intuitively, it is possible to anticipate the sharpest manifestations of stratification in the marginal ice zone in spring and summer after the beginning of solar heating and ice melting. The fairness of the latter is confirmed by all the data published in the literature. The presence of thin (10–20 m) water layers freshened by 2.5–3‰ at the very surface in the marginal ice zone has been recorded during summer observations in the regions of the East Greenland current and Spitsbergen [165, 235]. According to the graphs in ref. [235], the vertical salinity gradient at the lower boundary of the 10 m freshened layer could reach 0.5–1.0‰/m, which gives Väisälä–Brunt frequencies up to $10^{-1}\,\text{s}^{-1}$, i.e. conditions that are rarely found in the ocean! These sharp density jumps can create a 'blocking effect' (see Section 4.4 and ref. [81]) which effectively prevents the transmission of an impulse to the underlying layers from wind–wave turbulence generated in the near-surface freshened layer by a wind of force 3–4, or by interaction of the waves and swell with the lower boundary of the ice cover. No wonder that according to ref. [235] extremely fine thermal layering of the intrusive type could be observed at a depth below 20–30 m. One of its characteristic features was the presence of cores of relatively warm water with a temperature up to $+4.3°C$, i.e. 5.5°C higher than the water temperature in the freshened layer, presupposing isopycnic exchange of lenses at a distance of at least 40–60 km by the normal to the ice edge.

More sophisticated measurements near by the surface, which are still reported very rarely, give grounds for assuming from the inversion temperature distribution in the near-surface layer that the water salinity in the upper 1 m layer in calm sunny weather and under intensive ice melting does not exceed 5–7‰, and a jump to salinities of the order of 30‰ and higher occurs on horizons not deeper than 2–3 m [10]. However, even in winter, according to the observations of Muench in the marginal ice zone of the Bering Sea [192], a typically two-layer stratification was observed with a jump layer on the horizon 50–60 m and vertical gradients of $2 \times 10^{-1}°C/m$, $5 \times 10^{-2}‰$ per m, and 4×10^{-2} units of σ_t/m which corresponds to high values of the Väisälä–Brunt frequency ($\simeq 2 \times 10^{-2}\,\text{s}^{-1}$).

Let us now consider the horizontal gradients of the physical characteristics which are typical of the near-surface layer in the marginal ice zone. According to ref. [78], the temperature contrast in the underlying surfaces (water–ice) can reach 8.7°C in May, and the corresponding air temperature contrast is 6.9°C at a distance of about 80 miles. Vize [18] reports that the salinity contrast can reach 3‰ at the same distance, and in the given case the most freshened waters were under ice cover less than 10 miles from the edge. A zone of the highest horizontal gradients of temperature and salinity is found within the marginal ice zone, which, as a matter of fact, is the marginal frontal zone. Practically from data recorded even in winter [18, 78, 165, 192] the width varies from 20 to 70 km at average gradients of 5×10^{-2}°C/km and 1×10^{-2}% per km. It is natural that the horizontal gradients of temperature and salinity may be much greater at fronts that can form within this frontal zone under the influence of local frontogenetic factors. In particular, the data in ref. [130] indicate the presence of horizontal temperature gradients of 0.5–1.0°C/km at the frontal interfaces in the marginal ice zone even in winter. However, in the latter case they were due to ice-edge upwelling when the warm and salt Atlantic waters emerged at the surface from a depth of 150–160 m. We have here a clear case of the complicating influence of the large-scale water-exchange processes between the Atlantic and Artic oceans on the structure of the near-surface layer in the marginal ice zone.

The presence of a relatively thin layer of lighter water on the surface in the marginal ice zone, which is separated from the lower layers by a sharp density jump, makes the near-surface layer very mobile and highly reactive to all the local impulse effects as a result of the causes discussed in Section. 5.1. Apparently, jet currents in any direction can develop in this layer as a result of short-term ice motions, local inhomogeneities of the wind field, rapid changes in atmospheric pressure, instability of the frontal interfaces, breaking and emptying snow patches, and due to ice-edge upwelling, which will be discussed below in greater detail. Of particular interest in this respect are the narrow concentrated flows of floating ice in the form of protuberances with cyclonic or anticyclonic vortices at the end, beginning from the ice edge and directed transverse the marginal ice zone to a distance up to 100 km toward the open sea, which were discovered on the basis of satellite information. The best example of this structure was found by us in the marginal zone of the fast shore ice in the Tatar Strait (Fig. 5.4). Unfortunately, there are no contact measurements which would illustrate the hydrological situation in the marginal ice zone at the time of observation (4 April 1979). However, satellite information on such structures was supplemented in two other cases with detailed shipboard surveys in respective marginal ice zones [165, 235]. One of these cases deals with four such structures with horizontal scales of 5–15 km which were observed at a distance of 25–40 km from one another on a 100 km section of the ice edge north-west of Spitsbergen [165]. In another case, a larger structure ($\simeq 50$ km) of this type with a vortex pair at the end (judging by the data in ref. [235]) was found at the edge of East Greenland current and was the object of special investigations [235]. The authors of both these cited works considered the discovered structure to be vortex formations and attributed their origin to frontal instability. Not in the least refuting this possibility, it is necessary to note that the cited works do not make clear what concrete fronts could manifest this type of instability. The assumption of Johannessen et al. [165] that the observed vortex formations were manifestations of barotropic instability contradicts their own conclusion on the point that the frontal zone recorded near the edge was 'weak' and that the 'substantial horizontal velocity shear' was not connected with it. In another case [235], baroclinic instability is attributed to the East Greenland polar front. An assumption was made later [220] that the frequently recurring almost in the same place ($\simeq 79°30'$ N, $2°$ E) vortical disturbances of the type described in

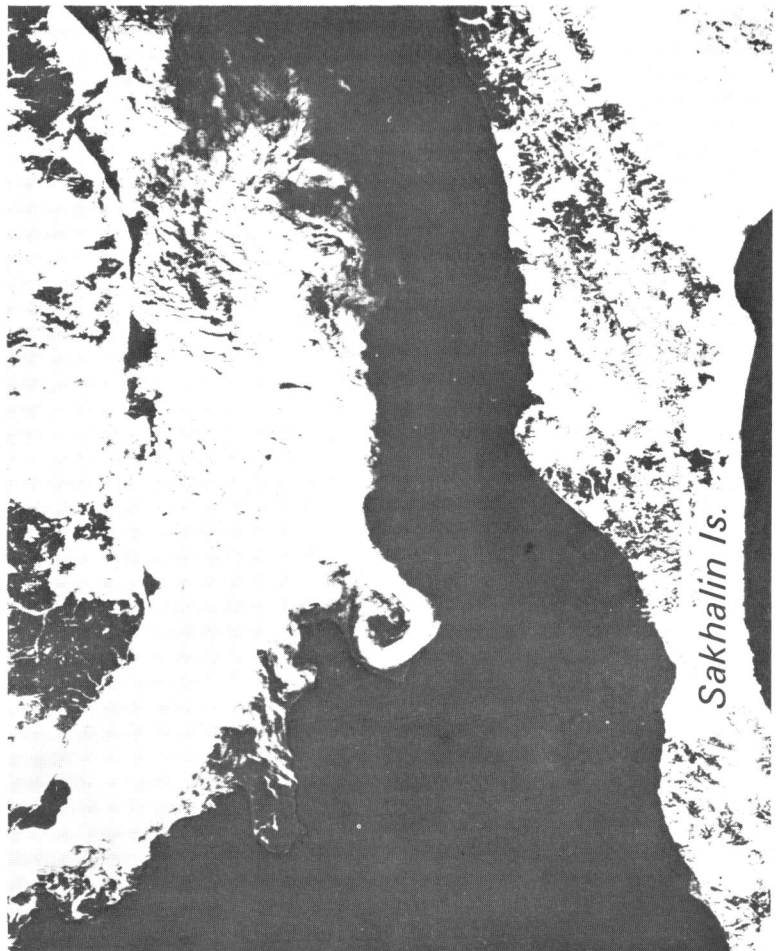

Figure 5.4. Transversal jet with an anticyclonic vortex (diameter $\simeq 30$ km) at the end observed in the marginal zone of the fast shore ice in the Tatar Strait. Fragment of image after 29 th launch of the 'Meteor' satellite (medium resolution scanner, 0.5–0.7 μm band), 4 April 1979, 0336 GMT.

ref. [235] are a consequence of the topographic effect connected with depressions and rises of the bottom in the given region (see Fig. 1 in ref. [220]). If this is so, then the observed vortical disturbance of the ice edge is due to the barotropic component of the current, which has no relation to the local regime of the marginal ice zone. The conclusion should be drawn that many aspects of the origin of marginal structures, similar to those in Fig. 5.4, and the associated water and ice motions are still not cleared up. What can be considered today as known for certain? The following is known:

- floating ice is transferred in the discovered structures by jet currents of ordered character and convergent nature;
- jet currents begin at the very ice edge and, in the majority of cases, are

directed along the normal from the ice edge toward the open sea;

- all the cases of similar transversal filaments have been discovered in light wind weather;
- their lifetime reaches several days, during which the structure preserves its position in relation to the edge.

If it is taken into consideration that these structures attain a size of 10–50 km in 1–5 days, the rate of ice transfer by the transversal jet currents should reach 40–50 cm/s. It is reasonable to conclude in this case that these currents are concentrated in the near-surface freshened layer. This assumption is not in discordance with the data in ref. [165]. It is difficult to say today whether the vortices found in connection with transversal jets develop as secondary formations, or vice versa, a vortex or vortices, developing in the marginal ice zone, form a convergent jet, sucking the floating ice from the marginal ice zone away to the open water. Indirect evidence of the fact that the currents forming the marginal ice zones are of vortical character is given in ref. [78], where the appearance of near-surface water lenses with clear marginal ice zone T, S characteristics at a distance from the ice edge is pointed out. There may be many reasons for the origin of eddies in the marginal ice zones. For example, the lenses of anomalously warm and salt water recorded in ref. [235] near the ice edge on horizon 40 m were quite obviously cores of intrathermocline eddies [104], because their influence was traced in the thermohaline structure down to 600 m. In this case, their disturbing vortical influence on the near-surface layer of the marginal ice zone should also have been manifested. The very exchange of lenses of alien waters through the marginal ice zones, having a width from 40 to 60 km, merits attention and explanation. The connection is not quite clear yet between the transversal jets and the jet current which commonly develops alongside the ice edge and the fronts which may form in the marginal ice zone for various reasons, including ice edge upwelling. However, the combination of a jet current along the ice edge with transversal jet currents, and the regularities in the behaviour of the latter, as aforesaid, resemble greatly the situation in a coastal upwelling (see Section 5.3.2). Therefore, it seems to us that jet currents with eddies at the end, perpendicularly directed to the ice edge, can be a manifestation of the front instability at the edge connected with an upwellling.

The comparatively recently discovered phenomenon of ice edge upwelling [130] deserves special discussion in the context of this section. A theoretical analysis of water motions induced by the wind close to the ice edge, adopted for a solid immobile cover, was done for the first time by Gammelsrod et al. [147] for homogeneous water and by Clarke [131] for stratified conditions. An upwelling, accompanied by a jet current along the ice edge, should be one of the main elements of the developing circulation in both cases owing to the conversion of the wind tangential stress rotor ($\mathrm{rot}_z \tau_y$) on the ice edge ($x = 0$) into infinity. The comparison made by Clarke [131] of the dynamic conditions close to the edge of a common ice cover (where the Ekman layer in the water is thicker than the ice cover) with the conditions close to the coast or the boundary of shelf glaciers (where the Ekman layer is thinner than the solid side boundary) established essential differences in the character of Ekman transport as well as in the factors inducing upwelling. In the case of a common ice edge, the upwelling is generated only by divergence of local Ekman transport, directed by the normal to the ice edge, while long shelf waves, increasing the vertical velocities as well as the velocity of the boundary jet current, are added to this factor at the boundary of shelf glaciers and at the coast. However, in all cases with an immobile ice edge the most advantageous for the upwelling is a wind directed along the edge so that the open water remains to the right-hand side (in the northern hemisphere).

It would seem that these results are a principal solution of the upwelling problem in the marginal ice zone. However, the fact that they were obtained under the assumption of immobility of the ice and a fixed position of its edge gave rise to doubts in their trustworthiness. Ice researchers know well the exceptional mobility of the ice edge. According to Buckley *et al.* [130], the ice edge to the north of Spitsbergen can be displaced dozens of kilometres in a few days. Displacement of the ice edge to 35 km in 1 day was observed in the Bering Sea during storms [201]. The question arose as to whether the ice motion alone in relation to the water can induce upwelling in the absence of any direct local effect of the atmosphere. Another question sprang up as to whether the influence of the wind was more effective on the ice than on the water surface from the point of view of impulse transmission, as a result of which it would be possible to anticipate different relationships between the wind direction and the development of upwelling than in the models cited above [131, 147]. The work by Røed and O'Brien [208] is dedicated to these and some other questions. An analysis of the results of the numerical simulation of the water–ice system has demonstrated that the tangential stress transmitted by the moving ice to the water is sufficient to generate divergence and upwelling in the marginal ice zone. And if the ice motion is induced by local wind, upwelling should develop at a wind directed along the edge toward the side opposite to the one at which upwelling should occur in the case of immobile ice. This is so because the ice moves faster than the near-surface layer of the water under the effect of the same wind. As it is easy to imagine displacement of the entire ice mass by some resultant of all the applied forces, a variety of local situations can be anticipated wherein displacement of the ice edge is not necessarily bound with the local wind. Hence, the problem of upwelling in the marginal ice zone should inevitably permit various formulations and different solutions, respectively. The possibility of these local situations, when ice motions and displacement of the edge can occur in the absence of wind, apparently explains the frequent observations of jet currents, vortices, and fronts in the marginal ice zone in calm weather.

A few words should be said about fast ice in regions where it forms only in winter and disappears completely in summer, e.g. in the northern part of the Sea of Japan and in the Tatar Strait, the Sea of Okhotsk, the eastern coast of Kamchatka, etc. If there is open water outside the fast ice limits, then a marginal ice zone with all the specific features described above develops there in winter and early spring. In particular, this is demonstrated in Fig. 5.4 by the transversal jet current with an anticyclonic vortex at the end in the marginal zone of fast ice in the Tatar Strait. But after the beginning of melting and breaking up of the fast ice,* the formed floating ice rapidly moves away from the coast and extends along the front that forms on the outer boundary of the coastal water band which is freshened by the ice melting. Apparently, this front has many similar features with run-off fronts [103], but it is usually extended along the coast to the entire length of the former fast ice zone. Judging by satellite images, it is at this front where characteristic manifestations arise as described by us in refs [33, 34] (see also Section 5.3.1). The question is of jets propagating from the front toward the sea almost parallel to one another with an average distance of several tens of kilometres in-between and well visible owing to the outlining bands of finely crushed ice. The evolution of these jets is traced well in Fig. 5.5, where four successive images (with an interval of about 1 day) of an area in the northern part of the Sea of Japan (Tatar Strait) were obtained by a medium resolution scanner during the 30th and 31st launches of the 'Meteor' satellite in the visible range of the spectrum on 28–31 March 1983 during break-up and intensive melting of the fast ice: (a) 28 March 1983, 2348, 31st launch of 'Meteor'; (b) 30 March, 0200, 30th launch of 'Meteor'; (c) 30 March, 2320, 31st launch of 'Meteor'; (d) 31 March, 2345, 31st launch of 'Meteor' (GMT).

Characteristic structures (clusters of floating ice) are designated in Fig. 5.5 by numbers. The first image (Fig. 5.5a), demonstrating the initial stage of fast ice destruction at the western coast of the sea, shows no indications of a front and, all the more, indications of its instability. Only the initial outlines of the future structures 1 and 2 are slightly visible. Several jet currents (1–3) with parallel axes in the south-east direction and with a distance of ≃ 35 km in-between are distinguished in the second stage (Fig. 5.5b) of the region of Krestovozdvizhensky Cape and southward of it, and structures 2 and 3 have a well-expressed mushroom-like shape. A day later structure 2 (Fig. 5.5c) took the shape of a symmetrical 'mushroom' with eddies at the end of the jet (it is magnified in Fig. 5.6b), while structure 3 practically disappeared. Structure 2 disappeared one more day later, although the meandering front itself, surrounding the water freshened by melting ice, remained, which is visible by the accumulation of ice along the convergence line.

There is very much in common between the structures discussed above and those discussed earlier in connection with arctic marginal ice zones. It seems to us that we are dealing in all cases with specific ordered forms of local non-stationary currents representing a response of the rotating fluid (ocean) to most different impulse disturbances of the equilibrium state. In our opinion, the basic elements forming these ordered currents are jets, vortices, and vortex dipoles (mushroom-like currents). Certain regularities, associated with the formation of such ordered currents in nature and in a laboratory experiment, are discussed in Section 5.3.

---

*End of March to June in the Far East seas.

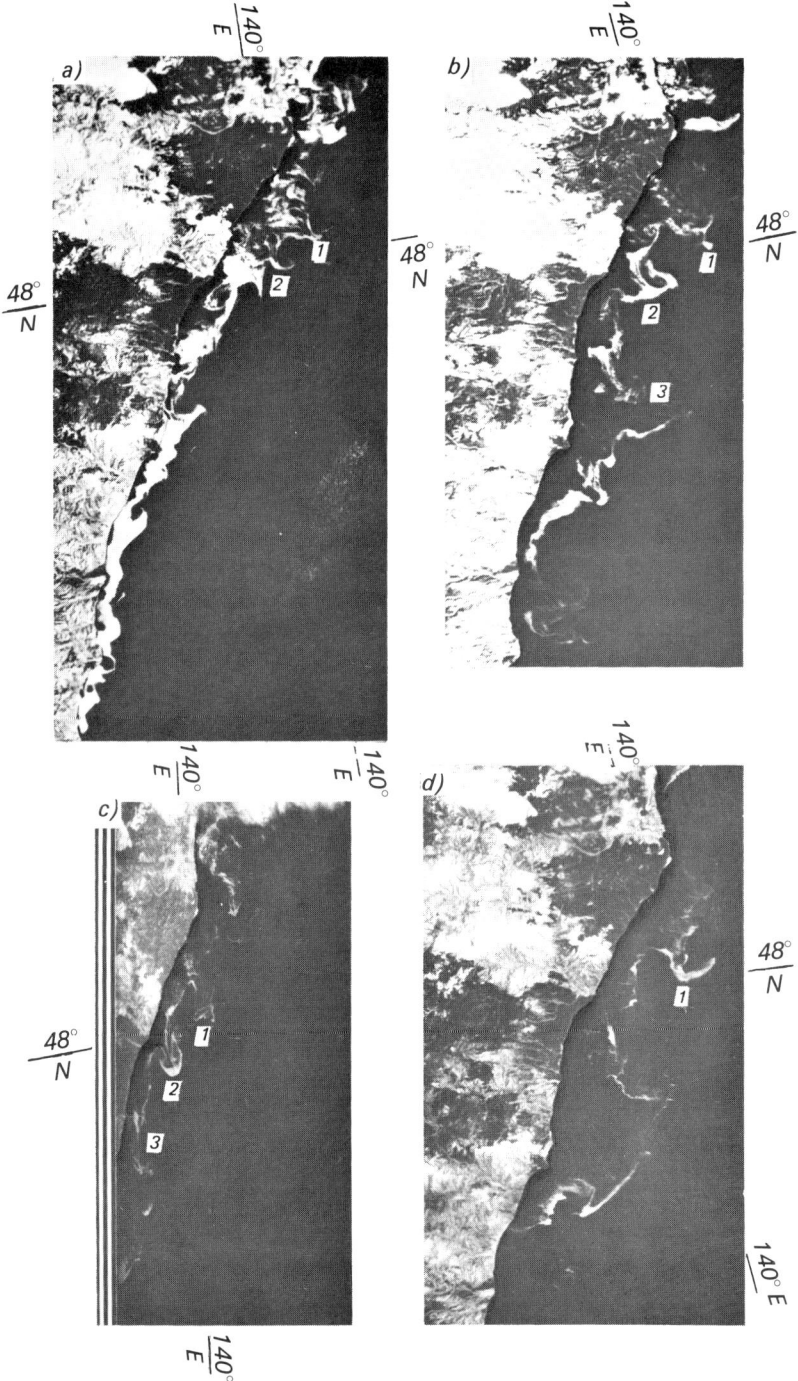

Figure 5.5. Evolution of fast ice melting in the Tatar Strait (explanations in the text).

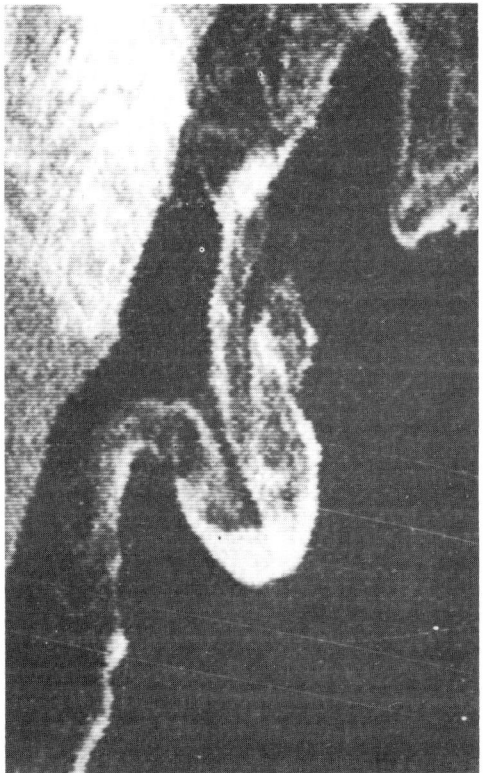

Figure 5.6. Mushroom-like currents in the ocean. (a) Diagram; (b) structure (2 in Fig. 5.5c) forming during intensive melting of fast ice in the Tatar Strait. Magnified fragment of image obtained after the 30th launch of the 'Meteor' satellite (medium resolution scanner, 0.7–1.0 μm band), 30 March 1983 [33].

5.3. Coherent (ordered) forms of non-stationary motion in the near-surface layer of the ocean

Two strong factors in the ocean induce ordering of turbulent motions of various scales, arising as a result of most different unordered disturbing effects. These factors are the earth's rotation and stratification. As stratification is frequently expressed most sharply in the near-surface layer of the ocean (see Section 4.3), and various atmospheric effects of different scales are applied to it, it is logical to anticipate in it from a chaos of initial disturbances systematic origin of various coherent (ordered), regularly developing non-stationary forms of motion existing for a long time. Their presence can be considered today as determined for sure on the basis of numerous satellite observations and measurements. Beyond all question, these forms of motion could be described on the mathematical basis of statistical hydrodynamics as some random velocity fields against a background of mean ordered motion in the scales of overall circulation of the ocean. Nevertheless, we prefer to describe them as ordered motions, having quite specific physical causes and conditions of generation, and a quite natural evolution. We believe that this approach makes it possible to learn more about the physics of the near-surface layer of the ocean than automatic attribution to turbulence of everything non-stationary and not quite understandable. The non-turbulent nature of the laboratory analogues of these motion forms is obvious, as their typical final Reynolds number is not more than 100, even in the most extreme cases [21, 27], regardless of the Reynolds number of the initial disturbances. Apparently, the availability of mechanisms for ordering motions in the ocean can ensure the necessary division of the motions by scales, so that it becomes possible to discuss the small values of the final 'turbulent Reynolds number' under field conditions (see below). We will describe here, mainly on the basis of satellite information, some of the ordered motion forms observed in the near-surface layer of the ocean.

5.3.1. *'Mushroom-like' currents in the near-surface layer of the ocean*

In the past decade a thorough and systematic analysis of numerous satellite visible band and infrared (IR) images of the ocean has revealed many interesting features of the ocean currents, eddies, and fronts. Totally unknown and quite unexpected among these features are the well-organized, recurrent, quasi-symmetrical patterns of some $\simeq 10$ to $\simeq 200$ km in size consisting of a narrow jet which ends up in a pair of vortices of opposite sign, so that the whole structure is strongly reminiscent of the cross-section of a mushroom (Fig. 5.6). This resemblance prompted us [33–35] to call this pattern a 'mushroom-like' current. The jet length L and the size of the vortical part B are of the same order in the majority of cases, and the width of the jet is usually not more than 10–25% of L. Patterns of this kind are 'visible' mainly due to the presence of one or another natural tracer (e.g. floating ice, plankton, suspended matter), or thermal contrasts on the ocean surface or in the near-surface layer. In some cases, similar patterns are formed by surface roughness contrasts (most likely by a combination of surface slicks and rippled patches) which are quite visible near the edges of the sunglint.

An analysis of several series of successive satellite images of the same pattern (see, for example, Fig. 5.5) has made it possible to establish that the currents creating such peculiar patterns in the field of respective tracers or contrasts over periods of 3–7 days (sometimes more) have been of non-stationary nature [33]. The current velocity, determined approximately by the increment of the pattern size at the beginning of development, reached 1 knot and even more. Judging by the momentum needed to produce such fast motions, it was logical to consider that these currents encompass only a comparatively thin near-surface layer of the ocean and their penetration into the depth is limited by density stratification due to temperature or salinity. Seasonal conditions, advantageous for the formation of such stratification, could be anticipated practically in all cases of observing mushroom-like patterns. Therefore, a logical conclusion was made that mushroom-like currents are produced by localized, short-term application of momentum to the surface or near-surface layer of the water because local sources of kinetic energy of various nature are always available there. In this case, the peculiar combination of the jet and the eddy motion could be the consequence of conservation of the amounts of angular momentum of both signs contained in the original disturbance. Then the joint effect of inertia, viscosity, and continuity of motion should lead to vorticity concentration in the form of a pair of vortices.

By now, a considerable body of knowledge has been accumulated concerning mushroom-like currents in the ocean, so that it is possible to make certain conclusions on the subject matter and to point out a number of consistencies in the processes of their formation and evolution. Most of this knowledge is derived from satellite information, and practically none comes from traditional shipborne means of measurement and observation. This gap in knowledge has been filled up by laboratory simulation of vortex dipoles (see below), having much in common with mushroom-like currents. Let us consider first the facts that are directly related to the ocean.

Mushroom-like currents (or vortex dipoles) may be considered as a horizontal plane version of vortical rings which in the case of nuclear explosions or intensive forest fires have a peculiar mushroom-like shape with a vortical ring propagating upward on the top of a highly localized and powerful vertical jet. In both instances, the mushroom-like patterns appear as a consequence of an extremely highly localized concentration of energy. Since for the ocean a sharp space and time inhomogeneity of the external forcing (e.g. atmospheric; see Section 4.5.1 and 4.5.2) and of the internal structure of the currents is rather common, the mushroom-like currents apparently may represent an extremely unique form of horizontal non-stationary motion of the near-surface waters of the ocean. Although not strictly two-dimensional in the ocean or in a laboratory tank, these currents most definitely do not possess any symmetry in the third (vertical) dimension, so that the continuity equation has to be ultimately satisfied in a relatively thin layer of water. This kind of argument leads to the conclusion that for complicated situations with multiple locations of forcing a dense

Figure 5.7. Diagram of two types (I and II) of compact 'packing' of mushroom-like currents in the ocean.

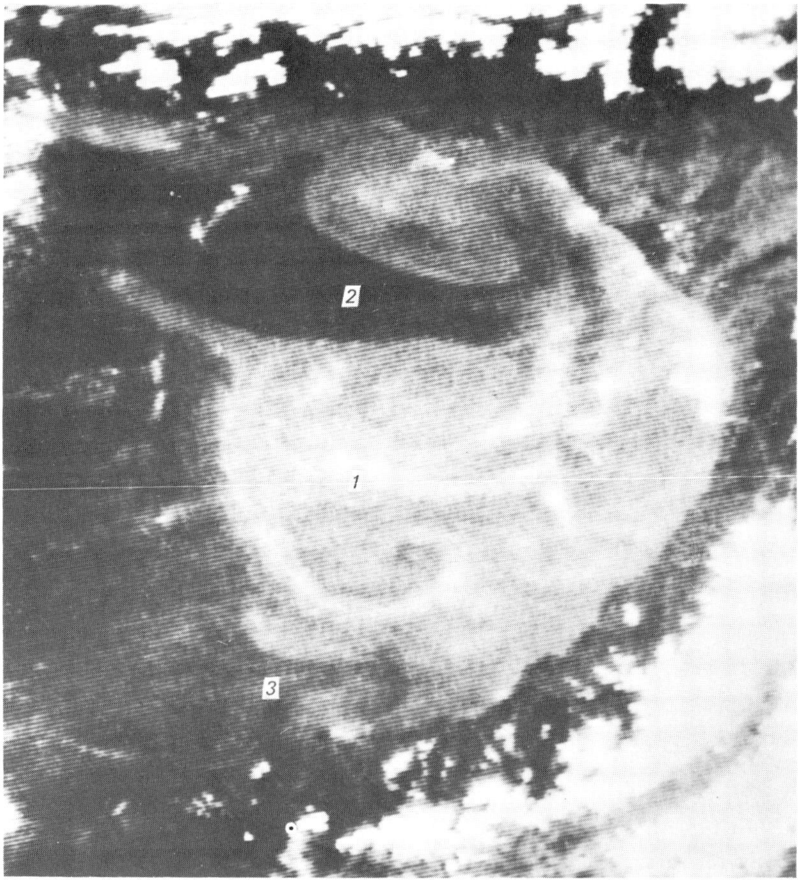

Figure 5.8. Example of the formation of type I compact 'packing' of mushroom-like structures in the eastern part of the Black Sea. Magnified fragment of image after the 30th launch of the 'Meteor' satellite (small resolution scanner, 0.5–0.6 μm band), 8 June 1981. See full image in ref. [54].

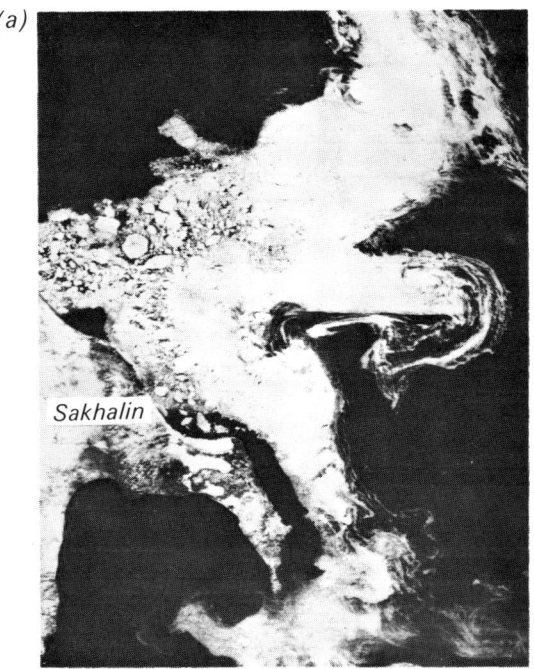

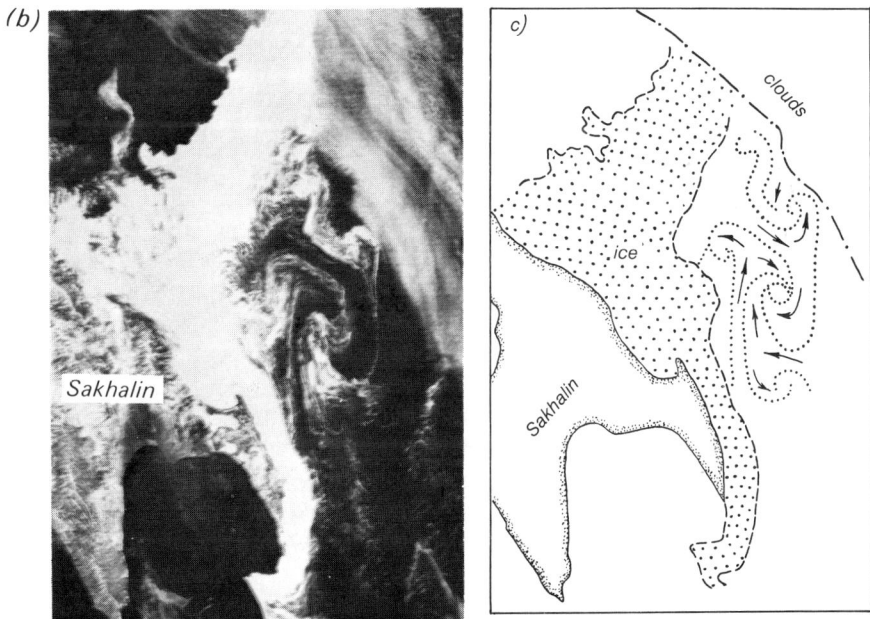

Figure 5.9. Formation of type II compact 'packing' at the edge of ice cover near the eastern coast of Sakhalin. (a, b) Fragments of images after the 30th launch of the Meteor' satellite (medium resolution scanner, 0.5–0.7 μm band), 10 and 15 April 1984, respectively; (c) interpretation diagram of fragment (b), demonstrating dense 'packing' of mushroom-like structures.

'packing' of vortex dipoles may be typical in which neighbouring mushroom-like structures are interlocked on a common vortex shared by them (Fig. 5.7). A similar situation may be observed when compensating motions of secondary nature are produced by local pressure gradients generated by an original strong disturbance. This situation is most likely in enclosed or semi-enclosed ocean areas close to coastal or frontal boundaries. One of the best examples of dense packing of type I (Fig. 5.7) is provided by a fragment of the image of mushroom-like currents obtained from a 'Meteor' system satellite in the eastern part of the Black Sea (Fig. 5.8). The visibility of the mushroom-like patterns is due to the presence of suspended matter brought in abundance by numerous rivers during the snow-melt season in the Caucasian Mountains. In this particular case, the jets of the two secondary mushroom-like currents (2 and 3) are directed in the opposite sense relative to the jet of the main one (1), which is directed westward. One of the vortices of the latter is shared by the secondary current (2) in such a way that there are only three vortices involved in two mushroom-like patterns. Other known cases, e.g. one near Sakhalin Island (Fig. 5.9, the tracer is floating ice), demonstrate type II (Fig. 5.7) packing of mushroom-like currents whose jet sections are normal to each other, so that one vortex is shared by each pair of perpendicular structures (Figs 5.9b and 5.9c). This pattern of floating ice distribution originated from a solid ice sheet of some 80 km long and 25 km wide which 5 days earlier (Fig. 5.9a) occupied the same position as the central mushroom-like structure on 15 April 1984 (Figs 5.9b and 5.9c). It is of interest to note that the type II perpendicular packing of mushroom-like structures is often observed in the disintegrating Von Karman vortex streets frequently visible in clouds over oceanic islands, e.g. over the Canary Islands (see Figs 1c and 1d in ref. [35]).

From the arguments advanced in this section it follows that mushroom-like currents should occur in the ocean rather frequently. Yet, the number of observed occurrences on systematically analysed satellite images of some selected areas of the ocean over 5 years amounts to several tens—a hundred cases only. Apparently, this is associated with the inconstant presence of the necessary natural tracer in the near-surface waters of the ocean. Besides, both the tracers and the near-surface stratification are greatly dependent on the seasonal conditions. Therefore, in certain seasons mushroom-like currents either cannot be observed because of the absence of a tracer, or may not exist at all.

Local disturbances in the motion of the ocean waters may result from a variety of kinetic energy sources available in the ocean–atmosphere system [34]. In the open ocean, these are jet-like air streams, unbalanced sea level and atmospheric pressure differences, and local dynamic instabilities of fronts and currents. In the coastal areas, these are river discharges,* melting of ice, local winds focused by coastal geometry and orography, bypassing of capes and large

---

*A mushroom-like form of a discharge lens is more likely to occur with relatively small rivers and lagoon outlets when their openings are protected by dikes and jetties, ensuring a directed discharge.

ice sheets, water exchange through straits as well as the same local dynamic instabilities of motion as in the open ocean. Apparently, the latter factor is the cause of the generation of a mushroom-like structure, whose evolution is observed in Fig. 5.5. Examples of the generation of mushroom-like currents in the case of water exchange through straits can be found in satellite IR images of the Baltic Straits [159], where such currents are clearly visible. Similar currents are frequently observed near the Kuril Straits in the Pacific (Fig. 2 in ref. [35]).

Perhaps of greater interest are the rapidly accumulating examples of the appearance of mushroom-like currents in the process of local dynamic instability of synoptic scale motions: eddies, frontal jet currents, and coastal upwelling frontal systems. Cold transversal jet streams which develop in the near-surface layer of the coastal upwelling zones during the relaxation phase of intensive upwelling events (see Section 5.3.2 and ref. [36]) often have mushroom-like or hammer-like structures at their offshore extremities. Perhaps, the occurrence of such structures would have been even more frequent if these jet streams carried water of higher positive buoyancy. As it is, the buoyancy gradient is only marginal, and the water easily mixes down while losing momentum on its way offshore. Contrary to this, the opposing or compensating influxes of warm waters near the surface into the areas of cold upwelled water more often have the appearance of large mushrooms.

Mushroom-like currents are accompanied by a strong deformation field because of their small spatial dimensions (10–200 km) and high velocity (up to 1 knot). From our estimates based on satellite data, the local deformation rate reaches values of $10^{-5} - 5 \times 10^{-4} \, s^{-1}$, which is 2–3 orders of magnitude higher than the rate of deformation typical of climatic frontal zones [103]. As a result, considerable redistribution of the concentration of passive scalers or temperature occurs in the near-surface layer under the influence of the intensive deformation field of a specific shape. When contrasts in the concentration of such scalars and temperature exist in the area of generation of these currents, the sharpest fronts are formed on the leading edge of the growing mushroom hat and on both sides of its jet part. Therefore, mushroom-like currents which form in the frontal zones (e.g. of upwelling or climatic origin) should create very complicated and contrast frontal structures that are well seen in satellite images in the visible or infrared ranges of the spectrum (see Section 5.3.3 and ref. [37]). It is also evident that mushroom-like currents are a very efficient mechanism of horizontal (in particular, transfrontal) exchange of heat, mass, and momentum in the near-surface layer of the ocean.

It is possible that the internal friction at the sharp frontal boundaries, accompanying mushroom-like currents, contribute to preservation of their coherent (ordered) character, ensuring maintenance of relatively low values of the 'turbulent Reynolds numbers' (Re_* about 100) in the scales of the entire current. For a thermal front of width $B_f = 100$ m to be in an equilibrium state [103] at a deformation rate D_d of $5 \times 10^{-4} \, s^{-1}$, the horizontal turbulent heat diffusivity on it should be

$$K_l = 2B_f^2 D_d = (2 \times 10^4) \times (5 \times 10^{-4}) = 10 \, m^2/s.$$

If the horizontal momentum exchange coefficient is of the same order of magnitude, then at a mean current velocity value $\bar{U}$ of the order 0.1 m/s and a typical width d of the jet part of the order 10 km = 10^4 m, we obtain $Re_* = Ud/K_l = 100$. At smaller deformation rates (order of $10^{-5}\,\text{s}^{-1}$), more diffusive fronts are observed ($B_f \simeq 1$ km) [103], which gives $K_l \simeq 20\,\text{m}^2/\text{s}$, and ensures $Re_* \simeq 100$ for mushroom-like currents of larger size.

Sharp thermal contrasts, which in frontal zones define boundaries of mushroom-like currents, can be seen in the IR images of the coastal Norwegian current, Leeuwin current (along the southern coast of Australia), East Australian, etc. A noticeable anticyclonic Gulf Stream ring on a NOAA-5 satellite IR image (p. 16 in ref. [194]) is accompanied by a warm mushroom-like offshoot with sharp frontal boundaries directed radially away from the outer edge of the ring toward the open ocean. This pattern is reminiscent of the manifestation of density front instability in the laboratory experiments carried out by Griffiths and Linden with a rotating fluid [153]. Numerous mushroom-like structures were observed by us during 1980–1985 in the visible band images (tracer—floating ice) in the area of a quasi-stationary anticyclonic gyre existing in the southwestern part of the Sea of Okhotsk [35] (Fig. 5.10). A similar situation is frequently observed in the background of the general anticyclonic circulation in the Alboran Sea [175]. It seems that mushroom-like currents are particularly frequent in regions with a general anticyclonic character of motion. Besides, they are connected, apparently, with any unstationary or unstable situation associated with the major ocean currents. For example, an ERTS-1 image of 4 July 1973, published in ref. [213], shows a series of mushrooms, 20–30 km long, directed across the Gulf Stream around 35°30′ N, 72°30′ W towards a cyclonic disturbance near its northern boundary. In this case, the whole pattern is visible owing to surface roughness contrasts.

The series of images demonstrating the evolution of mushroom-like currents in the Tatar Strait (see Fig. 5.5) has shown that the rate of their longitudinal

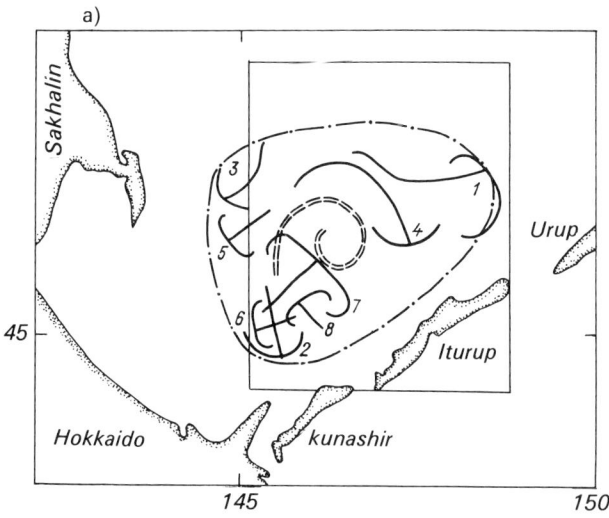

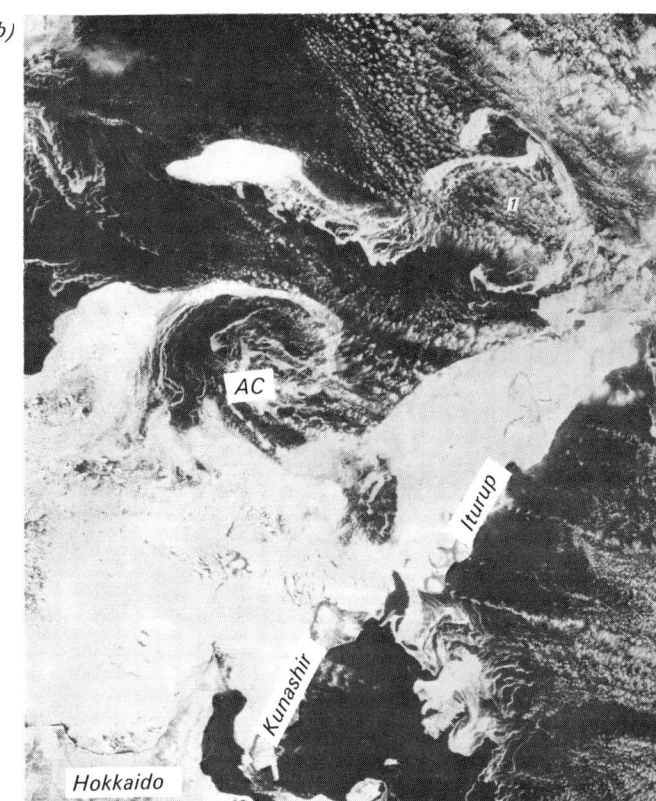

(b)

(d)

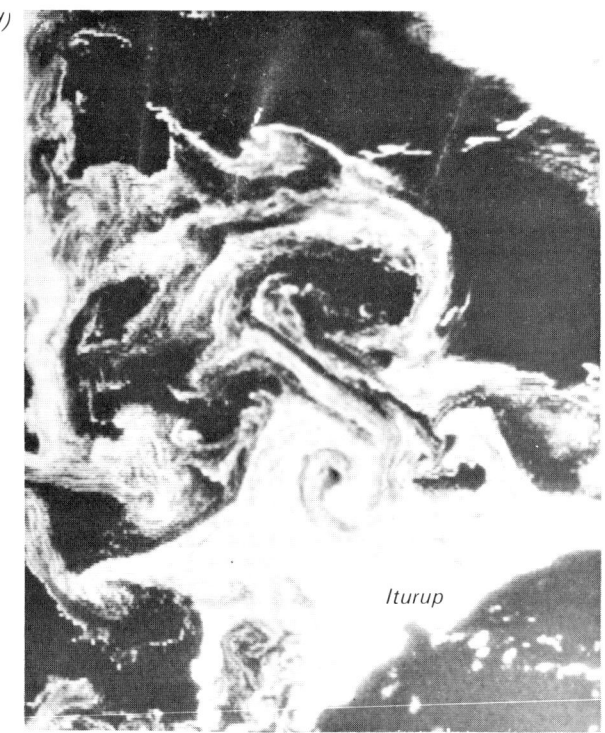

Iturup

Figure 5.10. Examples of mushroom-like currents observed in 1980–1985 south-west of the Sea of Okhotsk. (a) Diagram of structures 1–8 (to scale) discernible in the images after the 29th, 30th, and 31st launches of the 'Meteor' satellite (medium resolution scanner, visible range) in 1980–1984: (1) 31st launch, 29 March 1984 [see (b)]; (2) 30th launch, 14 March 1984; (3,4) 31st launch, 30 March 1983 [see (c)]; (5, 7, 8) 29th launch, 16 April 1980; (6) 30th launch, 19 February 1982. The double dotted line shows the position of an anticyclonic eddy, 29 March 1984, which itself was probably part of a mushroom-like structure at the ice cover edge (b); the dash-dotted line is the boundary of the anticyclonic gyre [35]. The thin line shows the boundary of the area whose image is given in (d). (b) Mushroom-like structure I [see (a)], whose jet part is directed by the normal to the boundary of the anticyclonic gyre (fragment of image after 31st launch of the 'Meteor' satellite, medium resolution scanner, 0.6–0.7 μm band, 29 March 1984) [35]. (c) Two mushroom-like structures 3 and 4 [see (a)] with an arc-shaped jet part (fragment of image after 31st launch of the 'Meteor' satellite, medium resolution scanner, 0.6–0.7 μm band, 30 March 1983) [35]. (d) Complicated combination of at least four mushroom-like structures of various space scales from ≃ 70 to ≃ 140 km (magnified fragment of image after 30th launch of the 'Meteor' satellite, small resolution scanner, 0.6–0.7 μm band, 2 April 1985).

increase (L) drops sharply from 15 to 8 cm/s between three successive images separated by an interval of about 24 h. Retarded growth of the longitudinal size is caused by the growth of the vortex pair, absorbing the continuously increasing amount of water transferred in the jet part. Hence, it is logical to assume that the current velocity along the axis of the jet should be essentially greater than $\partial L/\partial t$. For example, since the size of structure 3 (Figs 5.10a and 5.10c) increased during the first interval (28–30 March 1983) at a rate of $\overline{\partial L/\partial t} \simeq 37$ cm/s ($\Delta L = 35$ km, $\Delta t \simeq 26.2$ h) and during the second interval (30–31

March 1983) at a rate of $\overline{\partial L/\partial t} \simeq 19\,\mathrm{cm/s}$ ($\Delta L \simeq 15\,\mathrm{km}$, $\Delta t \simeq 21.3\,\mathrm{h}$), the true velocity in the jet should be more than 1 knot, and probably remained so for the first 24 h or longer. An analysis of the evolution of this and some other mushroom-like structures has shown that they all increase to $\simeq 70\text{–}80\%$ of their final size in the course of $\simeq 24\,\mathrm{h}$ and later grow insignificantly. This prompts us to assume at least two phases in the development of mushroom-like currents generated by an impulse effect, namely, the first (rapid) one which lasts for about 24 h, or even less, and the second (slow) one which lasts for several days during which the structure does not practically grow (except for the continuous slow enlargement of the vortex pair) and ends with complete breakdown of the structure. The situation as a whole resembles the evolution of a mixed patch in the process of collapse in a stratified fluid during which the patch goes through several stages with a variable dependence of the increasing radius on time. A dependence of this type for mushroom-like currents will probably be found either in laboratory experiments or theoretically, as it is hardly possible to ensure a sufficient frequency of observations of one and the same structure of this type from a research vessel or satellite. Some simple laboratory experiments performed by us on generating mushroom-like currents in a homogeneous fluid without rotation have shown that the differences in the dependence $L(t)$ for the initial and subsequent stages are quite obvious and essential.

An analysis of satellite images also reveals other specific features in the formation and behaviour of mushroom-like currents that do not demonstrate vortical asymmetry* of a definite sign although anticyclonic asymmetry is observed in many cases. They can have a rectilinear or a curved jet (Fig. 5.10a). Both vortices can be formed simultaneously, but primary formation is possible of one anticyclone with subsequent generation of a cyclone (see the discussion in ref. [34]).

As mentioned before, practically all the information on mushroom-like currents is currently based on satellite data. The only instrumental measurement known to us, which apparently can be interpreted as measurement in the area of a mushroom-like current and which confirms the intensity of concomitant frontogenesis, was made in the cold transversal jet in the Californian upwelling system [228]. The thermal sensor, towed on horizon 2.5 m, crossed the jet and one of the vortices in the pair. The recorded temperature drop and gradient reached 2°C and 0.25°C/km, respectively, on the third day after the formation of the current. Another instrumental measurement, which is probably related to the type of motion discussed, was made in the anticyclonic ring of the Gulf Stream with a diameter of $\simeq 100\,\mathrm{km}$ [166]. Measurements in the radial cross-section have shown a narrow and more rapid jet of warm and less salty (as compared to the surrounding) waters, propagating clockwise in the azimuthal direction against the background of common orbital motion of the waters in the ring. The jet with boundaries of frontal character encompassed a layer with a thickness not more than 45 m (above the seasonal thermocline), had a width

---

*Vortical asymmetry here means prevalence by intensity of one of the vortices in the pair.

of $\simeq 4.1$ km, and was situated at a distance of 5–6 km from the ring centre. The velocity in the jet, measured by a Doppler current meter with a vertical resolution of 6.4 m, reached 1.1 m/s. Unfortunately, there was no satellite information that would allow us to determine whether the jet had a closed circular character or ended with a vortex pair. However, it is possible that the jet had a shape similar to that shown in Fig. 5.10c. The very existence of such jets in both cases (anticyclonic Gulf Stream ring and anticyclonic eddy in the Sea of Okhotsk) is apparently evidence of the non-solid-body character of water rotation in the eddies.

It follows from all the results cited above that mushroom-like currents in the ocean resemble to a certain extent pairs of point vortices in an ideal fluid as discussed by Lamb [62]. However, these vortex pairs correspond to an idealized case of strictly two-dimensional motion when the closed lines of the flow coincide with lines of equal vorticity, or equal relative vorticity, while the external field is characterized by zero or homogeneous vorticity, respectively. Such vortex pairs are possible in a homogeneous rotating or non-rotating layer of non-viscous fluid of infinite or finite depth in the absence of friction against the bottom or surface. Most of these conditions are not observed in the real ocean, and continuous stratification does not permit strict two-dimensionality of the phenomenon. As mentioned previously, its initial stage is definitely non-stationary. The space scales and velocities of the observed mushroom-like currents [Rossby number $Ro = U/(Lf)$ of about 10^{-1}] presupposes a substantial degree of geostrophic adaptation. In its turn, it is impossible without at least weak vertical motions. Besides, on a rotating earth we deal with the β-effect which exerts an influence on the vortex motions of the type and size considered here, while the latter effect is not considered in the classical description by Lamb. Nevertheless, as follows from the above-cited satellite information, ordered (coherent) vortex motions of the dipole type exist in the ocean despite all the stated differences between the real ocean and Lamb's ideal fluid.

What does theory say on the potential existence of vortex dipoles in the ocean on a rotating earth? As theoretically presented by Larichev and Reznik (see the discussion in refs [27, 34]), a stationary solution of the soliton type in the form of vortex dipoles is possible on a beta-plane. However, these idealized dynamic structures should have a strict orientation in relation to the axis of rotation, i.e. they should propagate only in the zonal direction which is the only one possible in the presence of the β-effect. The above oceanic mushroom-like currents do not manifest such a selective behaviour and are clearly non-stationary. It is easy to see from the satellite information that they can propagate in any direction.

Theoretical predictions stimulated certain laboratory experiments in order to obtain vortex pairs of the soliton type in the presence of the β-effect. Although at approximately the same time vortex dipoles were more easily reproduced in the laboratory experiment without the β-effect [144] and even without rotation [21], theoretical consideration of the potential existence of various isolated vortex structures (dipoles or monopoles) in the ocean was always done taking into account the β-effect. However, an analysis [27] has shown that non-linear

(a) (b)

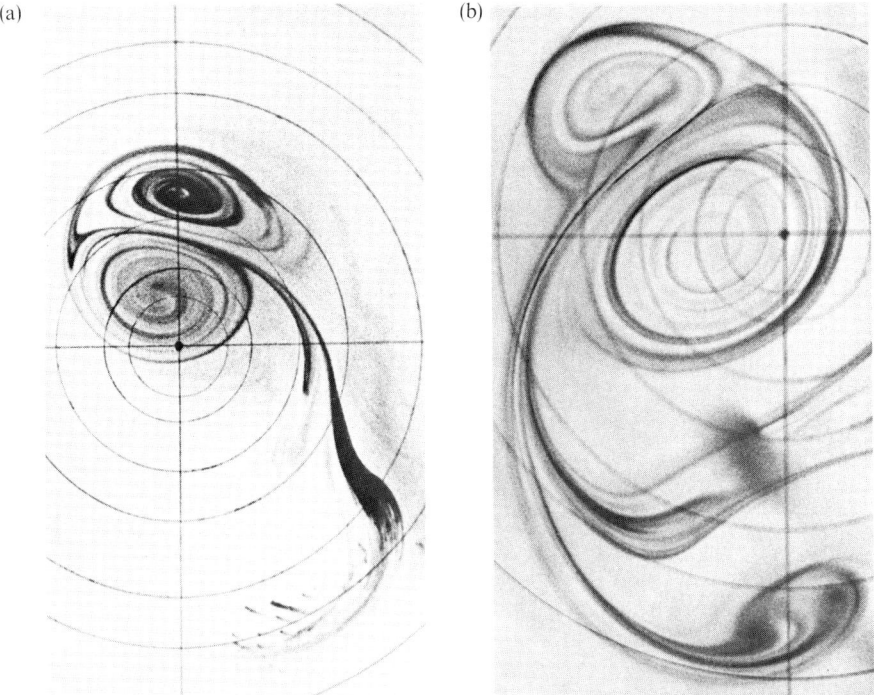

Figure 5.11. Examples of vortex dipoles in a laboratory experiment. (a) Dipole with AC asymmetry forming under the effect of a symmetrical air jet on the surface (two-layer fluid, $\Delta S = 10\permil$, $H_1 = H_2 = 3.5$ cm, time after introducing disturbance $= 56$ s); (b) dipole with C asymmetry forming 18 s after the effect of an axial symmetric air jet on the water surface (two-layer fluid, $\Delta S = 5\permil$, $H_1 = H_2 = 4$ cm). Cyclonic asymmetry developed as a result of interaction with residual motion of anticyclonic character from a vortex dipole which developed for 200 s in the previous experiment.

effects in the formation and evolution of vortex dipoles in the ocean and in the laboratory experiment are most likely more important than the β-effect, and that the behaviour of the vortex dipoles in the laboratory experiment on an f-plane is very similar to their behaviour in the ocean. Laboratory experiments [27] made in homogeneous and stratified fluids with various methods of disturbing their solid-body rotation have shown that the vortex dipoles represent a universal response of a solid-body rotating fluid to any local impulse disturbance, introducing finite quantities of positive or negative relative angular moment into the system. This gives additional weight to the conclusion, made on the basis of analysing satellite information, on the universality of the mushroom-like shape of non-stationary currents in the near-surface layer of the ocean where these disturbances are most frequent. The experiments [27] made it possible to give answers to the questions formulated above. The forming vortex dipoles may have arbitrary vortical asymmetry. The character of asymmetry, anticyclonic (AC) (Fig. 5.11a) or cyclonic (C) (Fig. 5.11b), depends on the degree of prevalence of locally introduced angular moment of one or the other sign in the case of disturbance. Imparting equal quantities of negative or positive angular moments (axially symmetric jet of water and air) to the water always induced in a rotating system (unlike the case of a non-rotating system) an AC-asymmetric dipole as in ref. [144]. The behaviour of asymmetric vortex dipoles in the experiment is very similar to the behaviour of Lamb's vortex pairs, i.e. they move along circular trajectories, rotating around the stronger vortex of a pair. The centre of rotation can be found from the classical formula given in ref. [62] under the condition that the velocity field in the vortex pair is known. The character of vortex asymmetry can change during interaction with other dipoles. The initial development of the dipole is determined by the intensity of disturbance, i.e. comparatively weak and short-term disturbances induce the initial formation of AC with the development of C several seconds later; C forms simultaneously with AC in the case of more intensive and longer effects. It is possible to note also interesting consistencies in the behaviour of dipoles in the laboratory tank which were also observed in the real ocean, namely, (a) one of the vortices of the dipole gives rise to an independent mushroom-like current (this system is discussed in ref. [37] and in Section 5.3.3); and (b) the dipoles form compact 'packings' of various configurations of the type presented in Fig. 5.7. The conclusion on rapid suppression of small-scale turbulence, generated in the case of local disturbance of any type, by rotation is a very important result of the experiments. The vortex dipole or dipoles develop in each case from chaos as a result of this suppression and subsequent ordering of motion. Coherent motion in the form of a vortex dipole rapidly propagates to the entire water layer (to the upper layer in a two-layer system) in absolute conformity with the Proudman–Taylor principle. The vortex motion is also transmitted to the lower layer in a two-layer system through the mechanism of geostrophic adaptation in the case of sufficient force and duration of the exciting effect.

The experimental results considered give hope that a well thought-out combination of laboratory, field, and theoretical studies, in conjunction with analysis of available satellite information, will make it possible in the near future to determine finally the main consistencies of the origin and evolution of mushroom-like currents (vortex dipoles) and their role in the formation of quasigeostrophic motions of larger scales in the ocean.

5.3.2. *Systems of transversal filaments in coastal upwellings*

The phenomenon to discussed here consists in the development in coastal waters under the influence of upwelling of breakthroughs of relatively cold surface waters rich in biogenic elements from the coast toward the open ocean in the form of narrow frontal filaments [127] (or 'jets' [143, 187]). Such filaments (Fig. 5.12) were observed in many satellite images [136, 164, 194, 228, 231], obtained by means of scanning IR radiometers and optical scanners (CZCS, NIMBUS-7 satellite), on the background of a clearly expressed and well-known 'patch' structure of temperature and chlorophyll distribution in the zones of upwelling.

The uncommonness of this phenomenon is associated with the following circumstances:

- the detected filaments propagate transverse the shelf and the continental slope to a distance of several hundred kilometres from the coast [127, 194, 231] (Fig. 5.12) which exceeds greatly the local baroclinic Rossby radius of deformation R_N (for zones of coastal upwelling $R_N < 50$ km [127]) determining the typical width of the coastal band of upwelling waters;
- for the relatively cold water of upwelling origin to flow on the surface in the form of filaments, the latter must be characterized by a deficit of salinity and density* whose origin should be explained in each concrete case;
- no existing hydrodynamic model of coastal upwelling predicts the origin of transversal filaments.

The physical nature of the observed cold filaments has not yet been established, though various hypotheses have recently been advanced [36, 164, 190, 231] in regard to their origin. They will be discussed below in connection with the physical picture that is outlined from satellite and ship-borne observations. This picture can be presented in the following generalized form.

(1) Transversal filaments are almost always detected on satellite images of the Oregon, California [127, 143, 194, 228], Benguela [231], and Canaries [136] upwellings and coastal upwelling of waters in the Gulf of Lions [187], but they have never been observed in the Peru upwelling [127].

(2) Sometimes systems of transversal filaments, arranged at some characteristic distance from one another when following the coast line, are observed (Fig. 5.12). This distance is 50–100 km in the case of the California upwelling, while in the Benguela upwelling it reaches 200–400 km [231].

(3) Filaments or systems of filaments form after termination or abatement of the wind, which is advantageous for the upwelling [228], and not in the period of intensive quasi-stationary upwelling. This circumstance is in good agreement with the fact established by aerial IR measurements [127] that intensive frontogenesis, return of the upwelling front to the coast, and general complication of space variability of the SST field commence in the period of upwelling relaxation.

(4) Filaments do not have a correct rectilinear shape. They meander and end sometimes with a vortex (cyclonic or anticyclonic), or a vortex pair, forming a mushroom-like or T-shaped [164] structure (Fig. 5.12).

(5) The filaments may be of different characteristic length, from several tens of kilometres in the Oregon upwelling to 300 km in the California one (see Fig. 5.12), and even to 500 km in the Benguela upwelling. The width of the filaments ranges from 10 to 60 km. As far as it is known from the few direct ship-borne measurements [143, 190, 207], their penetration into the depth is restricted to a thin near-surface layer of the ocean (not deeper than $\simeq 50$ m).

---

*In relation to the surrounding or underlying waters.

(a)   b)

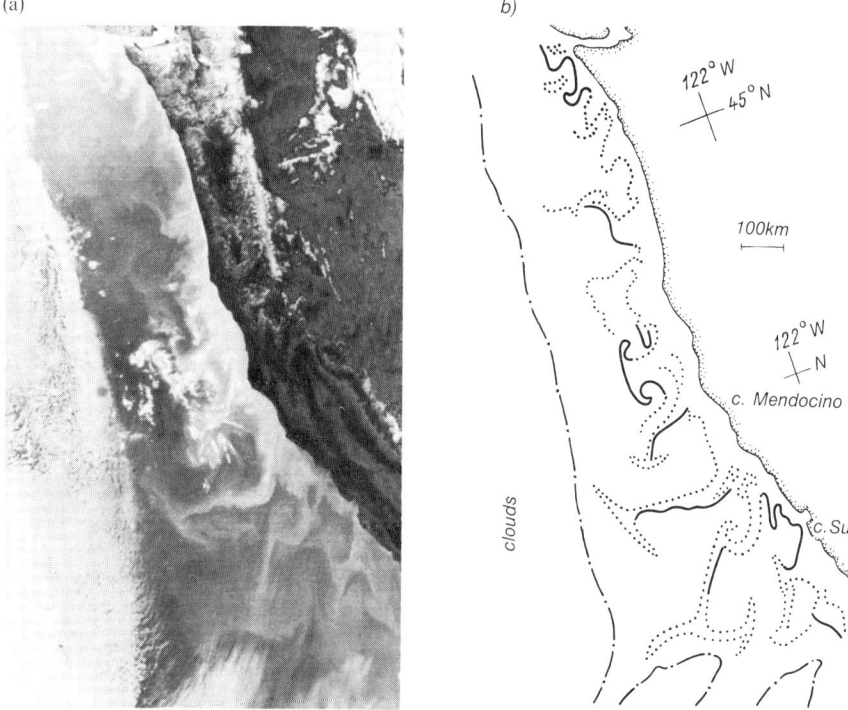

Figure 5.12. IR image of the California–Oregon coastal upwelling (a), obtained on 8 September 1976 from the NOAA-5 satellite [194, p. 22], and the interpretation diagram of the image (b). Dotted line: boundaries of colder waters; solid line: the sharpest fronts; dash–dotted line: boundaries of cloud cover.

According to the measurements made in the California upwelling, the thickness of one of the filaments was 30 m [190], and the lower boundary of another one on its southern side was at a depth of 30 m while on the northern side it was 60 m, i.e. in both cases, it was above the seasonal pycnocline [143]. The latter circumstance is in agreement with the established fact [127] that the 'patchiness' of the temperature field in the zone of the California upwelling is usually observed only to the horizon 25–30 m.

(6) The velocity of the current in the transversal filaments near the surface reaches 0.5 m/s in the California upwelling [143] and 0.4 m/s in the Gulf of Lions [187].

(7) The filaments transfer water which is only 1.5–2°C colder [190, 228], while the non-transformed waters of the upwelling directly in the spots of upwelling are 6–10°C colder than the surrounding waters. Publications [143, 207] give evidence on the reduced salinity of the waters in the filaments.

(8) The filaments have sharp frontal boundaries, and the thermal fronts on the left (if viewed in the direction of the current) boundaries of the filaments in the northern hemisphere, as a rule, are sharper than on the right ones. This is

seen clearly in Fig. 5.12 and indisputably follows from ship-borne measurements. A very sharp front (1°C over less than 100 m) was recorded in the California upwelling on the southern (left) side of the filament, and a more eroded front (1°C over several kilometres) on the northern (right) side [143].

(9) In some regions (e.g. California) the transversal filaments form quite often close to the same characteristic strips of the coast, i.e. capes, gulfs, canyons, and underwater ridges. It calls forth the assumption on the influence of the bottom topography or the shore outline on the origin of these filaments. However, in other regions (e.g. in the Benguela area), with the exception of one case, no obvious connection is observed of filament arrangements with the specific features of the bottom topography and with the outline of the shore [231].

(10) The lifetime of the transversal filaments may reach several weeks [207, 231], i.e. it often exceeds greatly the duration of a stable or quasi-stationary upwelling situation (2–5 days in the Oregon upwelling). The filaments practically never change their position in the course of their existence. The temperature drop at the boundary of the filament, recorded by crossing it with a thermal sensor on the third day of the structure's existence, was still 2°C [228].

It is necessary to mention the hypothesis, among all others making an attempt to explain the origin of transversal filaments in the coastal upwellings, which assumes the entrainment of cold coastal waters in the filament current, forming between two eddies, cyclonic and anticyclonic, close to the coast [164, 190]. This hypothesis in ref. [164] is developed to the state of a numerical model wherein the eddy pair is formed owing to the instability of the surface (0–150 m) California current in the southern direction and the subsurface (150–500 m) current in the northern direction. However, neither the time scale of the meandering process, inducing vortex separation (several tens of days), nor the depth where the anticyclonic eddy develops (deeper than 150 m) corresponds to the above-cited space and time characteristics of the filaments considered. Besides, the reasons for the propagation of the colder (and more dense) upwelling waters over the ocean surface are not explained in refs [164, 190]. The authors of other hypotheses [231] cast doubt on the filament character of this propagation, considering as alternatives either the bands of divergence or the effect of displacement of long shelf waves along the coast. In our opinion, it is impossible to refute the jet character of the given phenomenon. It is indicated by the frequent formation of mushroom-like (T-shaped) structures at the ends of the outbursts. Besides, the water motion velocity in the filaments, reaching 1 knot, is quite sufficient to ensure the above-mentioned time scale of the phenomenon even at a 500 km length of the filaments. Attention should be drawn to the fact, noted in ref. [231], of the absence of a noticeable displacement of cold bands that could have been anticipated in the Benguela current. Displacements of filaments are also not observed in other regions where there are constant currents along the coast. However, the typical average velocities of these currents, including the Benguela one, are of the order of 10–20 cm/s. In addition, these currents are not coherent unidirectional flows, and are characterized by a variable vortex structure. The more rapid transversal currents in the course of their existence (7–14 days) under these conditions can only gain bands, as is seen in Fig. 5.12. In addition, the above-cited hypotheses and assumptions are extremely abstract because they give no answers to quite natural questions that arise when learning the totality of known facts. The main questions are as follows:

(1) Which processes ensure the transformation of the deep cold waters lifted by the upwelling to the surface that forms the necessary deficit of salinity, giving them the possibility of keeping near the surface at subsequent propagation in the form of filaments?
(2) What role in the transversal filament formation does the pattern of slopes of the level surface and its evolution play, which is characteristic of upwelling zones, at termination or abatement of the wind which is advantageous for the upwelling?
(3) How do sharp fronts associated with filaments develop, and what is the role of the fronts, extending to great distances by the normal to the coast, when the main upwelling front is usually parallel to the coast and at a comparatively small distance from it (about R_N)?
(4) How are the transversal filaments of the cold surface waters connected with the 'patchy' space structure of the coastal upwelling?

Let us try to answer these questions on the basis of available information and current comprehension of the coastal upwelling physics and the nature of currents similar to the filament currents studied (see Section 5.3.1).

The actual data given above on the phenomenon considered here contain certain key information for comprehending its physical nature. First of all, it is necessary to consider the absence of transversal jet currents in the zone of the Peru coastal upwelling. Earlier the attention of one of us [103] was drawn to the absence of any essential variations of salinity by the horizontal and the depth in this region within 100–200 km from the coast. Hence, there are simply no sources in this region for the formation of the necessary deficit of salinity during the transformation of the upwelling waters which makes impossible their propagation on the ocean surface. During relaxation of the upwelling, the deep waters transformed close to the surface downwell into the depth without the formation of jets or sharp fronts. The latter circumstance was also mentioned in ref. [103]. Own sources of freshening are available everywhere in other regions of coastal upwelling. The waters of the Columbia and Fraser run-off, propagating southward along the coast down to South California, are the source in the Oregon and California upwellings, whose waters are colder and saltier than the surrounding waters [163]. The deep waters in the Benguela upwelling, rising to the surface, are 0.7–0.8‰ fresher than the waters in the upper layer in the open ocean [121], so that the T, S correlation of the space thermohaline variability is always positive in this region. A positive T, S correlation is observed at least some of the year in the Canaries upwelling and at the Morocco coast (from September until June in the regions more northward), and, as shown in the satellite IR images of October 1983 [136], transversal cold filaments are observed there at this time. Quite sharp thermal fronts (up to 0.5°C/km) are recorded even to the south of the Peru upwelling in the region between 35 and 42° S where the space variability of the salinity field is great in the upper layer of the ocean, and a very unusual variability of the SST field [50] is observed which could be due to the filament withdrawal of cold waters from the upwelling zone at the coast of Chile. More detailed information on the salinity and temperature variability in different regions of coastal upwelling is given in ref. [103], pp. 173–184.

The typical temperature and salinity values of the various water modifications in the region of the Benguela upwelling, given in ref. [121], make it possible to calculate easily that it is necessary for the fresh upwelling water with a temperature about 9–12°C and salinity 34.7‰ to warm up in the process of transformation to 16.5–17.0°C so that its density should reach (or could become slightly less than) the density of the surface waters in the open ocean that have a temperature of about 18.7°C and a salinity of 35.4‰. In this case, the resulting temperature difference is about 2°C, which is characteristic of cold transversal filaments. The deep water rising to the surface in the California upwelling with a temperature of 8–9°C must warm up by 6.0–6.5°C in the process of transformation to attain the density of the surrounding waters, but it must also freshen for this purpose from 33.7–33.8 to 33.2–33.3‰ and become 0.3–0.4‰ fresher as compared to the surrounding waters in the open ocean which have a salinity of about 33.6‰. The transformed upwelling waters will still remain 1.5–1.7°C colder than the surrounding waters which is observed in transversal filaments. This transformation requires the presence of freshened waters with a salinity of about 32.5–32.8‰ on the surface seaward of the upwelling band with which the waters of the upwelling would mix in a 1:1 relationship. According to ref. [163], the tongue of these waters goes down from the north along the California coast practically to the latitude of San Francisco.

The above estimate [36], made by us using climatic data, has been confirmed by direct measurements [207], and it turned out that typical values of the water salinity in the jets and surrounding background waters were still lower, i.e. 32.7–32.8 and 33.0‰, respectively. The reasons for this are the extremely low salinity of the waters flowing down from the north and freshened by the Columbia river discharge (32.4–32.5‰), and the fact that the mixing of the upwelling waters with the waters of the freshened lens does not occur in a 1:1 relationship, as we assumed for simplicity. Nevertheless, the most interesting fact was that the waters in the jet were of higher density than the surrounding near-surface waters by approximately 0.4–0.6 unit of σ_t, but were lighter than the underlying waters by 0.2–0.8 unit of σ_t. Dynamic calculations [207] demonstrated that the near-surface filament of cold and slightly more fresher waters with a thickness of about 40 m is carried away from the coast to the open ocean by the underlying geostrophic jet current, including a layer of about 300 m. An assumption is made in ref. [207] that the filaments could be the consequence of instability of the coastal system of currents, though the physical nature of this instability is still unclear. Let us indicate now the processes of transformation that can most effectively reduce the density of the upwelling waters. They are direct solar heating, wind–wave mixing, horizontal advection, and horizontal turbulent diffusion. The latter two processes in the upwelling zone are much more effective than the former two [206]. As a result of all the processes, an intermediate by its properties modification of waters, which separates from the waters of the open ocean by an increasingly sharper outer front as it gains homogeneity, forms seaward of the band of direct upwelling of waters in the near-surface layer. Andrews [121] pointed out this intermediate water mass, consisting of mixed waters, in his diagram of the Benguela upwelling. Its T, S characteristics, estimated by us above, are well within the limits indicated by Andrews. The transformed waters are distinguished well in many IR images of other zones of coastal upwelling by intermediate gradations of brightness. The typical width of the areas of such intermediate waters in the direction perpendicular to the coast is from 10–20 to 50–80 km. Sharp fronts at the phase of upwelling relaxation separate the intermediate waters from the neighbouring already on two sides, i.e. on the seaward and coastal sides. But if the intermediate waters formed in an upwelling, from the very beginning having a sharply expressed 'patchy' character, they themselves could have patches or lenses with a diameter of several tens of kilometres and could be surrounded with a frontal boundary on all sides. In this case, the process of upwelling relaxation inevitably includes the process of the collapse of these lenses.

Apparently, in case the instability of the outer front of regions and lenses of intermediate waters, inducing an increase of meander at the front, break of the front, and opposing inflow of warm waters in the direction of the coast should be the initial impulse for the origin of transversal filaments (Fig. 5.13). The developing jet of cold waters, flowing away from the coast, in the process of geostrophic adaptation forms on its left side (in the northern hemisphere) a sharper front, but it may also include, near the coast, the initial lateral front of the lens. The cold jet should inevitably become narrower in the direction of the open ocean owing to erosion of its right side by the opposing flow. It is possible that the intensive gradient barotropic currents, which should develop in the process of very rapid relaxation of the level surface slope during the phase of upwelling destruction, contribute to the development of outer front instability. The slope gradually grows during the phase of upwelling development owing to Ekman transport of the surface waters away from the coast, but, as follows from the observations in refs [163, 221], the level returns to its undisturbed position immediately after termination or abatement of the wind without any phase shift. The rapid sharpening and retreat to the coast of the main upwelling front on the coastal side of the intermediate waters, noted in ref. [127], are connected just with this process. It is quite likely that the returning water flow, directed to the coast *in its entire thickness*, is then reflected from the coast and is focused by the coastal bays and underwater canyons that cannot but contribute to break-throughs of the outer front and the formation of thick jets perpendicular to the coast and on which the transformed near-surface waters are carried out to the open ocean. In this, the given phenomenon qualitatively resembles a system of rip currents in the zone of intensive breakers. However, in the given case, there are additional complicating effects of the fronts and the earth's rotation. Correspondingly, a comparison suggests itself with convection

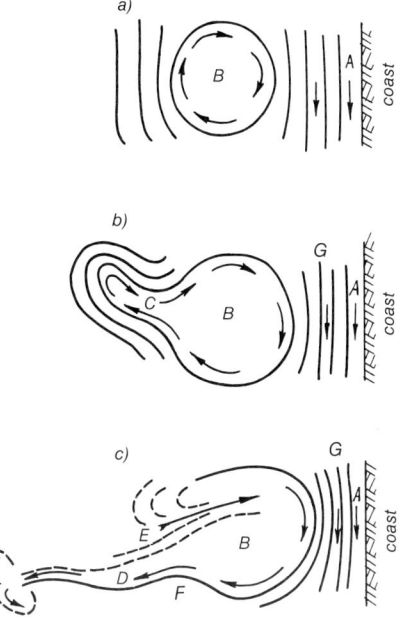

Figure 5.13. Diagrammatic representation of the development of instability and break-through of intermediate water lens outer front in the zone of coastal upwelling. (a) Formed lens of intermediate waters (arrows indicate the direction of motion); (b) moment of meander development due to instability of the outer front; (c) formation of transversal jet current and opposing inflow of warm waters. (A) Upwelling cold waters; (B) intermediate waters; (C) meander; (D) cold transversal filament; (E) opposing inflow of warm waters; (F) sharp frontal boundary; (G) main front of upwelling.

as the above-cited effects are similar to the effects of stratification and acceleration due to gravity in the development of convective instability.

The hypothetical pattern of the phenomenon suggested above meets all the questions set above and does not contradict the facts established from the observations. It would be good if this hypothesis together with the factual material discussed earlier stimulated the formulation of relevant theoretical and model problems, and the organization of subsequent field studies, which are unthinkable without most wide-scale application of modern aerospace techniques. In particular, these studies could be of great significance for industrial fishing, because as tuna fishing demonstrated to us and carried out simultaneously with satellite observations [174], the best catches were confined to the boundary of high-productive upwelling jets and fronts and frequently were more than 400 km away from the coast. Transversal upwelling filaments can be observed by spacemen because of their colour contrasts.

5.3.3. *On the near-surface circulation of waters in the sub-Arctic frontal zone (by satellite data)*

Having discussed separately the basic elements of ordered plane non-stationary motions of the oceanic near-surface layer, let us analyse now some cases of

(a)

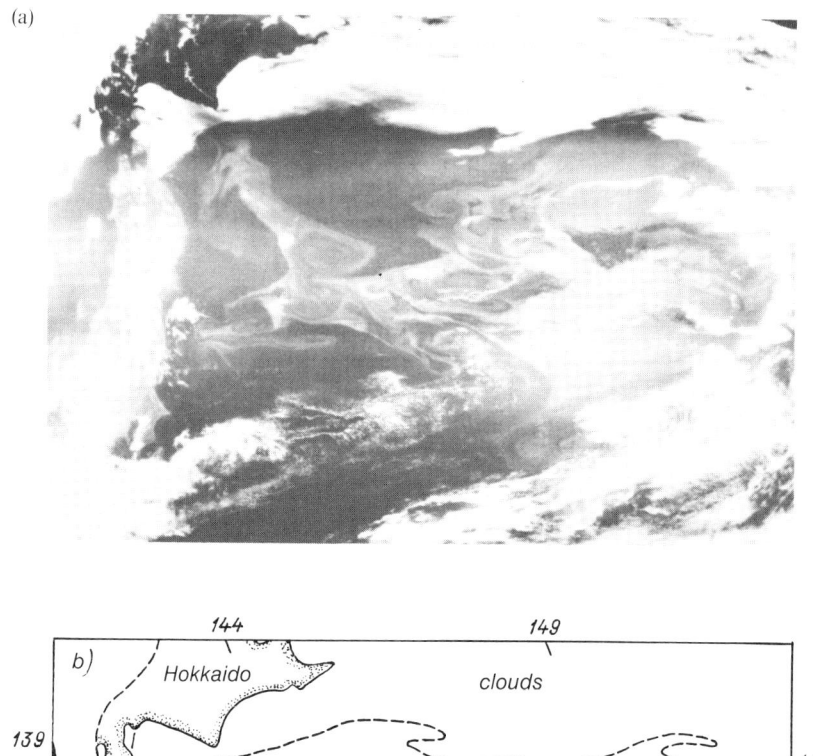

Figure 5.14. Satellite image (a) (30th launch of the 'Meteor' satellite on 19 May 1984, small resolution scanner, 0.5–0.6 μm band) and its interpretation diagram (b). The arrows indicate the direction of water motion. The letters and numbers designate eddies and jet currents, respectively (for explanation see text).

their joint manifestation against the background of large-scale general circula-tion. The sub-Arctic frontal zone to the east of Honshu, and the Kuril, Kuroshio, and Sangar currents may serve as a good example. Satellite images in the IR and visible ranges of the spectrum give a most visual synotic picture of the horizontal water circulation in this region and the possibility of observing the space–time evolution of its individual elements. An analysis of these images makes it possible to determine known and already described specific features of the given frontal zone as well as other interesting consistencies of the local interaction of the Kuril and Kuroshio currents and the structure of the region where this interaction is manifested most vividly [11, 37]. Situations are observed when the formation or motion of an eddy results in the formation or intensifica-tion of a whole series of jet currents in the meridional or zonal directions, and vortices accompanying them to a distance up to 300 km from the initial point of disturbance. The impression is created that similar disturbances arising under similar conditions result in similar structural specific features of the frontal zone.

Let us consider two images of the region eastward of Honshu relating to qualitatively similar situations and separated in time by a 3-year interval exactly, i.e. the image in the visible range after the 30th launch of the 'Meteor' satellite on 19 May 1984 (Fig. 5.14) and the IR image obtained from the NOAA-6 satellite [233] on 20 May 1981 (Fig. 5.15). The light-coloured areas in Fig. 5.14 (except the area of a sunglint patch in the right part of the image) correspond to warmer waters which are transformed Kuroshio waters [37].

It is possible to isolate in Fig. 5.14 a series of circulation elements that are easily discerned in the field of the tracer (apparently, phytoplankton [37]), namely, eddies A, B, C, and D; a mushroom-like structure (1) with a clearly

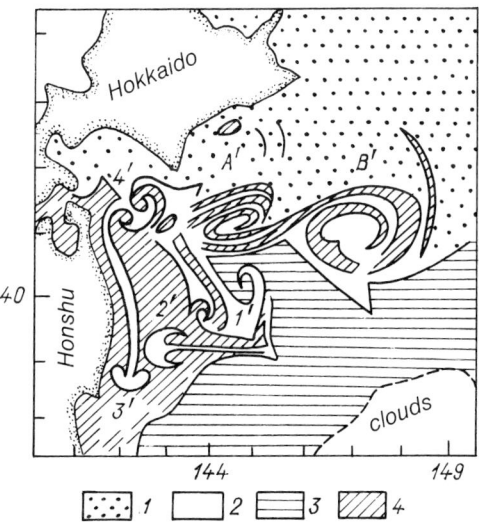

Figure 5.15. Diagram of the IR image obtained from the NOAA-6 satellite (20 May 1981), published in ref. [233]. (1) Cold sub-Arctic waters; (2) transformed cold waters; (3) warm waters of Kuroshio; (4) transformed warm waters.

expressed anticyclonic vortex and a less clear cyclonic vortex at the end of the jet indicating propagation of the flow (mushroom-like current) in the meridional southward direction down to $37°15'$ N; jet (2) propagating along $37°04'$ N in the westward direction, which is indicated by the anticyclonic element of vorticity at the end of the jet; and mushroom-like structure (3). Eddies A and B and jets 1–3 can be related to the mixing zone with consideration of the known average positions of the Kuroshio and Kuril currents. Eddy C is most likely positioned in the Kuroshio current proper, and the cyclonic eddy D southward of Kuroshio.

The northern boundary of Kuroshio can be well discerned in the IR image of 20 May 1981 (see Fig. 5.15) where two anticyclonic eddies A′ and B′ and mushroom-like structures 1′–4′ are also clearly visible. Structures 2′–4′ are a reflection on the satellite image of respective jet flows of relatively cold waters: the flow along the northern edge of Kuroshio, but in the opposite (westward) direction (2′); the waters of the Kuril current propagating to the south seaward of the narrow band of warm waters in the Sangar current along the Honshu coast (3′); and the flow in the SE direction near the Sangar Strait (4′).

Practically all the eddies of different sizes and directions of rotation in both images (Figs 5.14 and 5.15) are manifested on the surface in the form of a spiral of alternating bands of warm and cold water with a width of 10–20 km. Apparently, this spiral-like structure forms in the upper layer of the ocean as a result of entraining waters with different temperatures from the north and the south into orbital vortical motion [103].

Interesting specific features of circulation in the given zone are associated with anticyclonic eddies A and A′, having cold water in the centre on the surface. It is easy to notice that the location and configuration of the combinations of the three principal elements of circulation, namely, A(A′), mushroom-like structure 1(1′), and jet 2(2′), are almost identical in both situations (Figs 5.14 and 5.15). The jet part of the mushroom-like current 1(1′) with a width of 20–35 km in both cases is as if a tangent to the vortical ellipse was applied to a point on its western periphery. The direction of the jet is opposite to the direction of water motion in the eddy. The length (L) and the size of the vortical part (B) of the mushroom-like structures 1(1′) in the images are equal to $L_1 \simeq 140$ km, $L_{1'} \simeq 220$ km, $B_1 \simeq 80$ km, and $B_{1'} \simeq 150$ km, respectively. In both cases, the jet currents 2(2′), propagating westward to a distance of $L_2 \simeq 250$ km and $L_{2'} \simeq 215$ km, begin close to their 'peaks'.

The succession in the formation of circulation elements 1′–3′ can also be determined in the series of IR images [233]. The surface waters in eddy A′ started to whirl into a spiral in the interval between 28 April and 4 May 1981; the origin of jet current 1′ was observed on 13 May; and the mushroom-like structures 2′–4′ appeared in the image on 20 May when the formed mushroom-like current 1′ reached the anticyclonic meander of Kuroshio ($39°$ N). Thus, the formation of jet currents 1′–4′ coincided in time with a certain stage of development of the spiral-like structure of eddy A′ conditioned by entraining waters from the Kuril current. Lacking a similar series of images for May 1984, it can be assumed that the time succession was similar. The most likely reason for the origin of current 1(1′) was entrainment of the Kuril current waters from the

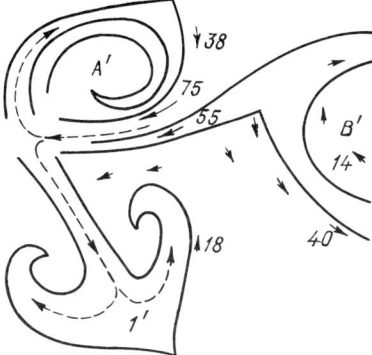

Figure 5.16. Hypothetical diagram of mushroom-like current formation 1(I'). The dotted lines indicate the assumed directions of motion. The small arrows and numbers correspond to the directions and velocities of the currents (in cm/s) obtained on 20 May 1981 by Vastano and Borders (see the discussion in ref. [37]).

East along the southern periphery of anticyclone A(A'), forming an intensive jet flow (its velocity on 20 May 1981 reached 75 cm/s, Fig. 5.16). Inevitable retardation of this flow in the direction of the western boundary of the region (Honshu Island coast) probably in combination with an impulse directed NW to SE (mushroom-like current 4', Fig. 5.15) from the Sangar Strait in a closed space of the given area most likely led to the formation of a gigantic mushroom-like structure. Anticyclone A(A') became its right-hand anticyclonic branch, and the left-hand cyclonic part degenerated into the mushroom-like current 1(1')* (Fig. 5.16). This current 1(1'), reaching the Kuroshio front, in turn resulted in the formation of jets 2(2'), exerting an influence on the flow of the Kuril current waters along the Honshu coast (3'). The propagation velocities of the mushroom-like structures 2' and 3', estimated by us in successive IR images for 20 and 21 May 1981 in ref. [233], are about 25–30 cm/s. Conformably, the velocities of the currents in these structures should be even higher (see Section 5.3.1 and ref. [35]).

Thus, the initial impulse [entrainment of the Kuril current waters on the periphery of anticyclone A(A')] induced in both cases the formation of several interconnected jets of zonal and meridional direction 1–2 (1'–3') with a length of 150–250 km in combination with a series of additional vortices in a closed area between Hokkaido in the north, the Kuroshio front in the south, and Honshu in the west. The 'impulse' nature of their origin is evidenced by the obviously passive character in the redistribution of heat and the plankton concentration in the developing jets and eddies 1–2 (1'–2'), consisting of alternating bands of warm and cold waters (Figs 5.14a and 5.15).

---

*Similar degeneration of one of the 'mushroom' branches into an independent mushroom-like current was observed by us in the 'Meteor' satellite images of the region eastward of Kamchatka (tracer—floating ice) and in a laboratory experiment (see Section 5.3.1).

What typical conditions could be favourable for the repeated formation of such a characteristic combination of jet streams and eddies that was observed in May 1981 and 1984? To note, in both cases (Figs 5.14 and 5.15) the centres of the main anticyclonic eddy were at 144°30′ E, i.e. in the region of the quasi-stationary anticyclonic meander of Kuroshio where the number of observed anti-cyclonic eddies is maximum as confirmed by statistical processing of available data by different authors (see the discussion in ref. [37]). In both cases, eddy A(A′) was close to the boundary between the transformed waters of Kuroshio and the inflowing waters of the Kuril current. Both images were obtained in the second half of May when, on the one hand, the temperature contrasts of the waters in the mixing zone were still very great, and, on the other hand, the seasonal thermohaline began to form. The latter circumstance probably creates conditions for the propagation of jets in a comparatively thin near-surface layer of the ocean, whose thickness is not more than 100–150 m. A similar pattern of interconnected jet streams was not observed in May 1976 when the warm anticyclonic eddy was in the transformed waters of the Kuril current [11].

Chapter 6

The near-surface layer and the effectiveness of remote sensing of the ocean

Modern aerospace techniques of observation and measurement, which are designed mainly for the acquisition of mass information on the ocean, are characterized by direct 'vision' which is restricted to the sea surface or to several metres beneath it. The scientific effectiveness of these means depends on the development of methods for obtaining with their aid information on the in-depth processes and phenomena in the ocean which would be cheaper, more operative, and comprehensive than the information collected by traditional ship-borne methods, and which would possess a number of principally new properties, i.e. a wide field of view, all-weather and systematic character. The contradiction between the tempting technological potentialities and the afore-cited natural restrictions set a large number of complicated methodological problems in the way of wide-scale application of aerospace techniques and means in oceanology. Their great number is connected in one way or another with the near-surface layer. From the point of view of a scientist who wishes to 'look into' the thickness of the ocean waters by means of aerospace techniques, the near-surface layer is the medium through which various representations of the indepth processes are 'projected' on the ocean surface, like on a screen. In a certain sense, the near-surface layer with its specific structure plays the role of a complicated information converter which in some cases amplifies or attenuates this information, and is sometimes absolutely impenetrable or 'opaque' for the information flows that are associated with the phenomena, processes, and effects which we are interested in. This last chapter considers certain properties of the near-surface layer in this context. The range of problems that should be discussed in this connection is so wide that, generally speaking, it should be presented in a separate monograph. Therefore, in this chapter we set the objective of formulating only a series of important questions that would then assist essentially in organizing studies of the near-surface layer of the ocean in the necessary directions.

It would be tempting to write down here a certain transfer function of the near-surface layer of the ocean in a generalized form, as is done in the case of transmitting radiation, which was radiated and reflected by the earth's surface, through the atmosphere (see ref. [79]). In the simplest cases, when the question is of direct visibility of underwater objects, or parts of the water body bottom through the thickness of the water and free surface disturbed by waves (see, for example, ref. [68]), the latter approach is possible and is even dictated by the physical nature of the effects occurring. However, the actual situations are much more complicated in the majority of other cases, and the similarity with an information converter or transmission of a useful signal, despite its clearness, is rather of metaphoric than of profound physical sense. First of all, practically all the kinds of radiation recorded by remote aerospace devices, with the exception of radiation in the visible range of the spectrum (the colour of the sea), are formed not by the water thickness, wherein the processes and phenomena that we are interested in develop, but in a very thin water layer very close to the surface. Conformably, a correct formulation of the problem should consist, in all these cases, in studying the effects that the near-surface layer of the ocean with its structure and variability renders not on the transmission, but on the formation of a signal from which it is intended to derive useful information. Secondly, the formation of a useful signal is most often directly influenced not by the process that occurs in the thickness of the ocean waters and which should be studied, but by some other process in the near-surface layer, having some well or poorly known bonds with the process being studied. Thirdly, the problem becomes additionally complicated by the differences in many of its bonds that prevent a generalization and often make it necessary to consider each concrete situation independently.

Examples of all the above-cited difficulties are vividly demonstrated by analyzing here such a well-known process as sea waves. Wind waves of force more than 5–6 effectively mask, for example, visible surface manifestations of fronts, Langmuir circulations, and internal waves [103]. In contrast to strong waves, light waves and ripples often stress the visible contrasts of roughness associated with the surface manifestations of the ocean water dynamics mentioned earlier. When visually observing objects on the bottom through the water thickness and a free surface, any waves and even ripples represent a natural obstacle. On the other hand, it would be practically impossible to detect the emergence of internal waves to the sea surface if there were neither ripples nor waves. It is the surface waves in a certain range of wavenumbers, modulated by internal waves penetrating into the near-surface layer from a deeper stratified thickness of the ocean waters, which exert an influence on the character of light reflection and scattering in the visible range of the spectrum, or on the radio emission of the surface, and the resulting optical or radio brightness is then perceived as signals from which it is necessary to isolate useful information on the internal waves [85, 122, 213]. At least two levels of ties are found in this case between the recorded brightnesses and the internal waves studied, namely, (1) interaction between the surface wind waves and the internal waves and (2) the influence of the characteristics of the surface waves on the properties of radio emission, or on the reflection and scattering of light from the ocean surface. Suspended matter in the waters of the near-surface layer is another interesting example. It contributes to the absorption and scattering of the light transmitted through the layer and, undoubtedly, is an obstacle for direct underwater vision. On the other hand, being redistributed by currents or waves (surface or internal waves), suspended matter of organic and inorganic origin can be a good passive tracer, making it possible to obtain from satellites, aeroplanes, and even ships useful information not only on the dynamics of the waters in the near-surface layer (see ref. [100] and Sections 2.3 and 6.2), but also on the in-depth topography of the bottom, exerting an influence on the internal waves and currents. The ice cover can also play a similar double role. It usually effectively prevents the recording of radiation from the water surface and any visual observations of phenomena hidden beneath it. However, in some cases the ice cover itself can carry information on these phenomena. Thus, the distance between the cracks in the ice cover, parallel to the ice edge in the zone where the ice is already weak due to melting, corresponds to the lengths of the internal waves propagating under the ice in the marginal ice zone [193]. At the same time, broken ice in the ocean is always a good tracer of currents in the near-surface layer.

Hence, it can be stated that the near-surface layer of the ocean, owing to its structure and variability and the processes occurring in it, can influence in many ways the fields of the radiated, reflected, and scattered brightnesses in the radio, thermal, and visible ranges of the spectrum. All these influences can be schematically presented as:

- weakening and suppressing effects;
- masking or noise effects;
- manifesting or amplifying effects.

Concrete examples of these effects will be discussed in the subsequent sections of this chapter when considering the problems of the oceanological information content of SST fields, fields of concentration of passive tracers, and characteristics of sea waves.

6.1. On the oceanological information content of the temperature fields of the surface and near-surface layer of the ocean

Perhaps the most effective mechanism of interconnection between the UQL and the underlying stratified thickness is the turbulent entrainment into the UQL and the near-surface layer of the horizontal temperature gradients due to the geostrophic relief of the seasonal thermocline. As shown in Section 5.1 [see equation (5.23)], in the process of geostrophic adaptation of the velocity and mass fields disturbances develop in the lower boundary of the UQL η'_g which are approximately $R_B^2/L^2 \simeq 100\text{–}500$ times greater than the disturbances ζ'_g in the free level surface. The lower boundary of the UQL should have a quite sharply expressed relief with an amplitude $\eta'_g = 10\text{–}50\,\text{m}$ on the same horizontal scales $L = 100\,\text{km}$ at $\zeta'_g = O|10\,\text{cm}|$ which is typical for the open ocean (beyond the limits of jet boundary currents). The depth of penetration of wind–wave and convective mixing from the surface into the thickness of the ocean waters is associated, first of all, with the flows of buoyancy and impulse through the free surface and does not depend directly on the formation of the UQL lower boundary geostrophic relief. As a result, where the lower boundary of the UQL is closer to the surface due to this relief, the cold waters of the seasonal thermocline are entrained more intensively into the UQL (see discussion on

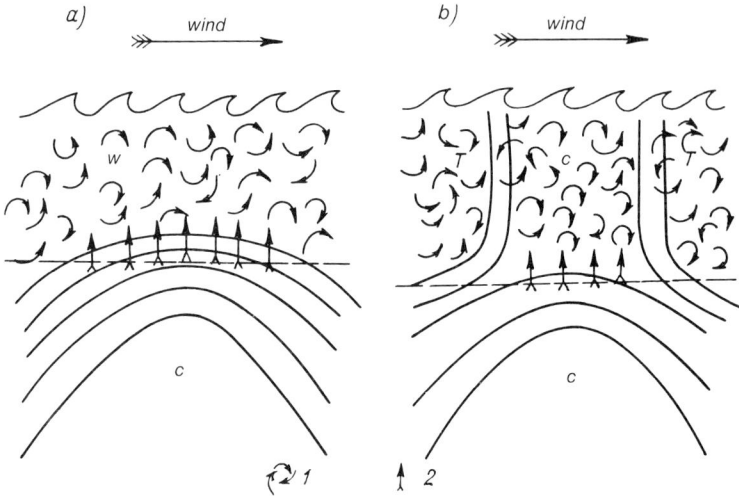

Figure 6.1. Diagrammatic representation of turbulent entrainment on the UQL lower boundary, having a considerable relief of geostrophic origin. (a) At the beginning of intensive wind mixing; (b) some time later. The dotted line shows the conditional lower boundary of the layer with intensive turbulence (turbocline). (1) Turbulence; (2) direction of entrainment.

turbulent entrainment [6, 14, 20] and Section 4.4), and horizontal temperature gradients develop in the UQL. If the horizontal scales of the mixing processes are considerably greater than L^*, then the space pattern of horizontal temperature inhomogeneities, developing in the UQL, should be quite an adequate plane reproduction of the seasonal thermocline geostrophic relief (in character resembling the reproduction of some locality relief on a plane topographic map). This process is shown diagrammatically in Fig. 6.1.

The inhomogeneity of the UQL by the horizontal in comparatively small space scales was revealed for the first time by Stommel and Fedorov [223], and the causes of its origin were discussed in the works of the same authors [94, 99]. In principle, the same combination of turbulent entrainment, induced by forcing from above with upwelling developing from below, is in the base of the development of a cold trace following tropical cyclones or typhoons on the ocean surface [113] (see also Section 4.5.1). The manifestations of all these inhomogeneities in the temperature field of the near-surface layer and in the SST field are connected with the diurnal cycle of convective mixing (see Section 4.3). Subsequent evolution of the developing horizontal inhomogeneities in the SST field is conditioned by the currents in the near-surface layer which are also under the influence of the diurnal cycle [243] and, generally speaking, does not coincide with the currents of the seasonal thermocline and deeper layers (see Sections 5.1 and 5.3). Sometimes, the waters in the near-surface layer are entrained by the geostrophic currents in the underlying gradients, which initially

*Which is most likely because atmospheric cyclones and anticyclones have characteristic diameters of 1000–3000 km.

could have no relation to the in-depth circulation, stratified waters; then near-surface horizontal temperature reconstruct under the effect of advection, as shown in Section 5.3.2 (see also ref. [207]). This is how the near-surface fronts of the shelf and slope waters, forming in the SST field characteristic tongues which create a visible image of vortical circulations, are entrained into the vortical circulation of the Gulf Stream rings [103].

At the same time, the formation of own space structures, e.g. inhomogeneities of kilometre scale connected with the internal waves, in the near-surface layer during the periods of solar heating can perform the role of 'noise', masking the useful signal from the geostrophic relief of the UQL lower boundary. Daily heating can in general screen off the useful thermal pattern, making it inaccessible for detection by photography in the IR range or by measurements with a towed thermal sensor close to the surface. Attenuation and distortion of the horizontal gradients of density, temperature, and salinity in the near-surface layer, associated with geostrophic motion in the deeper layers, may be induced by certain other effects characteristic only of the near-surface layer. Situations are possible when the near-surface variability (e.g. vast freshening areas) can fully eliminate manifestations of the in-depth geostrophic effects close to the surface and even result in a change of their sign to the opposite one. The effects of the different bonds of the individual specific features in the SST field with the character of motion in the thickness of the waters can also be of dynamic origin inherent of the geostrophic balance proper. This situation is characteristic of intrathermocline eddies, developing close to the surface, e.g. in the zones of coastal upwelling [57, 218] on lenses of transformed upwelling waters that displace away from the coast in the upper part of the seasonal pycnocline at the level of its density. They are manifested in the thermal field of the surface as cold patches, wherein the negative SST anomaly can reach $-1°C$. By analogy with common synoptic eddies and rings in the northern hemisphere it would seem possible in these cases to anticipate cyclonic rotation of the waters around a 'cold' core. However, the dome-shaped bend of the isopycs in the near-surface layer of the ocean, which is a mirror image to the cup-shaped bend in the underlying pycnocline, ensures only some diminishing of the orbital anticyclonic velocities, whose maximum values are observed on the intermediate horizon upon approaching the surface. This example demonstrates that it is sometimes possible to make a 180° mistake in the direction of the current when diagnosing the currents by the closed ring-shaped isotherms in the SST field.

A question of principle arises in connection with the above-discussed mechanisms, namely, how well does the SST field* reflect or transmit the most important details of the space thermal structure of the underlying layers? Certain qualitative ideas concerning this point have been given earlier (see p. 37 in ref. [103]). However, there are very few quantitative estimates in the literature

---

*Bearing in mind the objectives of this section and the obvious scarcity of adequate field data, we will consider below all the available types of information on SST as equivalent and comparable to one another despite the fact that we now know essential differences and uncertainties connected with the applied methods of measurement, both remote and contact (see Sections 3.4–3.6).

relating to the different aspects of the given question. In particular, it is not quite clear how the depth, where a correlation between the structural details of the SST field and the structural specific features of the temperature fields in the deeper layers practically disappears, depends on various hydrometeorological and other conditions. It is this dependence that should show how deep it is possible to 'look' into the thickness of the ocean waters under various conditions by analysing images or SST fields obtained by means of satellite IR radiometers.

An analysis of a large number of satellite IR images of the Alboran area in the Mediterranean Sea carried out in July 1980 showed [129], for example, that the well-known gyre in this basin was observed in 38 of 42 cases. There were no thermal manifestations of the gyre during days with a light wind up to 4 July 1980, and good 'visibility' started later when the wind intensified to 12 m/s. During days with a light wind, the Alboran gyre was not observed on the surface, but it was well manifested in the temperature field on the horizon 100 m. It is obvious that wind–wave vertical mixing, induced by intensification of the wind, results in sharpening of the horizontal thermal gradients in the UQL in conformity with the above-given diagram, which, in turn, ensured contrast 'visibility' of the gyre in the thermal field after 4 July 1980. If this is so, then it turns out that the cumulative effect of solar heating and layering of the near-surface waters in light wind summer weather at the scale of the natural synoptic period (3–7 days) can fully prevent the emergence on the surface of even quite strong and large-scale thermal contrasts when already on the horizon 100 m they are connected with the circulation of the water in an entire sea basin of diameter about 150 km. Hardtke and Meincke [154] studied a similar effect in the North Atlantic in June–September 1981 and connected it with the suppressing role of the seasonal thermocline located within the limits of the upper 50 m layer of the ocean. The results of their measurements showed that the specific features of the thermal structure in the deeper layers, conditioned by currents, are well manifested in the SST field and in satellite IR images respectively at vertical temperature gradients $\partial T/\partial z$ of about 0.05°C/m in the seasonal thermocline, and are practically not manifested at $\partial T/\partial z \simeq 0.25$°C/m. Apparently, the suppressing threshold is somewhere in between these two values. The diurnal thermocline, in which the vertical temperature gradient may exceed 1°C/m during the day in calm sunny weather, should have a very strong 'suppressing' effect in relation to the thermal and dynamic effects of the geostrophic currents that develop in the seasonal thermocline. On the other hand, any closed shallow circulation developing within the limits of the near-surface layer will modulate its thickness, redistributing the absorbed solar heat, and will create horizontal temperature contrasts which can act as indicators of the latter near-surface circulations. Thus, shallow near-surface currents are best seen in satellite IR images in the case of solar heating in calm or light wind weather.

As far as the layer thickness, wherein the masking or suppressing effect develops, is concerned, it may be different depending on whether there is the effect of a recurring diurnal cycle or heat is accumulated in the near-surface layer in the synoptic or seasonal time scale. It is possible to anticipate in summer the thinnest layers, which are arranged close to the surface, i.e. sources of thermal noise at a maximum space–time amplitude of the temperature noise fluctuations under conditions of precipitation or a river discharge freshening effect. Typical thicknesses and characteristics of heated layers in the diurnal cycle and under freshening conditions have been discussed in Sections 4.3 and 4.5. This question was the objective of special studies in the MARSEN experiment [168] carried out in the North Sea in August–September 1979 in the area of the Elbe discharge. A sharp diurnal thermocline and space patchiness of the temperature field in light wind weather developed above horizon 10 m, preventing the appearance of a sharp thermal front, bound with the Elbe discharge, on the satellite IR images. Only an increase in the wind velocity to more than 20 m/s, which occurred twice for 3–5 days in the 1-month period of the observations, improved the visibility of the front in the IR images.

The information content of the SST field from the point of view of manifestations of mesoscale and synoptic (as applied to the ocean) kinematics of the deeper water layers under common conditions of the open ocean and in the absence of freshening effects should depend on several factors, including the following:

- the presence (or absence) of strong geostrophic currents, fronts, eddies, or upwellings in the region being studied;
- the time of year; and
- the time of day.

The degree of influence of each of these factors can be estimated only on the grounds of mass analysis of the correlation bonds between the SST fields and the temperature fields on different deep horizons. There are numerous examples of the correlation analysis of SST variability in relation to the temperature variability in deeper layers. But analysis of the time fluctuations, recorded on one vertical and even if it repeats on several adjacent verticals, does not give an answer to the formulated question. Data are necessary on the correlation between the space structures of the temperature field observed on different horizons. In principle, correct analysis demands temperature fields that are averaged for each level for the same simultaneous periods of time of at least several hours to eliminate the noise signal of the internal waves from the cor-related data, which is especially strong on the horizons in the thermocline. However, oceanographers currently have no methods which would allow such data to be obtained. Hydrological surveys, performed from one or even several ships, are always asynchronous, and the required averaging of the SST fields is possible only on the basis of satellite data obtained from geostationary satellites of the METEOSAT type with quite poor space resolution, and even then an ideal cloudless sky is pertinent. Therefore, it is necessary to use the results of common ship-borne hydrological surveys without averaging, and to take either the data of the uppermost bathometer, relating to the same surveys, or the data of the upper horizon surveyed by an CTD probe, or the results of measuring

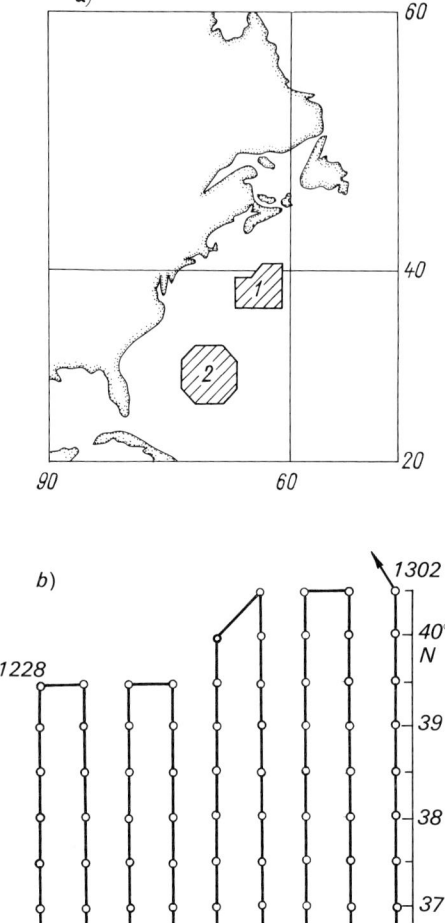

Figure 6.2. Regions of large-scale repeated hydrological or CTD probing 'RAZREZY' (1) and POLYMODE (2) in the North Atlantic (a); typical example of the arrangement of stations during one of the probings at the 'RAZREZY' polygon (b).

with a towed temperature sensor simultaneously with surveying, or instantaneous SST maps, recovered by one-time satellite data, for the SST fields.

The results of a correlation analysis of space ship-borne temperature surveys with a close arrangement of the stations are given below. The surveys were made during the POLYMODE (1977–1978) and RAZREZY (1984–1985) experiments (the arrangement of the surveyed regions 1 and 2 is shown in Fig. 6.2). The hydrological surveys in both regions covered areas of about 450×450 km with a distance of 20–30 miles (37–55 km) between the stations.

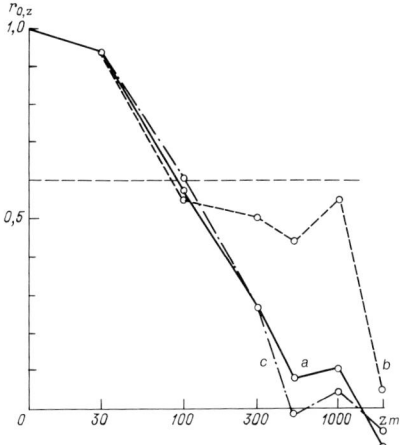

Figure 6.3. Variation of the correlation coefficient $r_{0,z}$ with the depth z. Region 2, Sargasso Sea, winter. (a) All data; (b) night-time stations 0000–0700 LST; (c) daytime stations 1100–1800 LST.

Naturally, each survey took several weeks. Neglect of the requirements of simultaneousness and averaging gives the character of estimates from below to our results owing to the inevitable distorting influence of the internal waves on all the horizons included in the diurnal, seasonal, and main thermocline. The results of the analysis (the values of the correlation coefficient $r_{0,z}$) are presented in Fig. 6.3 (Sargasso Sea, region 2: 119 stations, February 1977, 26th cruise of the research vessel *Akademik Kurchatov*), Figs 6.4a and 6.4b ('Gulf Stream' polygon, region 1: 119 stations, stage I, January 1984, and 143 stations, stage III, February–March 1984, fifth cruise of the research vessel *Vityaz*, and 38th cruise of the research vessel *Akademik Kurchatov*), Figs 6.5a and 6.5b ('Gulf Stream' polygon, region 1: 75 stations, August–September 1984, 40th cruise of the research vessel *Akademik Kurchatov*), and Figs 6.6a and 6.6b ('Gulf Stream' polygon, region 1: 81 stations, August–September 1985, tenth cruise of the research vessel *Vityaz*). The curves in each figure are divided on the basis of either the time of day (stations in the daytime at 1100–1800 LST, during the night from 0000 to 0700 LST, and all the stations from 0000 to 2400 LST), or the survey areas (where it is possible to consider separately the Gulf Stream jet, the region north of it, the Sargasso Sea region south of the jet, and the field as a whole, including the Gulf Stream together with the adjoining peripheral regions). The main conclusions, drawn on the basis of the clearest specific features of the graphs given in Figs 6.3–6.6, are reduced to the following.

(1) The coefficient of correlation $r_{0,z}$ diminishes with depth faster in the calm regions of the ocean, e.g. in the Sargasso Sea (Fig. 6.3) than in the same season in regions with strong jet currents of the Gulf Stream type (Fig. 6.4a). The values of $r_{0,z}$ for horizons below 100 m in the region of the 'Gulf Stream' polygon even in summer were substantially greater than in the Sargasso Sea. If the level of significant correlation $r_{0,z} = 0.6$ is adopted,

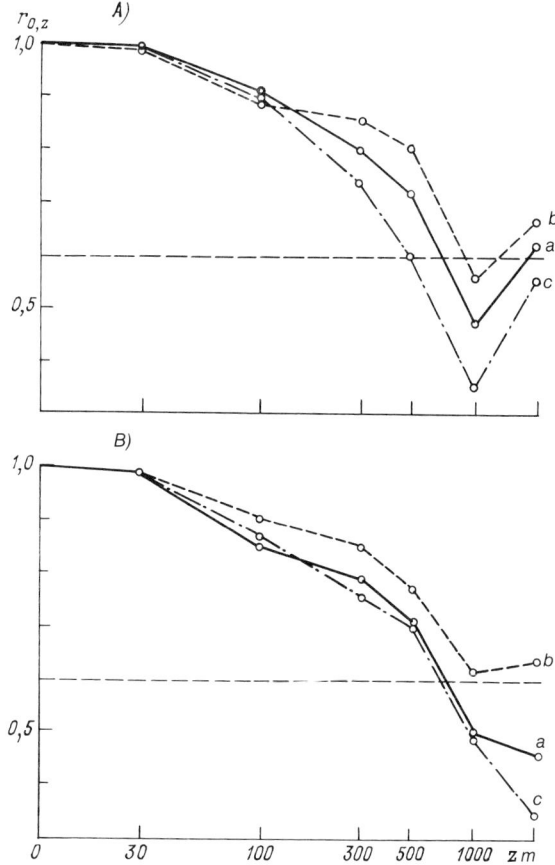

Figure 6.4. Variation of the correlation coefficient $r_{0,z}$ with the depth z. Region 1, Gulf Stream, winter. (A) January 1984; (B) February–March 1984. Designations of curves as in Fig. 6.3.

which covers with guarantee all the data samples at 95% provision, then far away from the Gulf Stream in the Sargasso Sea even in winter this level is on a horizon of about 90 m, while in the 'Gulf Stream' polygon in summer it occurs deeper (≈ 100 m in 1985 and $\simeq 380$ m in 1984) and in winter it decreases to the horizon $\simeq 700$ m and deeper.

(2) Conformably, the coefficient of correlation $r_{0,z}$ diminishes everywhere with depth in winter much slower than in the same region in summer.

(3) The diurnal heating of the near-surface layer qualitatively performs the same masking role as the seasonal one, though certainly the quantitative effect of diurnal heating is much less significant. The coefficient of correlation $r_{0,z}$ decreases with depth in one and the same region in summer and winter much faster during the day than during the night (Fig. 6.6a). It is evident that diurnal heating exerts a direct effect on the layers between 0 and 30 m which diminishes greatly the values of $r_{0,z}$ on all z beginning

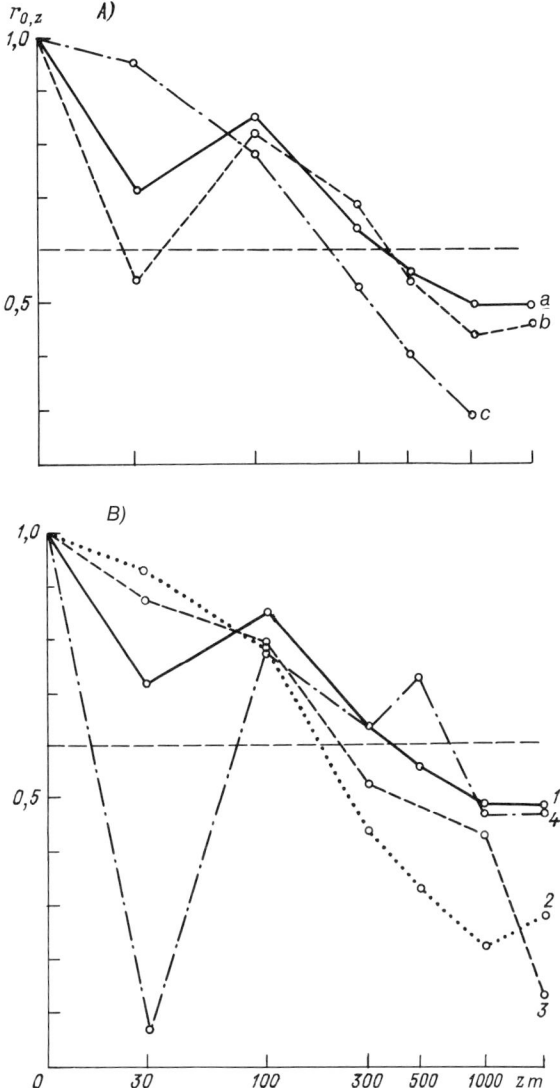

Figure 6.5. Variation of the correlation coefficient $r_{0,z}$ with the depth z. Region 1, Gulf Stream, summer (August–September 1984). (A) Differences in time of day (designation of curves as in Fig. 6.3); (B) differences in sub-regions: (1) whole field; (2) in the Gulf Stream; (3) south of the Gulf Stream; (4) north of the Gulf Stream.

with 30 m. In winter, as is shown in Figs. 6.4a and 6.4b, the values of $r_{0,z}$ differ between the day and night hours by not more than 0.05 within the upper 100 m, while the difference increases with depth, reaching 0.15–0.20 for $r_{0,300}$, $r_{0,500}$ and $r_{0,1000}$, and the diurnal correlation is always smaller than the nocturnal one. The role of the UQL, whose thickness in winter exceeds 100 m, is clearly observed here. The conditional threshold of a

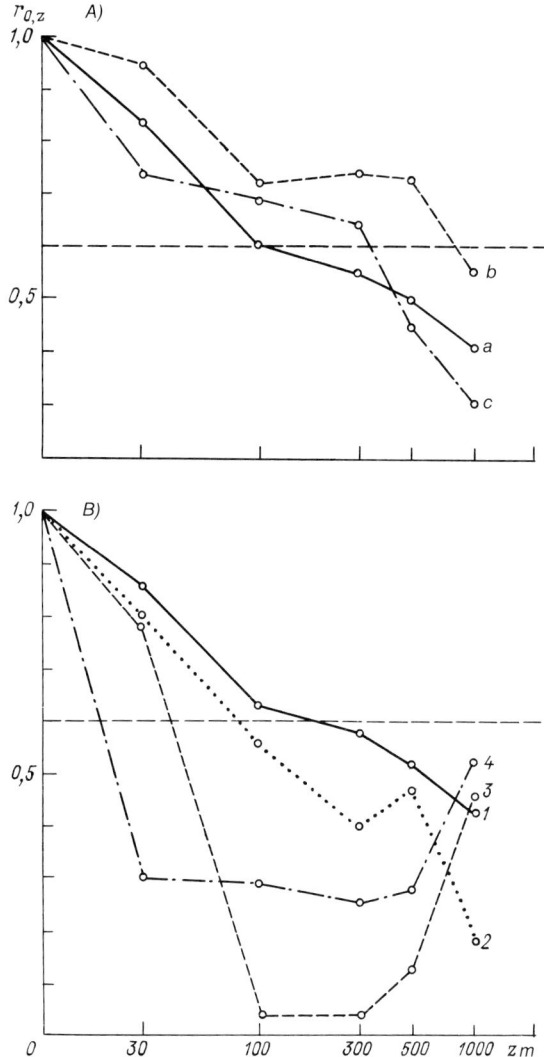

Figure 6.6. Variation of the correlation coefficient $r_{0,z}$ with the depth z. Region 1, Gulf Stream, summer (August–September 1985). (A) Same as in Fig. 6.5A; (B) same as in Fig. 6.5B.

significant correlation ($r_{0,z} = 0.6$) in summer is on the horizons between $\simeq 300$ m (in the daytime) and $\simeq 800$ m (at night), while in winter it lowers and fluctuates between $\simeq 500$ m (in the daytime) and 2000 m (at night). In the most extreme case for night stations, $r_{0,z} > 0.6$ even on horizon 2000 m in February–March 1984 (Fig. 6.4b).

The division of the areas covered by the surveys in the region of the 'Gulf Stream' polygon into subregions with different characteristics of the field of curents (Figs 6.5b and 6.6b) revealed additional consistencies in the links of the SST field with the temperature fields on the underlying horizons.

The values of $r_{0,z}$ decrease with depth most slowly in the Gulf Stream proper (as was to be expected) and when considering all the survey stations without division into subregions. The latter is also quite logical because the Gulf Stream is for the region as a whole the specific feature that creates a coherent trend in the space variability of the temperature. The sharp diminishing with depth of the $r_{0,z}$ values in the Sargasso Sea (in the south) and the North Atlantic (in the north) neighbouring the Gulf Stream leads to the fact that the conditional level of a significant correlation ($r_{0,z} = 0.6$) occurs there in summer on horizons 50 and 20 m, respectively (Fig. 6.6b). The latter confirms once more the first conclusion made by us. However, we encountered an interesting peculiarity during the summer survey of the 'Gulf Stream' polygon in 1984 (Figs 6.5a and 6.5b), i.e. local diminishing of the correlation on horizon 30 m and the recovery of the normal pattern of diminishing $r_{0,z}$ with depth on the lower horizons. Careful analysis of the $r_{0,z}$ behaviour in the subregions (Fig. 6.5b) demonstrated that the decorrelating contribution on horizon 30 m is introduced by the stations that were arranged on the northern periphery of the polygon inside the cyclonic meander of the Gulf Stream. There, on horizons close to 30 m, cold waters rising to the surface were observed, differing by 7–8°C from the waters of the Gulf Stream on the same horizons. Apparently, some of the stations included in Fig. 6.5a in the series of stations sampling the Gulf Stream were arranged too close to the sloping front and included also these cold waters on horizons close to the surface.

It should be added that the variability of the Gulf Stream northern boundary position in the region of the 'Gulf Stream' polygon with a period of about 2 weeks (from satellite data) is very highly correlated ($r \simeq 0.6$–0.7) with the variability of the current velocity (measured by instruments) not only in the entire thickness of the Gulf Stream, but also 2 m above the bottom on horizon 5000 m [170]. As certain changes in the SST are always associated with the variability of the Gulf Stream northern boundary position, it can be considered, in principle, that the SST field variability (especially in winter) in this region contains latent information on the variability of the currents in the entire 5 km thickness of the waters from the surface down to the bottom.

The results obtained above can be compared with other similar estimates in the literature [207, 232]. In a special experiment, FRONTS-80, carried out in winter in the region of subtropical convergence of the northern hemisphere in the Pacific Ocean [232], the SST field recovered by satellite IR data was compared with the temperature fields measured by contact methods on different horizons, including the ocean surface. The hydrological survey, used for comparison, was made in a 250×250 km square with a distance of 37 km (20 miles) between the stations. The coefficient of correlation between the satellite and contact patterns of the temperature distribution on horizon $z = 0$ m was 0.88 and remained just as high down to a horizon of 100 m (Fig. 6.7). The value $r_{0,z} = 0.55$ (95% provision and ten degrees of freedom) was adopted in ref. [232] as the threshold level of a significant correlation, because it was recognized that

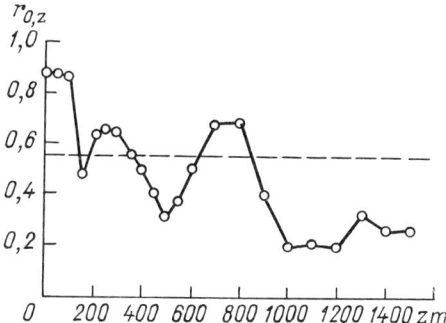

Figure 6.7. Variation of the correlation coefficient $r_{0,z}$ with the depth z in the region of subtropical convergence in the northern hemisphere in the Pacific Ocean according to the data in ref. [232]. The SST field recovered by satellite data was adopted for the correlated field on the horizon $z = 0$.

the temperature fields on the different horizons were not fully independent of one another owing to the presence of a determinate field of geostrophic currents (front meander) in the region surveyed. We took practically the same level of significance which was rounded to 0.6 because of the roughness of the estimates. Figure 6.7 shows several layers in which the coefficient of correlation falls below the level of significance. The first one is the depth of 150 m and is apparently bound with vertical displacements of the UQL lower boundary located just on this depth due to internal waves. In addition, warm intrusion with temperature inversion, which could contribute to decorrelation, was arranged often at the UQL base. The second minimum of correlation at a depth of 500 m was also connected with advection of alien waters from aside. The correlation disappeared completely below 800–850 m. In another case [207], the correlation of temperature fields connected with the transversal filaments of the cold waters in the California upwelling was studied (see also Section 5.3.2). The correlation of the SST field, obtained by contact measurements, was maintained at a level exceeding the minimum significant (in the given case 0.35 at 95% provision and 30 degrees of freedom) only down to horizon 40 m, i.e. by the thickness of the filaments. A high correlation of the SST field was observed at a greater depth with the field of the temperature horizontal gradients of the underlying stratified thickness down to a horizon $\simeq 300$ m, demonstrating that the coastal cold waters of upwelling origin are entrained towards the open ocean by the underlying jet geostrophic current which is well expressed in the seasonal thermocline. This example, as well as others given above, confirms once again the fact mentioned many times in this book that the SST field contains important information on the currents in the underlying layers. This information can be obtained with success in many cases, as will be shown below.

6.2. An estimate of the velocity field characteristics by a passive tracer observation in the near-surface layer of the ocean

Suspended matter, plankton (photosynthetic pigments), floating ice, and the temperature are natural tracers whose characteristics can be observed and/or measured by satellites. Spectral optical brightness measured by satellite instruments, certain derivative indices (e.g. 'colour index'), as well as the recovered values of concentrations and temperature can be quantitative measures of these tracers. In the general case, the variations of these values in time contain information of several hydrophysical fields simultaneously, namely, on the field of currents, performing advection of scalars; on the coefficient of diffusion, which scatters the particles or clusters of particles; on the field of admixture or heat sources; on stratification of the near-surface layer of the ocean; and, finally, on some initial distribution of the tracer concentrations that might not have been recorded, as could have occurred during the time period between respective satellite measurements (or images). The recovery of one or several of the stated fields by the succession of satellite measurements or images is a significant hydrophysical problem. This problem should be correctly formulated and correctly solved so that the interpreter does not adopt erroneously a random mixture of various effects for the effect of any one field, e.g. for advection of scalars in the field of currents.

Particular cases are certainly possible when the tracer possesses such properties that observing it makes possible estimation of the characteristics of a certain field. For example, the drift of a float with a radio transponder, an individual conspicuous ice-floe, or an iceberg makes it possible by the succession of positions over intervals of time Δt to obtain Lagrange fields of velocity. However, the situation is more complicated for the majority of natural tracers than this example, wherein the characteristic size of the single element of the tracer (buoy, ice-floe, iceberg), discernible from a satellite, was much smaller than the characteristic size of inhomogeneities in the field of currents. In the majority of cases, the single elements (particles) of the tracer cannot be distinguished from the satellite. In these cases, we deal with a random or ordered pattern of tracer element aggregates characterized by one or the other concentration (brightness), varying under the influence of elements (inhomogeneities) in the field of currents, whose characteristic size is commensurable to the size of the tracer aggregates. In another case, the situation is possible when the elements of the field of currents (e.g. turbulent eddies) are much smaller than the aggregates of tracer particles. In this case, we deal with a pure case of diffusion of the tracer aggregates or patches. Therefore, an adequate analytical model of a real field of currents can be obtained by presenting it in the form of superposition of the determinate component, performing transport of the inhomogeneities in the scalar field, and the fluctuating component responsible for turbulent diffusion. It is possible on this basis under certain assumptions to estimate the smoothed-in-time determinate components as well as (and in some cases only) the space statistical characteristics of the fluctuating component. The solution of this problem in a particular as well as in a general formulation is also possible in the presence of other factors, inducing a variation of the tracer concentration. The vector of the surface current was recovered, for example, in ref. [211], by successive maps of SST isotherms and neglecting the heat flux at the ocean–atmosphere interface. The question is discussed in ref. [178] on the correspondence of the slope of the space spectrum of the quasi-geostrophic velocity field to the slope of the space spectrum of plankton concentrations, the plankton being a passive tracer. The problem is considered in ref. [86] in a general formulation as applicable to problems of remote sensing of currents in the ocean on the basis of describing the evolution of the passive scalar (tracer) concentration field $C(t, \mathbf{x})$, $\mathbf{x} = (x_1, x_2)$ with the help of the stochastic differential equation

$$\partial C/\partial t + \mathbf{u} \cdot \nabla C + \lambda C = \chi \nabla^2 C + I(t, \mathbf{x}), \tag{6.1}$$

where $\mathbf{u} = \mathbf{u}(t, \mathbf{x})$ is the vector of the current velocity on the ocean surface, ∇ is the gradient operator in the horizontal plane, ∇^2 is the Laplacian operator, χ is the coefficient of molecular or turbulent diffusion, $I(t, \mathbf{x})$ is the field of distributed sources, and λ is the constant coefficient (feed-back factor). When $C(t, \mathbf{x})$ is the SST field, the value I is the total flux of latent, sensible, and radiating heat normed by the UQL thickness, and λC is the linear parametrization of the heat-exchange rate through the boundaries of the area considered. In the case of a biological population (plankton), I is the birth rate or the rate of supply of new specimens from the deep layers and λC is their death rate. In the general case, the fields $\mathbf{u}(t, \mathbf{x})$ and $I(t, \mathbf{x})$ are assumed to be random, and the problem is to estimate the statistical characteristics of vector $\mathbf{u}$ by observations of the field $C(t, \mathbf{x})$ and some *a priori* information on the source. Hence, the matter is reduced to the solution of an inverse problem of scalar transport. To note, in reality the values λ and χ change in space and in time. Consideration of this dependence may be the objective of subsequent studies.

The given scheme includes the aforementioned simple situation when only single tracers are observed (buoys, ice-floes, etc.). In this case, there is neither a source nor dissipating terms in equation (6.1), which saves us from having to transfer to equations for scalar fluctuations, as is required in more complicated cases [86].

It is useful to consider as a simple example of applying the approach developed in ref. [86] the case when only one component of the velocity, e.g. u_1, differs from zero, and the field of concentrations reaches a stationary state (i.e. ceases to vary in time). We will consider the non-random source I and the diffusion term as assigned, which means that the right-hand side of equation (6.1) is known. In this case, the analogue of the autoregression equation, written for the observed values (because a determinate case is considered), after all the transformations [86] is as follows:

$$u_1 \nabla_{x_1} C = I + \chi \nabla^2_{x_2} C. \tag{6.2}$$

Integrating (6.2) by x_2 within the limits of the stream of width D, and assuming that the transversal gradient of concentration ∇_{x_2} adopts values on the lateral frontal boundaries of the stream equal in absolute magnitude but opposite in sign, we obtain

$$\bar{u}_1 = \frac{\bar{I}}{\overline{\nabla_{x_1} C}} + \frac{2\chi \, \nabla_{x_2} C}{D \, \overline{\nabla_{x_1} C}}, \tag{6.3}$$

where $\bar{u}_1, I$, and $\overline{\nabla_{x_1} C}$ are values averaged by the stream width.

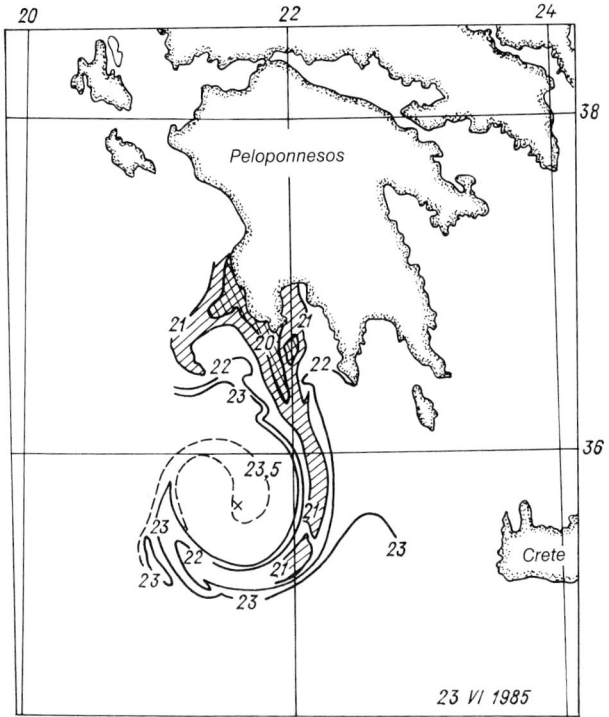

Figure 6.8. Interpretation diagram corresponding to the IR image in Fig. 4.7. The position of the isotherms corresponds to the gradations of brightness in the IR image. The actual values of the temperature variation limits are taken from the map published in the same issue of the *SATMER* bulletin [136].

Equation (6.2) and estimate (6.3) can be obtained also from the conservation equation. The objective of this reasoning is to show that the approach suggested in ref. [86] encompasses most different situations, beginning with determinate and ending with stochastic ones.

Let us apply formula (6.3) to estimate the velocity of the azimuthal current by the temperature field* in the quasi-stationary anticyclonic eddy in the Mediterranean Sea southward of Peloponnese by a series of satellite images obtained in June 1985 [136] (see Section 4.5.1). These images demonstrate that a stationary thermal regime characterized by stability of the isotherms (see Figs 4.7 and 6.8) formed after 19 June in the propagation of cold waters on the periphery of the eddy due to compensation of the inflow of cold waters ($\simeq 20°\mathrm{C}$) from the coast of Greece by mixing them with the surrounding warm waters ($\simeq 24°\mathrm{C}$) under conditions of diurnal solar heating. The images in ref. [136] clearly show that a jet of width $D \simeq 20\,\mathrm{km}$ with a longitudinal temperature gradient different from zero (here axis x_1 is directed along the current by the circumference of the eddy, and axis x_2 across the jet) extends approximately

---

*Below we will preserve the same symbol C for the temperature and the scalar concentration.

three-quarters of the circumference length whose diameter is about 100 km. Dividing the full temperature difference ($\simeq 4°C$) by this distance, we obtain $\overline{\nabla_{x_1} C} \simeq 1.7 \times 10^{-5}°C/m$. The estimates of ΔQ_D (see Section 4.1) for June at latitude $\varphi = 35°N$ with consideration of the heating of the upper layer of thickness 20–30 m from December to June from 14 to 24°C give an average diurnal accumulation of heat of about 6.7 MJ/m² (160 cal/cm²) due to solar heating. Adopting the thickness of the layer of cold waters entrained into vortical motion $h = 20$ m, we obtain $I = \Delta Q_d/(h\rho c_p) = 0.9 \times 10^{-6}°C/s$ ($\simeq 0.08°C/day$). The temperature gradient $|\nabla_{x_2} C|$ on the two lateral frontal boundaries of width $B \simeq 1$ km can be estimated on the average as $|\nabla_{x_2} C| \simeq 2 \times 10^{-3}°C/m$, whence we obtain $\nabla_{x_2} C/\overline{\nabla_{x_1} C} \simeq 118$. We take $\chi = 25$ m²/s as a typical value of the coefficient of lateral turbulent heat exchange at fronts with an effective width of 1 km (see p. 90 in ref. [103]). Then substituting all the above-mentioned values in (6.3), we obtain $\bar{u}_1 = 0.35$ m/s. This value exceeds greatly the 'visual' estimate of the velocity (0.17 m/s) which could be obtained directly by displacing the boundary of the coldest waters along the jet during the first 19 days of June when the arrangement of the isotherms was not yet stabilized. This was to be expected because it is impossible to consider 'by eye' the thermal processes associated with mixing and solar heating.

The algorithms for recovering the velocity field or its statistical characteristics (when it is impossible to determine the determinate part) by observations of the passive scalar field variations have as yet been only partially and indirectly verified. Full verification of the aforementioned procedures and their respective correction are possible only when results are obtained of lengthy simultaneous measurements of the velocity field and passive scalar concentration field. Apparently, it would be very useful to verify the suggested patterns of calculation by statistical simulation because it is very difficult to obtain such data in a field or laboratory experiment.

6.3. On the information content of the characteristics of surface waves

As follows from the materials discussed above (see Chapter 2), the characteristics of the surface waves can contain useful information on the currents close to the surface, fronts, eddies, internal waves, convective cells; on the specific features of the bottom topography; and apparently even on the stratification of the near-surface layer. Waves of different lengths experience different influences of the disturbing factors. The best information on currents, whose typical velocities in the ocean vary from 0.1 to 1.5 m/s, is contained in the characteristics, for example, of surface waves, whose phase velocities are within the same limits which is bound with the most noticeable modulation of the characteristics of just these waves by the current. Such phase velocities are peculiar to surface waves with lengths from 1 cm to 1 m. The same relates still more to the acquisition of information on internal waves and convective processes which typical velocities of about 0.1 m/s of water motion are connected with close to the surface. Respective manifestations can be anticipated in the most short-wave part of the above range of surface wavelengths. It is no coincidence that the search for ways of diagnosing the condition of the ocean water thickness by the condition of its surface is connected mainly with the short-wave range of the wave spectrum, i.e. with gravity-capillary waves. This circumstance is practically convenient, because the basic radiophysical methods of studying the ocean surface use the centimetre–decimetre range of radio waves which are reflected most effectively and are scattered by the short surface waves.

However, this does not mean that the characteristics of longer waves carry no useful information. Basic energy-carrying waves with lengths of tens and hundreds of metres experience refraction on space inhomogeneous currents, on shallow waters, and near the coasts. Diffraction effects can develop on submarine rises whose peaks are the closest to the surface. Correspondingly, the methods of detecting these effects should be different.

It is necessary to note here that the problem in all the above-mentioned cases is not to qualitatively diagnose correctly one or the other phenomenon in the water thickness by its visible manifestations on the surface. This problem can be considered as solved today. It is more important to perform a quantitative transfer from anomalous or modulated characteristics of surface waves to the characteristics of other phenomena which scientists are interested in. In regard to the aforesaid, it is just the situations wherein these anomalous or modulated characteristics are not easily detected on the basis of common visual observations that are of the greatest interest. Other considerations apart, the main difficulty is to avoid ambiguity in the solution of inverse problems.

Another circumstance should be stressed here. The methods of diagnosing the condition of the ocean water thickness by the surface manifestations should be operative. They should not be based on complicated and labour-intensive methods of measuring and analysing the field of surface waves because it is easier, faster, and more reliable to study the ocean water thickness in this case by common contact means, sending respectively equipped research vessels to the region of interest in the ocean. The operativeness of the methods of remote diagnostics should be still greater if they are intended for guiding research vessels to the objects under study. In all cases, any method of indirect diagnostics should ensure such advantages as compared to traditional methods of direct contact measurements of the phenomenon studied (e.g. currents) which would secure gains in material means, time, or quality of the data. All this imposes certain limits on the number and character of approaches to remote diagnosing of the ocean that are of any practical sense. Accordingly, it is possible to speak in this section of:

(1) the spectral approach, which, considering the points mentioned above, is applicable mainly to the short-wave range of the spectrum and may be very perspective when utilizing spectrum analysers on a ship [51]; and

(2) a small number of other practical techniques and methods of quantitative diagnostic estimates, including analysis of some selected statistical characteristics of the results of remote and contact measurements of sea waves.

6.3.1 *The spectral approach and its information content*

In principle, spectral presentation should essentially reflect in a certain generalized form the anomalous or modulated character of some spectral components in the complicated cascade of periodic oscillations that represent sea waves. Nevertheless, detection of the effects of modulation of anomalies on the basis of spectral representation of data on sea waves is possible only when the background undisturbed spectrum is known. Sometimes, the background undisturbed spectrum of the waves can be obtained in the neighbourhood of the area studied, i.e. where in conformity with *a priori* information the conditions of wave development (wind, time of its influence, and fetch) are identical, but the modulating factors (current, internal waves, disturbing specific features of the bottom topography, etc.) are absent. This approach usually cannot be applied because of the absence of *a priori* information, and it is necessary to rely on model ideas when analyzing the spectra. Considerable difficulties are encountered in this case too, because the spectra of wind waves undergo certain consistent changes with the development of the waves. The spectra proper and their changes should reflect in this case the local conditions and the evolution history of the wind field in each concrete case. Hence, it is also impossible to manage in this case without some minimal necessary hydrometeorological information. Otherwise, it is necessary to make assumptions, e.g. that the waves are fully developed while this does not always correspond to reality. Considering these and other feasible reservations, it is possible to indicate the following characteristics of the spectra wherefrom useful information can be obtained in regard to factors that disturb wind waves on the sea surface:

- the total level of spectral density characterizing the wave energy;
- the frequency and spectral density of the energy-transmitting maximum;
- the characteristic mean wave steepness;
- the availability of secondary or additional spectral peaks;
- the slope (or slopes) of the spectral curve in the selected interval (intervals) rightward of the main energy-transmitting maximum.

Most frequently spectra are analysed that describe the wave energy distribution by frequencies ω ['frequency spectra' $S(\omega)$] and by wavenumbers k ['space spectra' or 'spectra by wavenumbers' $\psi(k) = \psi(k, \theta)$] [112]. Here θ stands for the directions of the wave vectors of the spectral components in the chosen coordinate system. The frequency and space spectra should satisfy the condition of normalizing, so that for the frequency spectrum

$$\int_0^\infty S(\omega)\,\mathrm{d}\omega = \sigma_h^2 \tag{6.4}$$

and for the space spectrum

$$\int_0^\infty \mathrm{d}k \int_0^{2\pi} \psi(k, \theta)k\,\mathrm{d}\theta = \sigma_h^2, \tag{6.5}$$

where σ_h^2 is the full dispersion of level elevations h. When practically obtaining spectra by processing the measurements of sea waves, the values σ_h^2 in (6.4) and (6.5) may not coincide, generally speaking, because $S(\omega)$ can be obtained by processing the time series of level elevations $h(t)$ at a fixed point of observations, and the value σ_h^2 in (6.5) is obtained by averaging the instantaneous image of the level elevations in some restricted part of the space. This lack of coincidence can testify to non-stationy nature and space inhomogeneity of the wave field observed. Besides, (6.4) corresponds to an idealized quasi-one-dimensional pattern of $h(x)$ level elevations where the direction $x = ct$ coincides with the direction of wave propagation, and (6.5) describes the actual two-dimensional pattern $h(x, y)$.

But if the wave field is stationary and uniform, then in principle it is possible to transfer from one spectrum to another using the dispersion relationship, in particular

$$\omega^2 = gk. \tag{6.6}$$

It is easy to see that in this case

$$S(\omega) = \frac{2\omega^3}{g^2} \int_0^{2\pi} \psi(k, \theta)\, d\theta, \tag{6.7}$$

where $k = k(\omega)$.

From the methodological point of view, obtaining a frequency spectrum from a space one by formula (6.7), under the condition that (6.6) is observed, is equivalent to direct obtaining $S(\omega)$ from a time series of wave measurements at a point. In the presence of a current with velocity $\mathbf{u}(x, y)$, the dispersion relationship (6.6) transforms into

$$\omega = (gk)^{1/2} + \mathbf{u} \cdot \mathbf{k}. \tag{6.8}$$

It follows from (6.8) and (2.9), for example, that the results of remote aerospace measurements should be processed taking into account a correction for the current to compare the wave data, obtained in a point by direct measurements (e.g. by means of a wave graph), with the aforesaid results. This concerns not only the basic (energy-transmitting) range of the spectrum, but especially the centimetric–decimetric waves that were modulated greatly by the orbital velocities of the longer waves [215]. It also follows that the correct transition (6.7) from one spectrum to another is practically impossible with an unknown $\mathbf{u}$. On the other hand, if both spectra $S(\omega)$ and $\psi(\mathbf{k})$ are obtained for one and the same geographical point, energy-transmitting peak then comparing the values ω_{max} and k_{max} for one and the same that is easily discernible in both spectra it is possible under the condition $\omega_{max} \neq (gk_{max})^{1/2}$ to estimate the current velocity $\mathbf{u}$ or its component along the direction of the wave vector $\mathbf{k}$ from the dispersion relationship (6.8).

As stated in Chapter 2, when waves come from calm water to a spatially inhomogeneous current, their frequency ω remains invariable in a stationary system of coordinates, but the energy of the waves varies owing to interaction with the current. On a following current all the waves are suppressed ('smoothing'); on a contrary current the short waves are blocked (see Section

2.2), and the long waves become steeper owing to an increase in the amplitude and wavenumber [see Section 2.2 and expressions (2.10), (2.15), and (2.16)]. The blocking condition (2.14) is the boundary determining the character of wave transformation on the contrary current. All these changes in the wind waves on space non-uniform currents lead to essential variations in certain characteristics of the wave spectra [40, 161]. This was demonstrated experimentally in ref. [56], but in the latter case essentially greater effects of suppressing the amplitudes of high-frequency waves on a following current were obtained than those predicted by relationship (2.16). Thus, for example, the amplitudes of waves with frequencies 2.5–3.0 Hz decreased by almost two times in an experiment on a surface following current of 10 cm/s, while expression (2.16) gives only a 13–14% decrease of the amplitude. Kononkova et al. [56] explain this discrepancy by the 'smoothing' of the shortest waves due to subsurface turbulization of the water by the flow. Displacement of the energy-transmitting maximum towards lower frequencies and growth of the $S(\omega)$ values to the left of the energy-transmitting maximum by approximately 20% occurred in the experiments on the contrary current [56]. This behaviour of the frequency spectrum is connected with blocking of the short waves and partial pumping of energy into the long-wave range of the spectrum. Amplitude modulation of the waves, which is connected with the currents, is not the only reason for variations of the spectral density level of the recorded frequency spectra of the sea waves. According to the results of Khristoforov et al. [115] (see also the article by Khristoforov in ref. [85]), the lower levels of $S(\omega)$ should characterize short waves of smooth sinusoidal shape in slicks, while higher levels of $S(\omega)$ are characteristic of short waves of ripples, having a well-expressed trochoidal shape.

The above-cited examples demonstrate that even by measuring the characteristics of only the frequency spectra of wind waves it would seem possible to draw some conclusions on the character of wave modulation and to obtain estimates of the current velocity near the surface. To compare this, however, it is also necessary to have an undisturbed spectrum of waves, or at least reliable estimates of the undisturbed parameters of the most typical components of the waves, i.e. amplitudes a_0, wavelengths λ_0, and phase velocities c_0 for the main fixed frequencies ω_i. In the case of fully developed waves, it is possible to use for this purpose certain model spectral presentations that allow us to obtain some integral characteristics of the waves. The most universal dimensionless presentation of $S(\omega)$ obtained for fully developed wind waves at a wind velocity u_a at some height a over the sea surface is the Pierson–Moskovitz spectrum [203]

$$S(\omega)g^3/u_a^5 = b_1(u_a\omega/g)^{-5} \exp\left[-b_2(u_a\omega/g)^{-4}\right] \tag{6.9}$$

and its later modifications [156] (see also refs [40, 64]). This spectrum contains experimentally selected constants $b_1 = 0.4 \times 10^{-2}$ (or 0.8×10^{-2} with other normalizing of the spectrum) and $b_2 = 0.74$.

The Pierson–Moskovitz spectrum, which describes well the energy-transmitting components of the waves, makes it possible to determine by the wind velocity u_a the position of the spectral maximum of the waves $S_{max}(\omega_m)$,

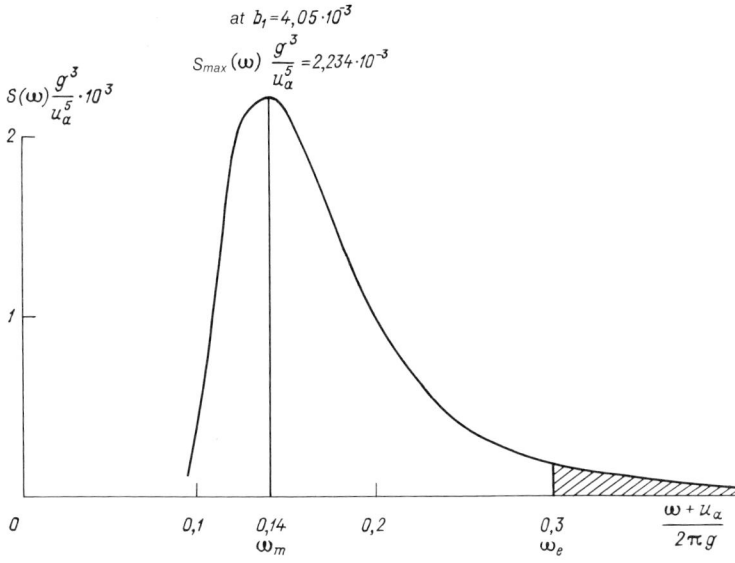

Figure 6.9. Dimentionless presentation of the Pierson–Moskovitz spectrum. The interval $S(\omega) \sim \omega^{-5}$ is dashed.

which, according to (6.9), takes place at the frequency $\omega_{\mathrm{m}}/(2\pi) \simeq 0.14g/u_{\mathrm{a}}$, at the dimensionless frequency $\omega_{\mathrm{m}}u_{\mathrm{a}}/2\pi g \simeq 0.14$, or at the dimensionless circular frequency $\omega_{\mathrm{m}}u_{\mathrm{a}}/g \simeq 0.88$ (see Fig. 6.9). The dispersion of σ_h^2 of the level elevations is determined from the shape of the dimensionless spectrum accordingly:

$$\sigma_h^2 \simeq 2.7 \times 10^{-3}u_{\mathrm{a}}^4/g^2. \tag{6.10}$$

The wind velocity u_{a} is implied in (6.9) and (6.10) at a height of $a = 20\,\mathrm{m}$. The following expression is commonly applied to consider the mean characteristic wave steepness ε_{m}:

$$\varepsilon_{\mathrm{m}} = a_{\mathrm{m}}k'_{\mathrm{m}} = (\sigma_h^2)^{1/2}k'_{\mathrm{m}}, \tag{6.11}$$

where $k'_{\mathrm{m}} = 1/\lambda_{\mathrm{m}} = k_{\mathrm{m}}/(2\pi) = \omega_{\mathrm{m}}^2/(2\pi g)$.

To consider the degree of 'normality' or 'abnormality' of the high-frequency part of the actual wave spectra, it might be useful to compare them in this part with the behaviour of $S(\omega)$ theoretically predicted by Phillips in the so-called equilibrium spectral interval [112, 202]

$$S(\omega) = \beta g^2 \omega^{-5}, \tag{6.12}$$

where the wave characteristics (and correspondingly such characteristics of the spectrum as dispersion of the surface elevations σ_h^2) should not depend on the wind velocity. The mean value of the dimensionless constant β, obtained by Phillips from many experimental determinations of different authors, is equal to 1.2×10^{-2}. The law $S(\omega) \sim \omega^{-5}$ is obtained for high-frequency wave components under the condition of energy saturation due to equilibrium between the supply of energy from the wind to the waves and the energy losses when they overturn and break. The space spectrum for the equilibrium interval is of the form

$$S(k) = \int_0^{2\pi} \psi(k)k \, d\theta = k^{-3} \int_0^{2\pi} f(\theta) \, d\theta = Bk^{-3}. \qquad (6.13)$$

In virtue of the normalizing conditions in (6.4) and (6.5) and relation (6.6), the constants β and $B = \int_0^{2\pi} f(\theta) \, d\theta$ in (6.12) and (6.13) correlate as

$$B = \beta/2. \qquad (6.14)$$

The limits of applicability of (6.12) and (6.13) have been determined by Phillips [112] as

$$\omega_m \ll \omega \ll 2g/u_* \qquad (6.15)$$

and

$$k_m \ll k \ll 2g/u_*^2, \qquad (6.16)$$

where u_* is the dynamic velocity or 'friction velocity' in the near-water layer of the air predetermining the velocity of surface drift in water, and ω_m and k_m are the frequency and wavenumber corresponding to the energy-transmitting maximum of the spectrum. These limits are bound with the fact that, generally speaking, $\beta = f(\omega u_*/g)$ and the exponent of the power at k can be considered as constants only at $\omega u_*/g \ll 2$ when the influence of the drift on the waves is negligibly small. Despite the determined limits, presented by Phillips himself [112], experimental data indicate that the high inequality in (6.15) is not fulfilled. Correspondingly, the section neighbouring ω_m and k_m exerts an influence on the value of constant β.

It should be noted that the interval with $S(\omega) \sim \omega^{-5}$ is also found in the Pierson–Moskovitz spectrum (6.9) where it begins (under moderate winds) from frequency ω_e, which greatly exceeds the frequency ω_m of the main energy peak ($\omega_e/\omega_m \simeq 2.1$–$2.3$ or $\omega_e u_a/(2\pi g) \simeq 0.29$–$0.32$). This remote from one another arrangement of ω_m and ω_e, as stated in ref. [40], has already long been recognized as characteristic for the conditions of fully developed waves on the basis of analysing a large number of wave measurements in ocean, seas, and lakes. It demonstrates that the Pierson–Moskovitz model approximates well the wave frequency spectrum just under these conditions. Nevertheless, spectra are found among real ones wherein $\omega_m u_a/g > 0.88$ (e.g. $\omega_m u_a/g \simeq 2.4$ [40]) and then $\omega_e/\omega_m \simeq 1.1$–$1.2$. The noted discrepancies are connected with the different nature of the series of field measurements of sea waves taken for analysis and testify only to the fact that the actual (especially the instantaneous) conditions of the sea surface can be quite far from being fully developed and, in general, are diverse in their spectral composition as well as in the degree of approach to energy saturation.

Experimental data have been accumulating in recent years demonstrating that the exponent of power n in the degree dependence of type $S(\omega) \sim \omega^{-n}$ differs greatly on most of the high-frequency slope of the spectrum ($\omega > \omega_m$) of common wind waves from the value $n = 5$ postulated by Phillips, and as a rule $n \leqslant 4$. The dependence $S(\omega) \sim \omega^{-4}$, for example, is substantiated in the presentations of sea waves as gentle turbulence [48].

When representing the frequency spectrum in the dimensionless form

$$\tilde{S}(\omega) = S(\omega)\omega^5/g^2 \tag{6.17}$$

and the experimental spectra as a function of the dimensionless frequency $\tilde{\omega} = \omega/\omega_m$ (with averaging of a large number of spectra of developed waves). Leikin and Rosenberg [66] pointed out three characteristic spectral intervals to the right of the main energy peak of the spectrum:

I— an interval adjoining the spectrum maximum (energy-transmitting area)

$$1.2 \lesssim \tilde{\omega} \lesssim 3.2; \quad \tilde{S}(\tilde{\omega}) = 4.0 \times 10^{-3}\tilde{\omega}^1; \tag{6.18a}$$

II— an equilibrium interval

$$3.2 \lesssim \tilde{\omega} \lesssim 10.5; \quad \tilde{S}(\tilde{\omega}) = 1.4 \times 10^{-2} = \text{constant}; \tag{6.18b}$$

III— a high-frequency interval proper

$$10.5 \lesssim \tilde{\omega} \lesssim 30; \quad \tilde{S}(\tilde{\omega}) = 4.5 \times 10^{-4}\tilde{\omega}^{1.5}. \tag{6.18c}$$

The experimental data for which the exponents of power n and the constants characterizing each interval in (6.18) were obtained were the result of careful measurements of surface waves from a pile foundation 20 km from the coast at a depth of 40 m. As shown in Fig. 6.10, the experimental points group quite well around the suggested universal three-interval representation which is called UTR below.

All the above-mentioned spectral models are considered currently by specialists as 'first generation models' [155] that only approach an adequate

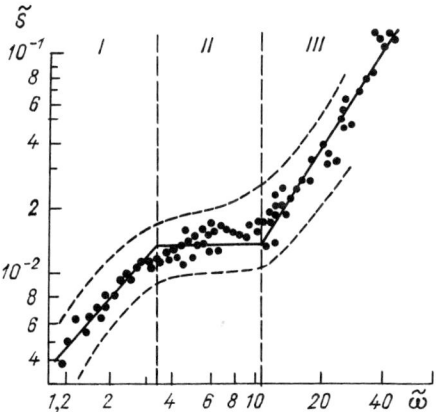

Figure 6.10. Universal three-interval representation of the frequency spectrum of sea waves by Leikin and Rosenberg [66] with experimental points obtained on the basis of measurements in the Caspian Sea.

spectral description of sea waves. The search for ways of constructing second generation models is connected with the realization of the following important facts:

(1)   overturn and breaking of waves cannot be considered as the only physical mechanism inducing the formation of an equilibrium interval in the high-frequency part of the spectrum. Non-linear interactions between the high-frequency components of the spectrum and the continuous transmission of wave energy towards the low frequencies from the pumping interval* play an important role in forming the low-frequency (energy-transmitting) part of the spectrum;

(2)   accordingly, if an equilibrium spectral interval exists under these conditions, then parameters of type β and B are not constants and they depend greatly in each case on the degree of development and the concrete conditions of wave development (wind velocity, duration of wind effect, fetch) which may not always be known *a priori*.

But even if all these conditions could be considered when constructing new spectral models, the basic difficulty still remains, preventing the application of spectral models as a norm. The difficulty is that the total level of spectral density and the full dispersion σ_h^2 in (6.4) and (6.5), as well as constants of type β and B for the isolated intervals, should depend greatly on the character of the two-dimensional structure of the field of wave elevations of level $h(x, y)$. For all that, it is still impossible to state with certainty which structure of $h(x, y)$ should be considered the norm.

It follows from the points mentioned above that it is currently not clear which model spectral representation of sea waves can be considered in each case as a reliable norm for determining anomalies in the express-analysis of real data (especially remote data) by comparison. Nevertheless, it is useful to determine just how the spectra of deliberately anomalous waves look like, and to compare them with some model representations, e.g. with UTR. With this purpose in mind, it is possible, first of all, to use spectral data on choppy water obtained by a group of workers of the Institute of Oceanology of the USSR Academy of Sciences while studying current rips in the White Sea in 1983 [8].

Figure 6.11 shows a frequency spectrum $S(\omega)$ of choppy water in current rips obtained during observations on 15 July 1983. Its basic energy-transmitting maximum corresponds to the frequency $\omega_m/(2\pi) = 0.48$ Hz. However, the striking property of this spectrum is the drop on the entire section to the right of ω_m by the law ω^{-5}. It turns out that the choppy water, which is mainly associated

---

*According to ref. [47], the pumping interval (the supply of energy to the waves) is located in fully developed waves in the part of dimensionless frequencies $\omega u_{10}/g = 2$–2.5, i.e. between the energy-transmitting maximum $\omega_m u_a/g$ and the high-frequency interval of dissipation at the extreme right-hand end of the spectrum. However, it should be noted that there is, as yet, no common opinion on the location of the pumping interval at the various stages of wave development (see the discussion in ref. [40]).

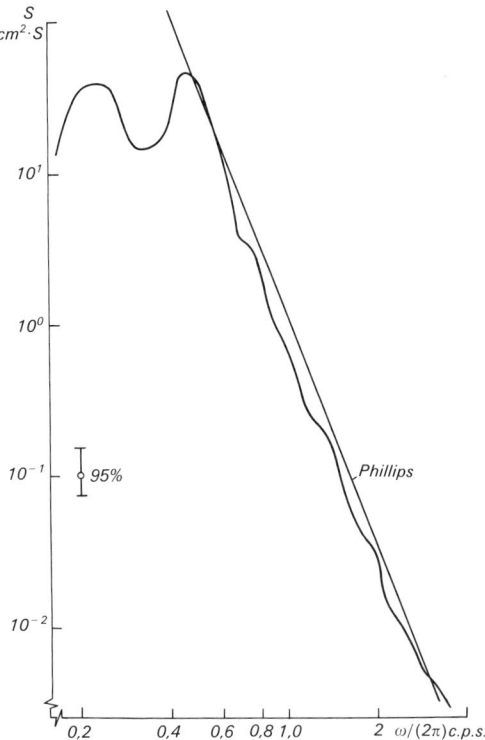

Figure 6.11. Frequency spectrum of waves (choppy water) in current rips from observations [8] on 15 July 1983 in the White Sea.

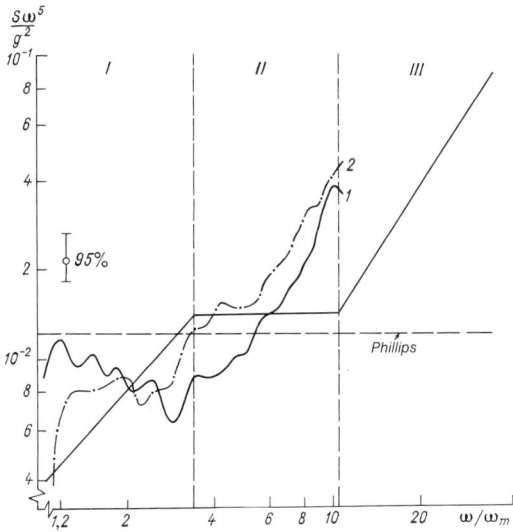

Figure 6.12. Comparison of choppy water spectra in a dimensionless form with a universal three-interval representation [66]. (1) 10 July 1983; (2) 11 July.

not with the wind, but with current rips of tidal origin on shallow waters (bottom rise to a depth of $\simeq 10$ m), corresponds better to the ideas of Phillips [112, 202] on surface waves under the conditions of energy saturation than common wind waves. When plotted on a graph in dimensionless coordinates $\tilde{S}(\tilde{\omega})$ (Fig. 6.12), the spectra of the choppy water demonstrate significantly large deviations from UTR. Their intervals $\tilde{S}(\tilde{\omega}) = $ constant, which are independent of the wind, are essentially more to the left ($1.2 \lesssim \tilde{\omega} \lesssim 3$–4) than in UTR, although the levels of the dimensionless spectral density of these intervals correspond to the values of constant β which are 1.5–1.8 times smaller than those cited in refs [66, 112]. And, vice versa, the most high-frequency ends of the spectra are characterized by a significantly greater (2–3.5 times) spectral density as compared to UTR. This means that the shorter decimetre and centimetre waves apparently have a steepness close to the limit conditioned by their intensive interaction with the tidal current which explains the extremely high values of their respective spectral densities and the fact that the areas of choppy water were clearly observed on the screen of the vessel's radar operating in the centimetre wavelength range. It is important to stress once more that the observed choppy water in the current rips was only partially due to the wind, which was light during the observations (from calm to 8 m/s [8]), hence the striking difference of its spectra from UTR, obtained with consideration of the wind. Apparently, the observed higher frequency of wave breaking ('white-caps') per unit of sea surface area is connected with the limit steepness of waves of the decimetre range.

Let us try to estimate by means of expression (6.11) the mean characteristic steepness of the waves corresponding to the choppy water spectra given in Figs 6.11 and 6.12. The value $k'_m = 0.147$ m^{-1} ($\lambda_m = 6.8$ m) corresponds to the spectral maximum at frequency 0.48 Hz. With the estimates in ref. [8] of full maximum dispersion of the level elevations in the zone of breaking ($\sigma_h^2 = 1.3 \times 10^{-2}$ m^2), obtained by integrating the spectra under discussion, expression (6.11) gives $\varepsilon_m = 0.016$, which is anomalously small and, in general, agrees qualitatively with the anomalously low values of β for sections $\tilde{S}(\tilde{\omega}) = $ constant and the low levels of spectral density of the low-frequency energy-transmitting sections of the choppy water spectra considered. But if the dispersion of the level elevations on sections $\tilde{S}(\tilde{\omega}) = $ constant is calculated by integrating (6.12) within the limits of the afore-cited boundary frequencies at $\beta = 0.9 \times 10^{-2}$, we obtain $\sigma_e^2 = 2.2 \times 10^{-3}$ m^2 ($\sigma_e = 0.047$ m). At a mean frequency $\bar{\omega}_e = 2$ Hz, we obtain $\bar{k}'_e = 2.56$ m^{-1} ($\bar{\lambda}_e = 0.39$ m), which gives $\bar{\varepsilon}_e = 0.12$, i.e. an estimate which is significantly closer to the limit steepness (see Section 2.2) than the previous one. This estimate confirms the aforesaid assumption on the substantial contribution of the decimetre waves to the frequency of wave-crest breaking.

The afore-cited estimates of steepness, as well as the earlier established fact of the anomalously low values of β in the interval $\tilde{S}(\tilde{\omega}) = $ constant, emphasize only that apparently it was necessary to introduce a correction factor, which took into consideration the difference in space structures $h(x, y)$ in the cases compared, into the actual spectra of the choppy water for comparison with (6.12) of Phillips [112] or with (6.18b) from UTR [66]. This can be explained as follows on a purely qualitative level. The structure $h(x, y)$, shown schemat-

Figure 6.13. Region of 'jumping' choppy water in the White Sea current rips (photograph by A. S. Kazmin).

ically in Fig. 6.14a, and the absence of progressive motion of the waves are characteristic of choppy water in current rips of the 'jumping waves' type (see Fig. 6.13). Common wind waves, whose idealized structure is shown in Fig. 6.14b and whose actual structure is close to that shown in Fig. 6.14c, are in the basis of (6.12) at $\beta = 1.23 \times 10^{-2}$ [112] or of (6.18b) at $\beta = 1.4 \times 10^{-2}$ [66]. It is quite obvious that in the case of choppy water (Fig. 6.14a) full dispersion σ_h^2 of level elevations, recorded at a fixed point, should be lower than in the case of common developed wind waves (Figs 6.14b and 6.14c) that regularly pass through the point of measurements A. Only high-frequency splashes of choppy water filling the entire space between the larger breaking crests, which are dashed in Fig. 6.14a

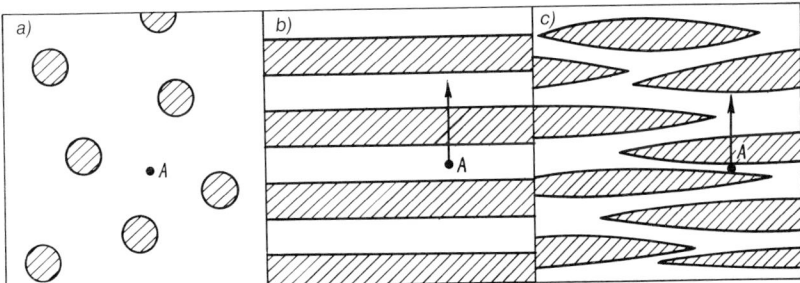

Figure 6.14. Diagrammatic representation of three types of two-dimensional structure of the sea surface elevations $h(x, y)$ (areas of maximum steepness are dashed). (a) For 'jumping' choppy water; (b) for idealized wind waves to which commonly considered frequency spectra correspond; (c) for real wind waves. The point of observation is designated by the letter A. The direction of wave displacement in cases (b) and (c) is indicated by an arrow.

(see also Fig. 6.13), can make a significant contribution to the characteristic mean steepness of the waves, which is estimated by the dispersion of the entire spectrum or its individual sections. The necessity of subsequent studies in this direction is quite obvious.

Attempts to estimate the characteristic steepness of the waves by the actual spectra of the waves show that this problem has been very poorly studied in the modern literature on sea waves. Although examples of the spectra of surface slopes can be found in the literature (see, for example, ref. [112]), they are still very few and their properties are poorly studied. The connection of the values β and B in (6.12) and (6.13) with the steepness of the waves in an equilibrium interval at a variable two-dimensional structure $h(x, y)$ has not yet been studied by anybody, and the problem of the limit values of β_{max} and B_{max} has not been discussed anywhere, although the existence of a limit wave steepness is an obvious fact for everyone. Besides, there is discordance in the expressions used in the literature for estimating the steepness of the waves, including limit steepness. Different authors have used as a steepness measure the values h/λ, a/λ, or ak which differ either by a factor of 2, as $h = 2a$, or by a factor of 2π, as $k = 2\pi/\lambda$. Apparently, it is necessary to introduce some uniformity and to use subsequently only the relation h/λ, which is equal to $1/7 \approx 0.142$ for the limit steepness of the Stokes two-dimensional wave [40, 73].

An express analysis of the space spectra of surface elevations is of special interest from the point of view of remote diagnostics of sea waves and the detection of anomalous states of the surface. In recent years, remote optical spectrum analysers of sea waves have been designed which can be used from on board research vessels, making it possible to obtain practically instantaneous data on the anomalies of the spectral distribution of wave energy in the range of small wavelengths. The available publications [51, 73, 85] are not numerous, but they illustrate that the energy of the spectral components with fixed wavelengths do not remain in the ocean at a constant level for long periods of time and they fluctuate quite significantly (5–7 times), in particular, under the influence of internal waves. It has been found, for example, that most appreciable modulation due to the interaction of surface waves with internal ones is characteristic of shorter waves ($\lambda = 3.5$ cm). Apparently, new results will soon be obtained on analysing the short-wave range of space spectra of sea waves. In the meantime, satellite radar systems (in particular, the synthetic aperture radar of the American SEASAT satellite) have already demonstrated the real possibility of obtaining space spectra of sea waves within the 50–400 m wavelength range in the region of the JASIN experiment [234]. It is possible to count in this range of wavelengths on detecting wave anomalies due to shallow water banks and refraction on currents, especially where the space inhomogeneities of the velocity field are most intensive (e.g. near oceanic fronts). Airborne measurements by means of a laser profilometer were used not long ago to study the differences in the sea surface state on both sides of sharp oceanic fronts. The results obtained are of great interest in the context of this section; therefore, they should be discussed in greater detail. The investigated fronts limited on both sides the extrusion of cold Labrador waters into the area of the warm North Atlantic current in the region of the Great Newfoundland bank [185]. The lengths and heights of the basic systems of waves and swells in the directions of their propagation were obtained by the energy-carrying peaks of the Doppler (apparent) one-dimensional space spectra of surface elevations along various flight tacks looking starwise. When approaching the cold side of a front (the SST jump at the front reached 5–7°C), there was increased turbulization of the near-water layer of the atmosphere up to the flight altitude (250 m above sea level) and gradual growth of the significant wave height* from 2 to 3 m in the near-frontal area, whose width varied in some cases from 30 to 60 km depending on the wind direction in relation to the front orientation. Turbulization of the near-water air layer was even more intensive on the warm side of the front. However, the flight tacks were too short to determine the width of the zone of the front's influence towards that side. Waves and swell changed the direction of propagation by 8–15° when crossing the front. The qualitative estimates of the changes in the heights and directions of wave and swell propagation when crossing the front determined in that way were then compared with the theoretical predictions by the formulae of Longuet–Higgins and Stewart [180, 181] taking into consideration the velocities and directions of the currents near the fronts using the data of buoy measurements. The coincidence was very good in some cases, while in others the discrepancies could be attributed to the sharp variations in stratification, the degree of turbulization, and the conditions of exchange by an impulse in the near-water layer of the atmosphere when crossing oceanic fronts of such intensity.

The above-cited examples make it possible to conclude that, despite certain obvious success, application of the spectral approach to diagnose the conditions of the near-surface layer by the surface waves is far from simple and far from being always the most effective method. The non-universal character of the spectral constants and slopes of the spectra, and the inevitability of idealization when simulating make it impossible to rely with confidence on the existing models of spectral representations when determining the anomalous character-istics of real spectra. The production of space spectra of waves, which, in principle, are more informative than frequency spectra, was beset until recently with many technological and methodological difficulties. On the whole, it is necessary to acknowledge that application of the spectral approach in this sphere is not more productive than, for example, studying the variability of the temperature field in the ocean [103].

6.3.2. *Other statistics and different approaches to the analysis of the characteristics of surface waves*

In our opinion, the most perspective way of determining informative anomalies in the sea surface state on the basis of analysing the statistical characteristics of waves consists in selecting such characteristics for analysis that are most

---

*Average height of one-third of the highest waves.

closely associated with the steepness and shape of the surface waves observed. Khristoforov *et al.* [115] (see also the article by Khristoforov in ref. [85]) have clearly shown that even very short waves under light winds have a different shape in areas of slicks and ripples, making different in these two cases such statistical characteristics of wavegrams as asymmetry, excess, etc.

In principle, many other characteristics such as the mean wave height $\bar{h}$ or the significant wave height $\bar{h}_s$ obtained as a result of direct measurements, as was done, for example, in ref. [185], can be used for comparative analysis. It is not absolutely necessary in this case to apply the spectral form of representing the results of measurements. The following can be used for comparison: (1) undisturbed characteristics measured in the neighbourhood (if there is such a possibility as in ref. [185]); (2) respective tabulated values of $\bar{h}$ or $\bar{h}_s$ for the known conditions (velocities and duration of wind action, fetch, sea depth), taken from available tables (see, for example, ref. [80]); and (3) respective climatic characteristics, whose sources and degree of truth are considered in ref. [70]. As mentioned before, estimation of the current velocity by the increment or decrease of the wave height when outletting a current can be the result of these comparisons. Estimates are given, e.g. in ref. [70], of the relative increments $\bar{h}_T/\bar{h}$ of average wave heights on a contrary current depending on the relationship $\bar{c}/u$ of the wave average phase velocity $\bar{c}$ to the current velocity u:

$$\bar{c}/u: \quad 5 \quad 6 \quad 7 \quad 10 \quad 15$$
$$\bar{h}_T/\bar{h}: \quad 1.9 \quad 1.5 \quad 1.4 \quad 1.2 \quad 1.1$$

These estimates, based on field studies [46], agree well with the experimental data [56] discussed above, and with the theoretical predictions of Longuet–Higgins and Stewart [180, 181]. Matushevsky [70] points out that in full accordance with the cited estimates the normal height of large waves can easily double on strong contrary currents (up to 2–2.5 m/s), e.g. in the axis of the Needle Cape current nearby south-east Africa, and the shape of the waves can become anomalous with a very steep front slope of the crests and with deep, almost flat troughs between them. So-called 'killer waves' develop with a height of 12–20 m. Moreover, the height of the crests proper, which is usually not more than 0.6–0.8 of the full height h_m of the maximum wave, can increase to 0.90–0.92 h_m. It is quite obvious that in the case of regions with high velocities of the constant or tidal currents and with frequent storms, the histograms of recurrence of waves with different heights should differ greatly under various combinations of the wind, wave, and current directions.

As visual estimates of the wave height are usually prone to significant errors, at waves up to force 6–7 it is possible to judge the wave height by the lengths of the most correctly shaped waves with overturning crests. The wave height can be approximately estimated in this case by the limit steepness from the known relationship $h \simeq 0.14\lambda$, and the most trustworthy estimate in this case can easily be the mean of several estimates. However, a still more useful parameter is the mean frequency of wave breaks per unit of surface area per unit of time which can be estimated by visual and by aerial photography. Local intensification of the frequency of wave breaks determined in this way can itself

testify to local anomaly of the waves which is associated either with the bottom topography or with the current, or with the internal waves [43]. On the other hand, physical models of various complexity are possible (see, for example, refs [38, 44, 188]) which allow us to connect quantitatively the specific frequency of wave breaks or its space distribution with the characteristics of wave disturbing factors, or, in some cases, with a density structure of the near-surface layer. In particular, if alternating bands are observed with a very different specific frequency of surface wave breaks, then, as discussed earlier in Chapter 2, the distance between the bands corresponds to length λ of the modulating internal waves. Repeated shipboard, aerial, or satellite observations with reliable coordinate referencing can give estimates of the direction of propagation and phase velocity c of the internal waves. It is possible to estimate the relative density drop $\Delta\rho/\rho$ on the jump layer within the limits of a two-layer model by the known phase velocity c and the UQL thickness, which is equivalent to estimating the enthalpy of the UQL [188]. If the internal waves, for example, are of non-linear character and propagate by well-expressed packets of the soliton type (i.e. they can be described, for example, by the equations of Korteweg–de Vries, as in ref. [197] or by other such equations), then it is possible to estimate also the thickness of the quasi-homogeneous layer by the non-linear additive to c and then its enthalpy [38]. The next step is to estimate the dispersion relationship for the internal waves on the basis of analysing repeated (or multiple) aerial surveys of the field of surface waves (with necessary frequency and high accuracy of referencing), or by antenna measurements of this field with a set of sensors arranged across the water area [13]. In principle, the required measurements could be made in the future also from satellites.*
In turn, it is possible to recover the parameters of near-surface layer stratification, e.g. vertical distribution of the Väisälä–Brunt frequency [13, 39], by the dispersion relationship for the internal waves by solving the inverse problem. The complicating or inhibiting factors in this case are the unknown constant currents in the near-surface layer of the ocean and the multimode composition of the internal waves.

The dispersion relationship for linear surface waves on deep waters makes it possible to determine the velocity of the currents by the Doppler shear ω_D of the frequency of the radio signal scattered in the reverse direction [222] (see also ref. [79]). Radar irradiation should be performed in this case under small (sliding) angles to the horizon from a stationary base (platform, coast) or a carrier moving at a known speed. According to the conditions of Bragg (resonance) scattering, the measured Doppler shear ω_D should occur on the sea wave component with a length $\lambda = \lambda_r/2$, which is equal to the half-length of the radiated radio waves λ_r, and should be different when a current is present or absent. In the latter case, it is simply equal to the frequency ω of short surface waves which scatters the signal. The increment of the phase velocity of these

---

*For certain technological reasons, satellite images of the ocean surface have not yet been obtained with the required high frequency and high accuracy of coordinate referencing (about ± 50 m).

waves in relation to the radar due to the current is

$$\Delta c = k^{-1}(\omega_D - \omega),$$

where $\omega = (gk)^{1/2}$ and k is derived from the Bragg condition $k = 4\pi/\lambda_r$. Accordingly, the current velocity is derived from (6.16) as $u \simeq \Delta c$. In experiments [222] performed with waves of $\bar{h}_s \simeq 1$ m, two radar signals of different directions transmitted from a shore station enabled the current vector to be obtained. The wavelengths λ_r of the radio signal used were within 14.7–123.5 m, which made possible observation of the radio signal scattering on the components of the surface waves with λ from 7.3 to 61.7 m. Currents in the 2–4 m thick near-surface layer were measured for control from drifters. Agreement between the velocities determined by the two methods was about ± 2 cm/s at velocities of 10–20 cm/s.

Finally, it is possible to use artificially generated monochromatic surface waves as a probing signal to map the near-surface currents on limited areas (lagoons, bays, and lakes). In this case, the base for estimating the velocities of the currents is expression (2.9), from which the following expression is obtained in the general case of a current directed at angle θ to the wave crest:

$$u \sin \theta = c_0 [\lambda_1/\lambda_0 - (\lambda_1/\lambda_0)^{1/2}].$$

It follows from this expression that the following current ($\theta = 90°$) with a velocity of 20 cm/s will increase the length of the surface waves at $c_0 = 1$ m/s by 37%, while the same contrary current ($\theta = 270°$) will decrease it by 45%. However, when $\bar{u}/c_0$ approaches 10^{-2}, modulation of the wavelength will be negligibly small. Unfortunately, exactly this order of $\bar{u}/c_0$ is characteristic of the interaction of a monochromatic swell with a field of weak currents in the ocean which hinders practical application of the given method under conditions when it is impossible to develop artificially monochromatic waves of predetermined frequency.

The given method was suggested and practically tested by Sheres [216] using two wave-makers and aerial photography in the Aqua-Hedionda lagoon off California. In all cases, the obtained estimates of the velocities were systematically below those measured by rotor current meters near the surface. This is because all the methods of estimating the current velocity, based on measuring the modulations of the surface waves, produce a mean value of velocity $\bar{u}$ for the water layer which is commensurable by thickness with the length of the waves. The estimation of the current direction was satisfactory.

A more detailed description of various methods for the acquisition of useful oceanological information from the results of measuring surface waves should be the objective of a special study.

6.4. Conclusion

Certainly, the previously discussed physical description of the near-surface layer and its specific characteristics cannot be considered exhaustive. The book presents only the principal features of the structures characteristic of this layer and estimates the limits of variability observed in it. Even if we really wanted to, it would be impossible to give a geographical description of the specific features of this layer in various areas of the World Ocean. Concrete situations, observed by us and other authors, were used only as examples to illustrate what differences from the traditional ideas could be associated with the near-surface layer, including in extreme situations. Apparently, it is impossible to map the properties of the near-surface layer as has been done, for example, for the typical UQL thickness in various seasons at the scales of whole oceans. The variety of combinations of the diurnal cyclic heating with the synoptic variability of the wind and other hydrometeorological conditions over the ocean leads to such irregularities in the variability of the near-surface layer that can be the object of only comparatively short-term prognosis. The only potential approach in this case, as mentioned in Chapter 1, is to elaborate comparatively simple prognosticating models that are typified by the different physical regimes of the near-surface layer. This approach should be the next step in the studies.

What useful applications will the information contained in this book find? In our opinion, it will, first of all, have an influence on the subsequent formation of modern notions on the physical nature of variability of the hydrophysical fields near the ocean surface. In particular, the material in Chapters 4 and 5 throws some new light on the origin of large-scale anomalies of the ocean surface temperature (ASST) under certain conditions. For example, vast positive ASST, observed by us for several weeks in the Sargasso Sea, were the result of accumulation of absorbed solar irradiance in a comparatively thin rain-freshened near-surface layer. Such anomalies cannot but be manifested in the average monthly SST values, but they have a very low correlation with the distribution of temperature in the deeper layers of the ocean. Another important application of the information on the near-surface layer should be its binding consideration when interpreting IR measurements of SST from satellites, especially when referencing IR data to ground-truth subsatellite SST measurements made by common contact means. The methodological aspects of this problem discussed in the book should prevent scientists from making typical errors which, unfortunately, are often found in publications of recent years on this subject. Apparently, the same logic is applicable to the estimates of the current vectors of the near-surface layer by the data of one or the other satellite measurements.

It follows from the data analysis presented in this book that the variability of the near-surface layer characteristics (in particular, salinity) can have far-reaching climatic consequences. This is associated with the fact that the suppressing role of near-surface density stratification in regard to turbulence and convection can tell not only in the diurnal and seasonal cycles, but also in the inter-year variability, preventing the formation of deep or intermediate waters in individual regions in the course of long-term periods of time. Qualitatively opposed effects may occur under anomalously intensive evaporation or anomalous cooling from the surface.

The ordered forms of near-surface layer water motions, determined on the basis of satellite information, e.g. mushroom-like currents (vortex dipoles) and transversal filaments in coastal upwellings and marginal ice zones, give an entirely new mental pabulum. On the one hand, we have obtained a concrete idea on the mechanisms and form of subsequent transmission and concentration in the ocean of the impulse imparted to the water in the course of numerous local and short-term atmospheric and other effects. On the other hand, it has become clear what forms can assume in the ocean the natural trend to ordering and self-organization of initially turbulent disturbances of motions under the influence of the earth's rotation and stratification. The intricate path of evolution, which at the end can lead to well-known synoptic-scale eddies, gradually becomes clearer. The practical consequences of understanding these processes (from the point of view of spreading biogenic elements, plankton, marketable species of fish and other animals, as well as pollutants in near-surfactants) will need to be addressed quickly. The causes of the origin of increased productivity 'patches' near the outer boundaries of the coastal upwelling zones are already clear. Apparently, the situation is the same in the large-scale climatic frontal zones of the ocean.

Expenditure allocated in any sphere of science for obtaining new knowledge is a long-term investment. New knowledge on the near-surface layer of the ocean is not an exception in this respect. We are unable to foresee as yet many potential practical uses of the new information. But we can state with confidence that we are far from possessing all the necessary knowledge on the near-surface layer of the ocean. An inquisitive researcher will meet with surprises and revelations. We would like to hope that the search will be continued.

References

1. Abramov, R. V. (1974). Studies of diurnal variations by hourly observations in a quasi-fixed point in the tropical zone of the ocean. In: *Atlantic Hydrophysical Polygon-70*. Nauka, Moscow, pp. 112–122.
2. Altberg, V. K. and Popov, E. A. (1934). Some results of measuring the water temperature in the surface layers and in the depth. *Izv. State Hydrological Institute* No. 67, 27–35.
3. Anisimova, E. P., Ivanov, V. N., Milekhin, L. I. and Speranskaya, A. A. (1985). On the temperature regime of the water surface in conditions of free convection. *Izv., Acad. Sci., USSR, Phys. Atmos. Ocean* **21**, 1113–1115.
4. Arsenyev, V. S., Galerkin, L. I., Kutko, V. P., *et al.* (1984). *Temperature Field in the Northern Area of the Atlantic*. Gidrometeoizdat, Moscow.
5. Arkhipova, E. G. and Rzheplinsky, G. V. (1949). Results of experimental observations of convection in natural conditions. *Tr. GOIN* **11**(23), 43–52.
6. Barenblatt, G. I. (1977). Strong interaction of gravity waves and turbulence. *Izv., Acad. Sci., USSR, Phys. Atmos. Ocean* **13**, 845–849.
7. Barenblatt, G. I. (1982). *Self-Similarity and Intermediate Asymptotics*. Gidrometeoizdat, Leningrad.
8. Barenblatt, G. I., Leikin, I. A., Kazmin, A. S., *et al.* (1985). Suloy in the White Sea *Dokl. Acad. Sci. USSR* **281**, 1435–1439.
9. Bezverkhny, V. A. and Solovyev, A. V. (1986). Experimental data on the vertical structure of the non-stationary near-surface thermocline. *Izv., Acad. Sci., USSR, Phys. Atmos. Ocean* **22**, 71–77.
10. Byzuyev, A. Ya. and Fedorov, K. N. (1973). On the similarity of the thermal structures in iceleads in Arctic ice and in freshwater lakes. *Probl. Arktiki Antarkt.* No. 41, 99–101.
11. Bulatov, N. V. (1982). Specific features of the cyclonic meander and eddy formation in the sub-Arctic front zone. *Issled. Zemli iz Kosmosa*, No. 3, 53–58.
12. Bune, A. V., Ginsburg, A. I., Polezhayev, V. I. and Fedorov K. N. (1985). Numerical and laboratory simulation of convection development in the surface-cooled water layer. *Izv., Acad. Sci., USSR, Phys. Atmos. Ocean* **21**, 956–963.
13. Burdyugov, V. M. and Grodsky, S. A. (1984). Method of estimating the dispersion relationship of internal waves by sea surface images. In: *Problemy Issledovaniya Okeana iz Kosmosa*, Nelepo, B. A. (Ed.). MGI, Acad. Sci. Ukrainian SSR, Sebastopol, pp. 5–18.
14. Varfolomeyev, A. A. and Sutyrin, G. G. (1981). Laboratory simulation of free non-stationary penetrating convection. *Dokl. Acad. Sci. USSR* **261**, 55–59.
15. Vasilkov, A. P., Kelbalikhanov, B. F. and Stefantsev, L. A. (1985). Effect of clarification of the thin surface layer of the ocean. *Izv., Acad. Sci., USSR, Phys. Atmos. Ocean* **21**, 1327–1330.
16. Vershinsky, N. V., Volkov, Yu. A. and Solovyev, A. V. (1981). On the vertical

microstructure of the thin surface layer of the ocean. *Dokl. Acad. Sci. USSR* **256**, 694–698.

17. Veselov, V. M., Korchashkin, N. N. and Lozovatsky, I. D. (1984). On the interaction of internal waves and small-scale turbulence in the thermocline with variability of the ocean surface state. *Izv., Acad. Sci., USSR, Phys. Atmos. Ocean* **20**, 78–85.

18. Vize, V. Yu. (1944). Hydrometeorological conditions in the marginal ice zone in Arctic seas. *Tr. Arkt. Antarkt. Nauchno-Issled. Inst.* **184**, 122–151.

19. Vinogradov, V. V. and Mironov, L. V. (1976). Water temperature time variability in the Tropical Atlantic. *Trudy Mezhvedomstvennoy Expeditsii TROPEX-74*, Vol. II. Ocean. Gidrometeoizdat, Leningrad, pp. 39–45.

20. Voropayev, S. I., Gavrilin, B. L. and Zatsepin, A. G. (1981). On the structure of the ocean surface layer. *Izv., Acad. Sci., USSR, Phys. Atmos. Ocean* **17**, 521–526.

21. Voropayev, S. I. and Filippov, I. A. (1985). The development of a horizontal jet in density uniform and stratified fluids. Laboratory experiment. *Izv., Acad. Sci., USSR Phys. Atmos. Ocean* **21**, 964–972.

22. Ginsburg, A. I. (1981). Methodological problems of measuring the temperature and salinity in the boundary layer of the ocean. *Meteorol. Gidrol.* No. 4, 65–70.

23. Ginsburg, A. I., Dikarev, S. N., Zatsepin, A. G. and Fedorov, K. N. (1981). Phenomenological specific features of convection in fluids with a free surface. *Izv., Acad. Sci., USSR, Phys. Atmos. Ocean* **17**, 400–407.

24. Ginsburg, A. I., Zatsepin, A. G., Sklyarov, V. E. and Fedorov, K. N. (1980). Effects of precipitation in the near-surface layer of the ocean. *Okeanologiya* **20**, 828–836.

25. Ginsburg, A. I., Zatsepin, A. G. and Fedorov, K. N. (1977). Laboratory study of the thermal boundary layer fine structure in water at the water–air interface. In: *Mesoscale Variability of the Temperature Field in the Ocean*, Fedorov, K. N. (Ed.). Institut Okeanologii AN SSSR, Moscow, pp. 109–125.

26. Ginsburg, A. I., Zatsepin, A. G. and Fedorov, K. N. (1977). Thermal boundary layer fine structure in water at the water–air interface. *Izv., Acad. Sci., USSR, Phys. Atmos. Ocean* **13**, 1268–1277.

27. Ginsburg, A. I., Kostyanoy, A. G., Pavlov, A. M. and Fedorov, K. N. (1987). Laboratory reproduction of mushroom-like currents (vortex dipoles) in conditions of rotation and stratification. *Izv., Acad. Sci., USSR, Phys. Atmos. Ocean* **23**, 170–178.

28. Ginsburg, A. I. and Fedorov, K. N. (1978). Cooling of water from the surface at free and forced convection. *Izv., Acad. Sci., USSR, Phys. Atmos. Ocean* **14**, 79–87.

29. Ginsburg, A. I. and Fedorov, K. N. (1978). Thermal state of cooled water boundary layer during transition from free to forced convection. *Izv., Acad. Sci., USSR, Phys. Atmos. Ocean* **14**, 778–785.

30. Ginsburg, A. I. and Fedorov, K. N. (1978). On the critical boundary Rayleigh number at cooling water through the free surface. *Izv., Acad. Sci., USSR, Phys. Atmos. Ocean* **14**, 433–437.

31. Ginsburg, A. I. and Fedorov, K. N. (1979). On the contribution of salinity and temperature to convective instability at evaporation of sea water. *Izv., Acad. Sci., USSR, Phys. Atmos. Ocean* **15**, 886–890.

32. Ginsburg, A. I. and Fedorov, K. N. (1982). Variability of salinity in the near-surface layer of the ocean. *Okeanologiya* **22**, 928–935.

33. Ginsburg, A. I. and Fedorov, K. N. (1984). Evolution of mushroom-like currents in the ocean. *Dokl. Acad. Sci. USSR* **276**, 481–484.

34. Ginsburg, A. I. and Fedorov, K. N. (1984). Mushroom-like currents in the ocean (by data of analysing satellite images). *Issled. Zemli iz Kosmosa* No. 3, 18–26.

35. Ginsburg, A. I. and Fedorov, K. N. (1984). Certain consistencies in the development

of mushroom-like currents in the ocean determined by analysing satellite images. *Issled. Zemli iz Kosmosa* No. 6, 3–13.

36. Ginsburg, A. I. and Fedorov, K. N. (1985). Systems of transversal filaments in coastal upwellings: satellite information and physical hypotheses. *Issled. Zemli iz Kosmosa* No. 5, 3–10.

37. Ginsburg, A. I. and Fedorov, K. N. (1986). On the near-surface circulation of waters in the sub-Arctic frontal zone (by satellite data). *Issled. Zemli iz Kosmosa* No. 1, 8–13.

38. Grishin, G. A. and Grodsky, S. A. (1985). Estimation of parameters of two-layer stratification of the ocean by satellite observations of internal waves. *Issled. Zemli iz Kosmosa* No. 4, 3–8.

39. Grodsky, S. A. and Kudryavtsev, V. A. (1983). Recovery of density profile by field dispersion relationships of short-period internal waves. In: *Metody Obrabotky Kosmicheskoy Okeanologicheskoy Informatsii* Nelepo, B. A. (Ed.). MGI, Acad. Sci. Ukrainian SSR, Sebastopol, pp. 59–66.

40. Davidan, I. N., Lopatukhin L. I. and Rozhkov, V. A. (1985). *Wind Waves in the World Ocean*. Gidrometeoizdat, Leningrad.

41. Dikarev, S. N. (1983). On the influence of rotation on the structure of convection in a deep homogeneous fluid. *Dokl. Acad. Sci. USSR* **273**, 718–720.

42. Dikarev, S. N. and Zatsepin, A. G. (1983). Development of convection in a two-layer, unstably stratified fluid. *Okeanologiya* **23**, 950–953.

43. Dulov, V. A., Klyushnikov, S. A. and Kudryavtsev, V. N. (1986). Influence of internal waves on the intensity of wind wave breaking. Field observations. *Morsk. Gidrophyz. Zh.* No. 6, 14–21.

44. Dulov, V. A. and Kudryavtsev, V. N. (1987). Influence of internal waves on the intensity of internal waves breaking. Breaking frequency per unit of observed surface area. *Morsk. Gidrophyz. Zh.* No. 6, 12–19.

45. Dykhno, L. A. (1985). Formation of mixed area structure under the effect of bypassing in a stratified fluid. *Izv., Acad. Sci., USSR, Phys. Atmos. Ocean* **21**, 528–536.

46. Zhevnovaty, V. T. (1971). Wind waves on tidal currents. *Priroda i Khozaistvo Severa* No. 3, 127–131.

47. Zakharov, V. E. and Zaslavsky, M. M. (1982). Intervals of pumping and dissipation in the kinetic equation for wind waves. *Izv., Acad. Sci., USSR, Phys. Atmos. Ocean* **18**, 1066–1076.

48. Zakharov, V. E. and Filonenko, N. N. (1966). Energy spectrum for stochastic oscillations of a fluid. *Dokl. Acad. Sci. USSR* **170**, 1292–1295.

49. Zatsepin, A. G., Kazmin, A. S. and Fedorov, K. N. (1984). Thermal and visible manifestations of large internal waves on the ocean surface. *Okeanologiya*, **24**, 586–592.

50. Zatsepin, A. G., Kazmin, A. S. and Fedorov, K. N. (1984). Hydrophysical conditions in the region of the sub-Antarctic frontal zone in the south-east Pacific. In: *Frontalniye zony yugo-vostochnoy chasti Tikhogo okeana*, Vinogradov, M. E. and Fedorov, K. N. (Eds) Nauka, Moscow, pp. 51–57.

51. Zuikova, E. M., Luchinin, A. G. and Titov, V. I. (1985). Determination of the characteristics of space–time wave spectra by an optical image of the sea surface. *Izv., Acad. Sci., USSR, Phys. Atmos. Ocean* **21**, 1095–1103.

52. Ivanov, R. N. (1965). Wave and drift surges in the sea. *Izv., Acad. Sci., USSR, Phys. Atmos. Ocean* **1**, 94–108.

53. Kazmin, A. S. (1983). On the influence of an oil film on the thermal structure of

the near-surface water layer. *Izv., Acad. Sci., USSR, Phys. Atmos. Ocean* **19**, 1075–1081.

54. Kazmin, A. S. and Sklyarov, V. E. (1982). Certain specific features of the Black Sea waters' circulation by 'Meteor' satellite data. *Issled. Zemli iz Kosmosa* No. 6, 42–49.

55. Kalmykov, A. I., Nazirov, M., Nikitin, P. A., *et al.* (1985). On the ordered mesoscale structures on the ocean surface detected by data of radar surveys from space. *Issled. Zemli iz Kosmosa* No. 3, 41–47.

56. Kononkova, G. E., Poborchaya, L. V. and Pokazeyev, K. V. (1975). Variation of the wind wave spectrum on the water flow. *Izv., Acad. Sci., USSR, Phys. Atmos. Ocean* **11**, 960–964.

57. Kostyanoi, A. G. and Rodionov, V. B. (1986). On the formation of inter-thermoclinic eddies on the Canaries upwelling. *Okeanologiya* **26**, 892–895.

58. Kraus, E. B. (1972). *Atmosphere–Ocean Interaction*. Clarendon Press, Oxford.

59. Kropotkin, M. A., Verbitsky, V. A., Sheveleva, T. Yu. and Tarashkevich, V. N. (1978). Radiation temperature of the water surface polluted with an oil product film. *Okeanologiya* **18**, 1107–1108.

60. Kudryavtsev, V. N. and Solovyev, A. V. (1985). On the parametric description of the cold film on the ocean surface. *Izv., Acad. Sci., USSR, Phys. Atmos. Ocean* **21**, 177–183.

61. Kuftarkov, Yu. M., Nelepo, B. A., Fedorovsky, A. D. (1978). On the cold temperature skin-layer of the ocean. *Izv., Acad. Sci., USSR, Phys. Atmos. Ocean* **14**, 88–93.

62. Lamb, H. (1945). *Hydrodynamics*. Dover, New York.

63. Lappo, S. D. (1940). *Oceanographic Handbook on the Arctic Seas of the USSR. General Pilot Book*. Glavsevmorput, Leningrad–Moscow.

64. Le Blon, P. and Maiseck, L. (1978). *Waves in the Ocean*, Vol. 2. Elsevier Oceanography Ser. No. 20. Elsevier, Amsterdam.

65. Leikin, I. A. (1987). Study of the wave spectra on a horizontally inhomogeneous current in the zone of current rips. *Izv., Acad. Sci., USSR, Phys. Atmos. Ocean* **23**, 52–58.

66. Leikin, I. A. and Rosenberg, A. D. (1980). On the high-frequency range of the wind wave spectrum. *Dokl. Acad. Sci. USSR* **255**, 455–458.

67. (1954). *White Sea Pilot Book*. Gidrographicheskoye Upravleniye VMS SSSR.

68. Luchinin, A. G. (1986). Signal/noise relationship in the image of a water body bottom observed through a wavy surface. *Izv., Acad. Sci., USSR, Phys. Atmos. Ocean* **22**, 195–201.

69. Malevsky-Malevich, S. P. (1974). Specific features of temperature distribution in the near-surface water layer. In: *Protsessy perenosa vblizi poverkhnosti razdela okean–atmosphera*. Gidrometeoizdat, Leningrad, pp. 135–161.

70. Matushevsky, G. B. (1985). Methods of determining climatic characteristics of wind waves and estimation of their truth. Review information, series 'Oceanology', No. 2, VNIIGMI, Mezhdunarodny Tsentr Dannykh, Obninsk.

71. Mikhailov, V. I. (1979). Certain reasons for the increase of ocean surface microlayer salinity in individual areas of the North Atlantic. *Tr. GOIN* No. 146, 99–102.

72. Kraus, E. B. (Ed.) (1979). *Simulation and Prognosis of Upper Layers of the Ocean* (translated from English). Gidrometeoizdat, Leningrad.

73. Monin, A. S. and Krasitsky, V. P. (1985). *Ocean Surface Phenomena*. Gidrometeoizdat, Leningrad.

74. Monin, A. S. and Fedorov, K. N. (1973). On the ocean upper layer fine structure. *Izv., Acad. Sci., USSR, Phys. Atmos. Ocean* **9**, 442–444.

75. Muromtsev, A. M. (1962). On the hydrology of the Suez Canal, Red Sea, and Gulf of Aden. *Meteorol. Gidrol.* No. 2, 42–45.

76. Nelepo, B. A., Sizov, A. A. and Panteleyev, N. A. (1982). Manifestation of internal waves on the surface in the open ocean. *Dokl. Acad. Sci. USSR* **266**, 225–228.

77. Nikiforov, E. G. (1956). On the ties of the wind current with wind waves. *Izv., Acad. Sci., USSR, Geophys. Ser.* No. 12, 1450–1461.

78. Nikolayev, Yu. V., Makshtas, A. P. and Ivanov, B. V. (1984). Physical processes in the marginal ice zones of drifting sea ice. *Meteorol. Gidrol.* No. 11, 73–80.

79. Novogrudsky, B. V., Sklyarov, V. E., Fedorov, K. N. and Shifrin, K. S. (1978). *Space Study of the Ocean*. Gidrometeoizdat, Leningrad.

80. (1975). *Oceanographic Tables*, 4th edn (reviewed and supplemented). Gidrometeoizdat, Leningrad.

81. Paka, V. T. and Fedorov, K. N. (1982). On the influence of the ocean upper layer thermal structure on the development of turbulence. *Izv., Acad. Sci., USSR, Phys. Atmos. Ocean* **18**, 178–184.

82. Panin, G. N. (1985). *Heat- and Mass-Exchange Between a Water Body and the Atmosphere under Natural Conditions*. Nauka, Moscow.

83. Pedlosky. G. (1984). *Geophysical Hydrodynamics*, Vol. 1 (translated from English). MIR, Moscow.

84. Pelevin, V. N. and Rutkovskaya, V. A. (1986). Volume absorption of solar energy by oceanic waters. *Okeanologiya* No. 6, 914–919.

85. Pelinovsky, E. N. (Ed.). (1982). *Influence of Large-Scale Internal Waves on the Sea Surface* (Coll. scientific works). Institute of Applied Physics, USSR Academy of Sciences, Gorky.

86. Piterbarg, L. I. and Fedorov, K. N. (1987). Estimation of the velocity field characteristics by observations of a passive scalar on the ocean surface. *Issled. Zemli iz Kosmosa* No. 1, 26–34.

87. Preobrazhensky, L. Yu. (1964). On certain specific features of water temperature distribution in the surface layer. *Tr. GGO* No. 150, 99–101.

88. (1974). *Manual on Hydrological Work in Oceans and Seas*, 2nd edn. (reviewed and supplemented). Gidrometeoizdat, Leningrad.

89. Ryanzhin, S. V. (1981). Langmuir circulations. In: *Termodinamicheskiye protsessy v glubokikh ozerakh*. Nauka, Leningrad, pp. 45–126.

90. Ryanzhin, S. V. and Mironov, D. V. (1985). On the recurrence and critical conditions of origin of Langmuir circulations. *Izv., Acad. Sci., USSR, Phys. Atmos. Ocean* **21**, 184–190.

91. Snopkov, V. G. (1974). Certain meteorological and aerological characteristics of trade circulation on a polygon. In: *Atlantic Hydrophysical POLYGON-70*. Nauka, Moscow, pp. 46–66.

92. Solovyev, A. V. (1979). Fine thermal structure of the ocean surface layer in the region of POLYMODE-77 polygon. *Izv., Acad. Sci., USSR, Phys. Atmos. Ocean* **15**, 750–757.

93. Solovyev, A. V. (1982). On the vertical structure of the ocean thin surface layer at light wind. *Izv., Acad. Sci. USSR* **18**, 751–760.

94. Stommel, G. (1969). Horizontal temperature variations in a mixed layer in S. Pacific (translated. from English). *Okeanologiya* **9**, 97–102.

95. Terner, J. S. (1973). *Buoyancy Effects in Fluids*. Cambridge University Press, Cambridge.

96. Fedorov, K. N. (1955). Wind currents in a sea of variable depth. *Izv., Acad. Sci., USSR, Geophys. Ser.* No. 3. 224–233.

97. Fedorov, K. N. (1956). Level and currents during a catastrophic storm in the North Sea in 1953. *Izv., Acad. Sci., USSR, Geophys. Ser.* No. 4, 437–445.
98. Fedorov, K. N. Coastal wind currents and Ekman theory. (1959). *Izv., Acad. Sci., USSR, Geophys. Ser.* No. 8, 1238–1241.
99. Fedorov, K. N. (1976). On one mechanism of forming mesoscale horizontal temperature inhomogeneities in the upper layer of the ocean. *Okeanologiya* **16**, 403–407.
100. Fedorov, K. N. (1976). Space observations of oceanic internal waves. *Okeanologiya* **16**, 787–790.
101. Fedorov, K. N. (1979). On the slow relaxation of a hurricane trace in the ocean. *Dokl. Acad. Sci. USSR* **245**, 960–963.
102. Fedorov, K. N. (1981). On the physical structure of the ocean near-surface layer. *Meterol. Gidrol.* No. 10, 58–66.
103. Fedorov, K. N. (1983). *The Physical Nature and Structure of Oceanic Fronts.* Gidrometeoizdat, Leningrad.
104. Fedorov, K. N. (Ed.) (1986). *Intrathermocline Eddies in the Ocean* (Coll. Articles). Institute of Oceanology, USSR Academy of Sciences, Moscow.
105. Fedorov, K. N., Varfolomeyev, A. A., Ginsburg, A. I., *et al.* (1979). Thermal response of the ocean to hurricane 'Ella'. *Okeanologiya* **19**, 992–1001.
106. Fedorov, K. N., Vlasov, V. L., Ambrosimov, A. K. and Ginsburg, A. I. (1979). Study of evaporating sea water surface layer by optical interferometry. *Izv., Acad. Sci., USSR, Phys. Atmos. Ocean* **15**, 1067–1075.
107. Fedorov, K. N. and Ginsburg, A. I. (1986). Phenomena on the ocean surface by visual observations. *Okeanologiya* **26**, 5–14.
108. Fedorov, K. N., Ginsburg, A. I., Zatsepin, A. G., *et al.* (1979). Experience of recording the temperature and salinity of the ocean surface layer by 'AIST' probe. *Okeanologiya* **19**, 156–163.
109. Fedorov, K. N., Ginsburg, A. I. and Piterbarg, L. I. (1981). On the physical nature of 'calm weather inhomogeneities' in the ocean temperature field. *Okeanologiya* **21**, 203–210.
110. Fedorov, K. N., Plakhin, E. A., Prokhorov, V. I. and Sedov, V. G. (1974). Specific features of thermohaline stratification in the region of the Tropical Atlantic polygon. In: *Atlantic Hydrophysical POLYGON-70*. Nauka, Moscow, pp. 236–282.
111. Phillips, O. M. (1973). On interaction of internal and surface waves (translated from English). *Izv., Acad. Sci., USSR, Phys. Atmos. Ocean* **9**, 954–961.
112. Phillips, O. M. (1977). *The Dynamics of the Upper Ocean*, 2nd edn. Cambridge University Press, Cambridge.
113. Khain, A. P. and Sutyrin, G. G. (1983). *Tropical Cyclones and Their Interaction with the Ocean.* Gidrometeoizdat, Leningrad.
114. Khristoforov, G. N., Zapevalov, A. S. and Shutov, A. P. (1985). Structure of temperature fluctuations in the near-surface layer of the ocean in calm weather. *Okeanologiya* **25**, 235–236.
115. Khristoforov, G. N., Smolov, E. E. and Zapevalov, A. S. (1979). Manifestation of nonlinear surface sea waves in statistical and spectral characteristics. *Morsk. Gidrophyz. Issled.* No. 3. 113–124.
116. Khundzhua, G. G., Gusev, A. M., Andreyev, E. G., *et al.* (1977). On the structure of the ocean surface cold film and on the ocean heat exchange with the atmosphere. *Izv., Acad. Sci. USSR, Phys. Atmos. Ocean* **13**, 753–758.
117. Shigayev, V. V., Druzhinin, S. N., Lebedev, V. L. (1982). Study of temperature surface film by results of sea observations. *Meteorol. Gidrol.* No. 5, 75–79.

118. Shtokman, V. B. (1946). Vertical distribution of thermal waves in the sea and indirect methods of determining the coefficient of heat conductivity. *Tr. Inst. Okeanol., USSR Acad. Sci.* **1**, 3–46.

119. Shuleikin, V. V. (1953). *Physics of the Sea*. USSR Academy of Sciences, Moscow.

120. Alpers, W. and Salusti, E. (1983). Scylla and Charybdis observed from space. *J. Geophys. Res.* **88**, 1800–1808.

121. Andrews, W. R. H. (1974). Selected aspects of upwelling research in the Southern Benguela current. *Tethys* **6**, 327–340.

122. Apel, J. R., Byrne, H. M., Proni, J. R. and Charnell, R. J. (1975). Observations of oceanic internal and surface waves from the Earth Resources Technology Satellite. *J. Geophys. Res.* **80**, 865–881.

123. Apel, J. R., Holbrook, J. R., Liu, A. K. and Tsai, J. J. (1985). The Sulu Sea internal soliton experiment. *J. Phys. Oceanogr.* **15**, 1625–1651.

124. Barnes, E. J. (1985). Eastern Cook strait region circulation inferred from satellite-derived, sea-surface, temperature data. *N. Z. J. Mar. Freshwater Res.* **19**, 405–411.

125. Bernstein, R. L. (1982). Sea surface temperature estimation using NOAA-6 satellite advanced very high resolution radiometer (AVHRR). *J. Geophys. Res.* **87**, 9455–9467.

126. Bourke, R. H. (1983). Currents, fronts and fine structure in the marginal ice zone of the Chukchi Sea. *Polar Rec.* **21**, 569–575.

127. Brink, K. H. (1983). The near-surface dynamics of coastal upwelling. *Prog. Oceanogr.* **12**, 223–257.

128. Bruce, J. G. and Firing, E. (1974). Temperature measurements in the upper 10 m with modified expendable bathythermograph probes. *J. Geophys. Res.* **79**, 4110–4111.

129. Bucca, P. J. and Kinder, T. H. (1984). An example of meteorological effects on the Alboran Sea gyre. *J. Geophys. Res.* **89**, 751–757.

130. Buckley, J. R., Gammelsrød, T., Johannessen, J. A., Johannessen, O. M. and Røed, J. A. (1979). Upwelling: oceanic structure at the edge of the Arctic ice pack in winter. *Science* **203**, 165–167.

131. Clarke, A. J. (1978). On wind-driven quasi-geostrophic water movement at fast ice edges. *Deep-Sea Res.* **25**, 41–51.

132. Csanady, G. T. (1984). The free surface turbulent shear layer. *J. Phys. Oceanogr.* **14**, 402–411.

133. Defant, A. (1940). Scylla und Charibdis und die Gezeitenströmungen in der Strasse von Messina. *Ann. Hydr. Met.* **68**, 145.

134. Dickey, T. D. and Simpson, J. J. (1983). The influence of optical water type on the diurnal response of the upper ocean. *Tellus* **35B**, 142–154.

135. Dietrich, G. and Kalle, K. (1957). *Allgemeine Meereskunde*. Berlin.

136. (1983–1985). Direction de la Meteorologie Nationale *SATMER*. Bulletin mensuel. Le Centre de Meteorologie Spatiale, Lanion, France, Nos 1–26.

137. Dunn, M. R. (1966). Line of rough water. *Mar. Obs.* **XXXVI**, 80–81.

138. Ekman, V. W. (1905). On the influence of the earth's rotation on ocean currents. *Ark. Mat. Astron. Fys.* **2**, 1–52.

139. Ewing, G. (1950). Slicks, surface films, and internal waves. *J. Mar. Res.* **9**, 161–187.

140. Ewing, G. and McAlister, E. D. (1960). On the thermal boundary layer of the ocean. *Science* **131**, 1374–1376.

141. Farmer, D. and Smith, J. D. (1978). Nonlinear internal waves in a fjord. In: *Hydrodynamics of Estuaries and Fjords*, Nihoul, J. C. J. (Ed.), Elsevier Oceanogr. Ser., No. 23, Elsevier, Amsterdam, pp. 465–493.

142. Fedorov, K. N. (1972). On the summer daily heating and diurnal heat budget of the upper ocean layer. *Studi in onore di Giuseppina Aliverti*, Napoli, pp. 27–40.

143. Flament, P., Washburn, L. and Armi, L. (1983). Observations of the subsurface structure of an upwelling filament coordinated with a sequence of satellite IR images. *EOS Trans. Am. Geophys. Union* **64**, 1059.

144. Flierl, G. R., Stern, M. E. and Whitehead, J. A. (1983). The physical significance of modons: laboratory experiments and general integral constraints. *Dyn. Atmos. Oceans* **7**, 233–263.

145. Foster, T. D. (1974). The hierarchy of convection. In: *Processus de formation des eaux oceaniques profondes en particulier en Méditerranée Occidentale, Colloq. Int. CNRS*, No. 215, Paris, pp. 237–241.

146. Foster, T. D. (1971). Intermittent convection. *Geophys. Fluid Dyn.* **2**, 201–217.

147. Gammelsrod, T., Mork, M. and Roed, L. P. (1975). Upwelling possibilities at an ice edge, homogeneous model. *Mar. Sci. Commun.* **1**, 115–145.

148. Gautier, C. (1978). Some evidence of cool surface water pools associated with mesoscale downdrafts during GATE. *J. Phys. Oceanogr.* **8**, 162–166.

149. Gonella, Y. (1971). The drift current from observations made on the Bouée Laboratoire. *Cah. Oceanogr.* **XXIII**, 19–33.

150. Gordon, G., Greenwalt, D. and Witting, J. (1984). Surface-wave expression of bathymetry over a sand ridge. In: *Remote Sensing of Shelf Sea Hydrodynamics*, Nihoul, J. C. J. (Ed.), Elsevier Oceanogr. Ser., No. 38. Elsevier, Amsterdam.

151. Grassl, H. (1976). The dependence of the measured cool skin of the ocean on wind stress and total heat flux. *Boundary Layer Meteorol.* **10**, 465–474.

152. Gregg, M. C., Peters, H., Wesson, J. C., Oakey, N. S. and Shay, T. J. (1985). Intensive measurements of turbulence and shear in the equatorial undercurrent. *Nature* **318**, 140–144.

153. Griffiths, R. W. and Linden, P. F. (1981). The stability of vortices in rotating, stratified fluid. *J. Fluid Mech.* **105**, 283–316.

154. Hardtke, P. G. and Meincke, J. (1984). Kinematical interpretation of infrared surface pattern in the North Atlantic. Oceanol. Acta **7**, 373–378.

155. Hasselman, K. (1982). The science and art of wave prediction—an ode to HO-601. In: *A Celebration in Geophysics and Oceanography—1982*. In honour of Walter Munk on his 65th birthday, 19 October 1982. SIO Ref. Ser. 84-5, 1984, pp. 31–37.

156. Hasselman, K., *et al.* (1973). Measurements of wind wave growth and swell decay during the joint North Sea wave project (JONSWAP). *Dtsch. Hydrogr. Z.* **8** (Suppl. A) 95 S.

157. Helland, P. (1963). *Temperature and Salinity Variations in the Upper Layer at Ocean Weather Station* M (66° N, 2° E). Norwegian University Press, Bergen, Oslo.

158. Höeber, H. (1972). Eddy thermal conductivity in the upper 12 m of the Tropical Atlantic. *J. Phys. Oceanogr.* **2**, 303–304.

159. Horstmann, U. (1983). Distribution patterns of temperature and water colour in the Baltic Sea as recorded in satellite images: indicators for phytoplankton growth. *Berichte aus dem Institut für Meereskunde an der Christian-Albrechts Universität, Kiel*, Bd. 1, N 106, 147 S.

160. Howard, L. N. (1966). Convection at high Rayleigh number. *Proc. 11th Int. Congr. Appl. Mech.*, Munich, pp. 1109–1115.

161. Huang, N. E., Chen, D. T., Tung, C. C. and Smith, J. R. (1972). Interactions between steady non-uniform currents and gravity waves with applications for current measurements. *J. Phys. Oceanogr.* **2**, 420–431.

162. Hughes, B. A. and Grant, H. L. (1978). The effect of internal waves on surface wind waves. *J. Geophys. Res.* **83**, 443–454.

163. Huyer, A. (1983). Coastal upwelling in the California current system. *Prog. oceanogr.* **12**, 259–284.

164. Ikeda, M. and Emery, W. J. (1984). Satellite observations and modeling of meanders in the California current system off Oregon and Northern California. *J. Phys. Oceanogr.* **14**, 1434–1450.

165. Johannessen, O. M., Johannessen, J. A., Morison, J., Farrelly, B. A. and Svendsen, E. A. S. (1983). Oceanographic conditions in the marginal ice zone north of Svalbard in early fall 1979 with the emphasis on mesoscale processes. *J. Geophys. Res.* **88**, 2755–2769.

166. Joyce, T. M. and Stalcup, M. C. (1984). An upper current jet and internal waves in a Gulf Stream warm core ring. *J. Geophys. Res.* **89**, 1997–2003.

167. Katsaros, K. B. (1980). The aqueous thermal boundary layer. *Boundary-Layer Meteorol.* **18**, 107–127.

168. Katsaros, K. B., Fiuza, A., Sousa, F. and Amann, V. (1983). Sea surface temperature patterns and air–sea fluxes in the German bight during MARSEN 1979, Phase 1. *J. Geophys. Res.* **88**, 9871–9882.

169. Katsaros, K. B., Liu, W. T., Businger, J. A. and Tillman, J. A. (1977). Heat transport and thermal structure in the interfacial boundary layer measured in an open tank of water in turbulent free convection. *J. Fluid Mech.* **83**, 311–335.

170. Kelley, E. A., Weatherly, G. L. and Evans, J. C. (1982). Correlation between surface Gulf Stream and bottom flow near 5000 m depth. *J. Phys. Oceanogr.* **12**, 1150–1153.

171. Kirwain, A. D., McNally, G., Pazan, S. and Wert, R. (1979). Analysis of surface current response to wind. *J. Phys. Oceanogr.* **9**, 401–412.

172. Krümmel, O. (1911). *Handbuch Der Ozeanographie. Bd. II (Wellen, Gezeiten, Strömungen).* Stuttgart.

173. Lange, P. and Hühnerfuss, H. (1978). Drift response of monomolecular slicks to wave and wind action. *J. Phys. Oceanogr.* **8**, 142–150.

174. Laurs, R. M., Fiedler, P. C. and Montgomery, D. R. (1984). Albacore tuna catch distribution relative to environmental features observed from satellites. *Deep-Sea Res.* **31**, 1085–1099.

175. La Violette, P. E. (1984). The advection of submesoscale thermal features in the Alboran Sea gyre. *J. Phys. Oceanogr.* **14**, 550–565.

176. Lazier, J. R. N. (1984). Arctic run-off in the North Atlantic Ocean. *Bedford Inst. Oceanogr. (BIO) Rev.* 21–24.

177. Leetmaa, A. and Welch, C. S. (1972). A note on diurnal changes in momentum transfer in the surface layers of the ocean. *J. Phys. Oceanogr.* **2**, 302–303.

178. Lesieur, M. and Sadourny, R. (1981). Satellite-sensed turbulent ocean structure. *Nature* **294**, 673.

179. Lighthill, M. J. (1967). Waves in fluids. *Commun. Pure Appl. Math.* **20**, 267–293.

180. Longuet-Higgins, M. S. and Stewart, R. W. (1960). Changes in the form of short gravity waves on long waves and tidal currents. *J. Fluid Mech.* **8**, 565–585.

181. Longuet-Higgins, M. S. and Stewart, R. W. (1961). The changes in amplitude of short gravity waves on steady non-uniform currents. *J. Fluid Mech.* **10**, 529–549.

182. Lynn, R. J. and Svejkovsky, J. (1984). Remotely-sensed sea surface temperature variability off California during a 'SANTA ANA' clearing. *J. Geophys. Res.* **89**, 8151–8162.

183. Madsen, O. S. (1977). A realistic model of the wind-induced Ekman boundary layer. *J. Phys. Oceanogr.* **7**, 248–255.

184. McAlister, E. D. and McLeish, W. L. (1965). Oceanographic measurements with airborne infrared equipment and their limitations. In: *Oceanogrphy from Space*, Woods Hole Oceanogr. Inst., Ref. N 65–10. pp. 189–214.

185. McClain, C. R., Huang, N. E. and La Violette, P. E. (1982). Measurement of sea-state variations across oceanic fronts using laser profilometry. *J. Phys. Oceanogr.* **12**, 1228–1244.

186. Miller, J. R. (1976). The salinity effect in a mixed layer ocean model. *J. Phys. Oceanogr.* **6**, 29–35.

187. Millot, C. and Wald, L. (1981). Infra-red remote sensing in the Gulf of Lions. In: *Oceanography from Space*, Gower, J. F. R. (Ed.). Plenum Press, New York, pp. 183–187.

188. Mollo-Christensen, E. and Mascarenhas, A. S. (1979). Heat storage of the oceanic upper mixed layer inferred from Landsat data. *Science* **203**, 653–654.

189. Montgomery, R. B. and Stroup, E. D. (1962). Equatorial waters and currents at 150° W in July–August 1952. *John Hopkins Oceanogr. Stud.* **1**.

190. Mooers, C. N. K. and Robinson, A. R. (1984). Turbulent jets and eddies in the California current and inferred cross-shore transports. *Science* **223**, 51–53.

191. Muench, R. D. (1983). The marginal ice zone experiment. *Oceanus* **26**, 55–60.

192. Muench, R. D. (1983). Mesoscale oceanographic features associated with the central Bering Sea ice edge: February–March 1981. *J. Geophys. Res.* **88**, 2715–2722.

193. Muench, R. D., LeBlon, P. H. and Hachmeister, L. E. (1983). On some possible interactions between internal waves and sea ice in the marginal ice zone. *J. Geophys. Res.* **88**, 2819–2826.

194. NOAA, US Department of Commerce (1979). *Oceanic and Related Atmospheric Phenomena as Viewed from Environmental Satellites.* Walter A. Bohan Co., Illinois.

195. Orlanski, I. and Polinsky, L. J. (1983). Ocean response to mesoscale atmospheric forcing. *Tellus* **35A**, 296–323.

196. ORSTROM/Meteorologie Nationale (1984). *Veille Climatique Satellitaire* (Bulletin bimensuel). Lannion, Centre de Meteorologie Spatiale, No. 1.

197. Osborne, A. R. and Burch, T. L. (1980). Internal solitons in the Andaman Sea. *Science* **208**, 451–460.

198. Ostapoff, F. and Worthem, S. (1974). The intradiurnal temperature variation in the upper ocean. *J. Phys. Oceanogr.* **4**, 601–612.

199. Owen, R. W., Jr. (1966). Small-scale horizontal vortices in the surface layer of the sea. *J. Mar. Res.* **24**, 56–65.

200. Paulson, C. A. and Simpson, J. J. (1981). The temperature difference across the cool skin of the ocean. *J. Geophys. Res.* **86**, 11044–11054.

201. Pease, C. and Muench, R. D. (1981). Cruise along ice edge in Bering Sea yields data on effects of gale. *Coastal Oceanogr. Climat. News* **3**, 43–45.

202. Phillips, O. M. (1958). The equilibrium range in the spectrum of wind generated waves. *J. Fluid Mech.* **4**, 426–431.

203. Pierson, W. J. and Moskovitz, L. (1964). A proposed spectral form for fully developed wind seas based upon similarity theory of S. A. Kitaigorodski. *J. Geophys. Res.* **69**, 5181–5190.

204. Price, J. F. (1979). Observations of a rain-formed mixed layer. *J. Phys. Oceanogr.* **9**, 643–649.

205. Reed, R. J. and Lewis, R. M. (1980). Response of upper ocean temperatures of diurnal and synoptic-scale variations of meteorological parameters in the GATE B-scale area. *Deep-Sea Res.* (Part A, Suppl. 1 to **26**, GATE) **1**, 99–114.

206. Reed, R. K. and Halpern, D. (1975). The heat content of the upper ocean during coastal upwelling: Oregon, August 1973. *J. Phys. Oceanogr.* **5**, 379–383.

207. Rienecker, M. M., Mooers, C. N. K., Hagan, D. E., Robinson, A. R. (1985). A cool anomaly off Northern California: an investigation using IR imagery and *in situ* data. *J. Geophys. Res.* **90**, 4807–4818.

208. Røed, L. P. and O'Brien, J. J. (1983). A coupled ice–ocean model of upwelling in the marginal ice zone. *J. Geophys. Res.* **88**, 2863–2872.

209. Saint-Guily, B. (1972). On the response of the ocean to impulse. *Tellus* **24**, 344–349.

210. Saunders, P. M. (1967). The temperature at the ocean–air interface. *J. Atmos. Sci.* **24**, 269–273.

211. Saunders, P.M. (1974). Tracing surface flow with surface isotherms. *Mem. Soc. R. Sci., Liege* **6**, 99–108.

212. Saunders, R. W., Ward, N. R., England, C. F. and Hunt. G. E. (1982). Satellite observations of sea surface temperature around the British Isles. *Bull. Am. Meteorol. Soc.* **63**, 267–272.

213. Sawyer, C. and Apel, J. R. (1976). *Satellite Images of Ocean Internal Wave Signatures* (Atlas). NOAA, Atlantic Oceanographic and Meteorological Lab., Miami.

214. Shay, T. J. and Gregg, M. C. (1984). Turbulence in an oceanic convective mixed layer. *Nature* **310**, 282–285.

215. Shemdin, O. H. (1977). Modulation of centimetric waves by long gravity waves: progress report of field and laboratory results. In: *Turbulent Fluxes Through the Sea Surface, Wave Dynamics and Prediction,* Favre, A. and Hasselmann, K. (Eds). Plenum Press, New York, pp. 235–255.

216. Sheres, D. (1982). Remote synoptic current measurements by gravity waves; a method and its test in a small body of water. *J. Phys. Oceanogr.* **12**, 200–207.

217. Simpson, J. J. and Dickey, T. D. (1981). The relationship between downward irradiance and upper ocean structure. *J. Phys. Oceanogr.* **11**, 309–323.

218. Simpson, J. J., Koblinsky, C. J., Haury, L. R., Dickey, T. D. (1984). An offshore eddy in the California current system. *Prog. Oceanogr.* **13**, 1–111.

219. Simpson, J. J. and Paulson, C. A. (1980). Small-scale sea surface temperature structure. *J. Phys. Oceanogr.* **10**, 399–410.

220. Smith D. C., IV, Morison, J. H., Johannessen, J. A. and Untersteiner, N. (1984). Topographic generation of an eddy at the edge of the East Greenland current. *J. Geophys. Res.* **89**, 8205–8208.

221. Smith, R. L. (1974). A description of current, wind and sea level variations during coastal upwelling off the Oregon coast, July–August 1972, *J. Geophys. Res.* **79**, 435–443.

222. Stewart, R. H. and Joy, J. W. (1974). HF radio measurements of surface currents. *Deep-Sea Res.* **21**, 1039–1049.

223. Stommel, H. and Fedorov, K. N. (1967). Small-scale structure in temperature and salinity near Timor and Mindanao. *Tellus* **19**, 306–325.

224. Stommel, H. and Schott, F. (1977). The beta spiral and the determination of the absolute velocity field from hydrographic station data. *Deep-Sea Res.* **24**, 325–329.

225. Strong, A. E. and McClain, E. P. (1984). Improved ocean surface temperatures from space-comparisons with drifting buoys. *Bull. Am. Meteorol. Soc.* **65**, 138–142.

226. Taylor, G. I. (1955). The action of a surface current used as a breakwater. *Proc. R. Soc. London., Ser. A* **231**, 466–478.

227. Thorpe, S. A. (1978). A near-surface ocean mixing layer in stable heating conditions. *J. Geophys. Res.* **83**, 2875–2885.

228. Traganza, E. D., Nestor, D. A. and McDonald, A. K. (1980). Satellite observation of a nutrient upwelling off the coast of California. *J. Geophys. Res.* **85**, 4101–4106.

229. Trump, C. L., Neshyba, S. J. and Burt, W. V. (1982). Effects of mesoscale atmospheric convection cells on the water of the East China Sea. *Boundary Layer Meteorol.* **24**, 15–34.

230. Unna, P. J. H. (1942). Waves and tidal streams. *Nature* **149**, 219–220.

231. Van Foreest, D., Shillington, F. A. and Legeckis, R. (1984). Large scale, stationary, frontal features in the Benguela current system. *Cont. Shelf Res.* **3**, 465–474.

232. Van Woert, M. (1982). The subtropical front: satellite observations during FRONTS 80. *J. Geophys. Res.* **87**, 9523–9536.

233. Vastano, A. C. and Bernstein, R. L. (1984). Mesoscale features along the first Oyashio intrusion. *J. Geophys. Res.* **89**, 587–596.

234. Vesecky, J. F., Assal, H. M. and Stewart, R. H. (1981). Remote sensing of the ocean waveheight spectrum using synthetic-aperture-radar images. In: *Oceanography from Space*, Gower, J. F. R. (Ed.) Plenum Press, NY, pp. 449–457.

235. Wadhams, P. and Squire, V. A. (1983). An ice–water vortex at the edge of the East Greenland current. *J. Geophys. Res.* **88**, 2770–2780.

236. Wesely, M. L. (1979). Heat transfer through the thermal skin of a cooling pond with waves. *J. Geophys. Res.* **84**, 3696–3700.

237. Witting, J. (1984). Wave–current interactions: a powerful mechanism for an alteration of the waves on the sea surface by subsurface bathymetry. In: *Remote Sensing of Shelf Sea Hydrodynamics*, Nihoul, J. C. J. (Ed.). Elsevier Oceanogr. Ser. No. 38, Elsevier, Amsterdam, pp. 187–203.

238. Woodcock, A. H. (1941). Surface cooling and streaming in shallow fresh and salt waters. *J. Mar. Res.* **4**, 153–161.

239. Woods, J. D. (1980). Diurnal and seasonal variations of convection in the wind-mixed layer of the ocean. *Q. J. R. Meteorol. Soc.* **106**, 379–394.

240. Woods, J. D. (1984). The upper ocean and air–sea interaction in global climate. In: *The Global Climate*, Houghton, J. T. (Ed.). Cambridge University Press, Cambridge, pp. 141–187.

241. Woods, J. D. and Barkmann, W. (1986). The response of the upper ocean to solar heating. 1: The mixed layer. *Q. J. R. Meterol. Soc.* **112**, 1–27.

242. Woods, J. D., Barkmann, W. and Horch, A. (1984). Solar heating of the oceans—diurnal, seasonal and meridional variation. *Q. J. R. Meteorol. Soc.* **110**, 633–656.

243. Woods, J. D., Barkmann, W. and Strass, V. (1985). Mixed layer and Ekman current response to solar heating. In: *The Ocean Surface: Wave Breaking, Turbulent Mixing and Radio Probing*, Toba, Y. and Mitsuyasu, H. (Eds). Reidel, Dordrecht, pp. 487–507.

244. Wu, J. (1983). Sea-surface drift induced by wind and waves. *J. Phys. Oceanogr.* **13**, 1441–1451.

245. Wu, J. (1985). On the cool skin of the ocean. *Boundary-Layer Meteorol.* **31**, 203–207.

246. Wunsch, C. (1978). The North Atlantic general circulation west of 50° W determined by inverse methods. *Rev. Geophys. Space Phys.* **16**, 583–620.

Subject Index